Principles and Applications of Domestic Animal Behavior

An Introductory Text

Edward O. Price
Professor Emeritus
Department of Animal Science
University of California
Davis, CA 95616, USA

www.cabi.org

CABI is a trading name of CAB International

CABI Head Office	CABI North American Office
Nosworthy Way	875 Massachusetts Avenue
Wallingford	7th Floor
Oxfordshire OX10 8DE	Cambridge, MA 02139
UK	USA
Tel: +44 (0)1491 832111	Tel: +1 617 395 4056
Fax: +44 (0)1491 833508	Fax: +1 617 354 6875
E-mail: cabi@cabi.org	E-mail: cabi-nao@cabi.org
Website: www.cabi.org	

A catalogue record for this book is available from the British Library, London, UK.

Library of Congress Cataloging-in-Publication Data

Price, Edward O.
   Principles and applications of domestic animal behavior: an introductory text /
Edward O. Price.
       p. cm.
   Includes bibliographical references and index.
   ISBN 978-1-84593-398-2 (alk. paper)
1. Domestic animals--Behavior. 2. Animal behavior. I. Title.

SF756.7.P75 2008
636--dc22

                                                           2008017353

ISBN-13: 978 1 84593 398 2

Typeset by SPi, Pondicherry, India.
Printed and bound in the UK by Cambridge University Press, Cambridge.

Contents

Preface

My interest in animal behavior began at a very young age. Being raised on a dairy farm and having cows, dogs, cats, a pony, sheep, goats and rabbits to care for provided me with many opportunities to observe animal behavior. We spent a lot of time with our cows, since they had to be milked twice a day. Each of the 60+ animals had a name and some had nicknames. We knew the cows that were dominant and those who were subordinate. We knew the tame cows and the more flighty ones, the ones that were first to enter the barn for milking and the ones that lagged behind. I will never forget the excitement of seeing the birth of calves and the sadness of seeing our animals depart from us later in their lives. I am thankful for the opportunity to observe the changes in their behavior as they grew up from calves to adult cows and had young of their own.

We also had wildlife on the farm, including woodchucks, skunks, raccoons, opossums, muskrats, rabbits, foxes, deer, pheasants and innumerable species of songbirds, reptiles and amphibians. The creek running through the fields held fish and attracted waterfowl on their migration in the fall. On one occasion, I captured a young red fox from a woodchuck den and kept it in our basement, hoping to tame it as a pet. I conscientiously fed it meat and tried to coax it to approach me, but to no avail. It remained fearful of me, in spite of my efforts to tame it. On another occasion, I was working in the fields and heard the cries of an animal in the nearby woods. I followed the sounds and soon spotted a very young dog trying to keep up with its feral (wild) mother. The mother ran off when I approached, so I took the pup home and we raised it. In spite of all the good care it received, it was never as friendly to us as our other dogs.

I never put these observations of animal behavior into any kind of scientific framework until I went to college and took courses in psychology and behavior. After receiving a BA degree in psychology, I was accepted into the Zoology Graduate Program at Michigan State University, where I worked under Dr John King, a new professor at MSU, who accepted me as his first graduate student. My coursework in animal behavior, my research experiences and daily discussions with professors and fellow graduate students at Michigan State laid a solid basis for the field I was to pursue as a career. In 1966, I accepted a position as Assistant Professor of Zoology at the State University of New York in Syracuse and was charged with developing an animal behavior program for my department. My research focus was the effects of domestication on the behavior of the Norway rat. Ten years later, I moved to the University of California, Davis, where I established a program of teaching and research in animal behavior in the Department of Animal Science. My research emphasis at UC Davis was the reproductive behaviors of cattle, sheep and goats.

I retired in 2002 after 36 years of teaching animal behavior and conducting research in this area. Two years ago, I decided to write this book. It is the text I envisioned for my introductory course at Davis but never had time to write. I firmly believe that students must be well grounded in the basic principles of animal behavior before they embark on advanced study or research. I also believe it is difficult to solve practical concerns associated with animal behavior if one does not understand the basic concepts underlying those issues. That is the purpose of this book, to provide a basic understanding of the principles and concepts of animal behavior, with an emphasis on domestic and captive wild animals. I have attempted, where possible, to illustrate the points I make with examples involving captive domestic and wild animals. A minority of examples is from the literature dealing with free-living wild animals. I have used these when I felt a point could be best illustrated by citing such examples. I have included no formal chapters on physiological mechanisms and animal welfare but have interspersed information on these topics throughout the book, as it seemed appropriate. The book contains very little on animal navigation, since it is of minor importance to captive animals.

This is basically a textbook for introducing college, university and veterinary school students to the field of animal behavior and the concepts underlying behavior. There are two major themes in the book, the development of behavior and social behavior. I have also included chapters on biological rhythms, human–animal interactions, animal handling and atypical behaviors. There are four chapters on the development of behavior, namely 'Interaction of Heredity and Environment' (Chapter 2), 'Behavior Genetics' (Chapter 3), 'Early Experience and Behavioral Development' (Chapter 4) and 'Learning' (Chapter 5). Eight chapters focus on social behaviors, 'Mating Systems and Reproduction' (Chapter 7), 'Male Sexual Behavior' (Chapter 8), 'Female Sexual Behavior' (Chapter 9), 'Maternal and Neonatal Behavior' (Chapter 10), 'Communication' (Chapter 11), 'Agonistic Behavior' (Chapter 12), 'Social Organization' (Chapter 13) and 'Personal Space and Social Dynamics' (Chapter 14). The chapters on social behavior can be thought of as following a general chronological progression from reproductive behaviors through to the integration of subsequent offspring into social groupings.

My students always seemed to appreciate the examples I used in my lectures to illustrate how the basic principles of animal behavior could be used to solve practical everyday problems faced by pet owners, farmers and others responsible for managing captive wild and domestic animals. Thus, a significant part of the book is devoted to this topic. In addition, my students told me they appreciated the numerous photos, drawings and personal stories that I included in my lectures. I have interspersed some of these throughout the book. I hope you will not only learn a great deal by reading this book, but that you will also enjoy it. I certainly have enjoyed writing it.

Acknowledgements

I am greatly indebted to my late father, C. Edward Price, for sharing his interest and love for domestic animals as I was growing up. Harold Musselman, my high school biology teacher, first introduced me to the formal study of animal biology and encouraged me to conduct extracurricular projects for his class. Dr John King, my mentor in graduate school, gave me the training and encouragement to pursue a career in zoology (animal behavior) and has been a lifelong friend. I am grateful to the many undergraduate and graduate students I have had in my animal behavior classes who have assisted me in my research program. Their enthusiasm and interest in animal behavior was an encouragement to me and probably was the single most important impetus for writing this book. The publications of my colleagues working in animal behavior research around the world have provided most of the material included in this text and those who I have known personally have been an encouragement to me over the years. Lastly, I am grateful to my wife, Mabell, who patiently allowed me to spend days and weeks on this book when we could have been doing other things together.

1 Introduction

Why Study Animal Behavior?

You have chosen to study animal behavior. Why? One reason is that you probably find animals interesting. Perhaps you have a dog or cat, or maybe a friend who does. Some of us grew up with pets or on farms where we were surrounded by animals every day. When you are exposed to animals on a regular basis, you cannot help but wonder why they do what they do. Have you found yourself asking why there are so many behavioral differences between individuals and species? Have you wondered how animals learn this or that and to what extent they think like humans?

Scientists interested in animal behavior often engage in basic research to answer these and other questions. Such research helps us understand animals, not only to satisfy our curiosity but also to determine how animals adapt to the environments in which they live. Nowadays, it is common knowledge that animal behavior is subject to natural selection and evolutionary processes in much the same manner as their morphology (structure) and physiology. Take, for example, animals living in cold climates. Many of these animals hibernate or assume a torpor-like state through the winter to ensure survival when food resources are in short supply or are inaccessible. Preparation for hibernation involves an intense period of feeding to build up fat reserves to get them through the winter. Some species hoard food in caches during the warmer months, which they can access during winter when food is less available. When I was a college student, I spent two summers working as a ranger naturalist in Denali National Park, Alaska. On my days off, I hiked the mountains and valleys in the park, observing and photographing wildlife. I remember watching the small, rodent-like pikas (*Ochotona princes*) gathering grasses and other vegetation into piles of 'hay' under rock ledges where their food supply would be accessible during the long winter months. As this book progresses, I will offer many examples of how domestic and some wild animals adapt to their environment via behavioral mechanisms.

Some animal behaviorists, like myself, not only conduct basic research to understand why animals behave as they do, but also investigate ways to use this basic knowledge to solve practical problems, such as how to improve reproduction, productivity (e.g. milk, egg and meat production) or the general well-being of domestic animals. Such applied research usually requires a thorough understanding of the basic biology of the animal so that reasonable hypotheses can be made prior to investigating applications. For example, recently we completed a study on weaning beef calves (*Bos taurus*) that was designed to reduce the stress experienced when calves were separated from their mothers. Beef producers in the USA traditionally wean beef calves at about 7 months of age, at a time when the calves are still obtaining milk from their mothers and are bonded strongly to them. Because calves react negatively to being separated from their dams at this age, the practice in the industry has been to separate them abruptly and at some distance, using an 'out-of-sight, out-of-mind' rationale that the calves

Fig. 1.1. Beef cows and calves separated by a fence, illustrating fenceline weaning. By eliminating the abrupt separation of mother and young, both show fewer behavioral signs of distress (Price *et al.*, 2003b).

would adapt to the separation more quickly if they could not see or hear their mothers. For several days after being weaned in this manner, a high percentage of calves spend much of their time pacing around the perimeter of their enclosures, vocalizing and ignoring feed placed in front of them. They are clearly stressed by the separation, even though they are in the company of other calves. In our research, we hypothesized that separating the calves and cows with just a fence between them would allow them to spend time in close proximity to each other and still prevent suckling by the calves (Fig. 1.1). When we conducted the research, we found our hypothesis was correct. Using the fenceline weaning technique, we noted our calves exhibited few signs of psychological stress. Pacing fencelines and vocalizations were uncommon, feeding was normal and the calves continued to grow at a normal pace through the weaning period. During the first few days of weaning, our experimental calves and cows spent less and less time adjacent to each other at the fenceline and, after 5 or 6 days, weaning appeared to be complete. Several months later, we found that the control group of calves in our study, which were weaned abruptly in the traditional manner, still had not regained the body weight they lost (i.e. not gained) during the weaning process. This research, and subsequent work by others (see Chapter 10), has convinced many beef producers that the traditional method of weaning by abrupt separation is not the most profitable and animal-friendly way to manage their animals.

Approaches to the Study of Animal Behavior

There are four general approaches to the study of animal behavior. These approaches are not mutually exclusive but rather arbitrarily describe the different emphases of scientists studying animal behavior.

1. Evolutionary or phylogenetic origins of behavior

Each species has its own unique behavioral adaptations to the environment in which it is found. These adaptations are attained in part by the expression of genes which have been selected in the species' evolutionary past. Thus, some scientists have focused their studies on how behavior has changed (over generations) in the recent or ancestral past. For example, I have been particularly interested in how domestication changes the behavior of animals (Price, 2002). Taking an animal from the wild and placing it in captivity constitutes a relatively major change in the animal's environment. In captivity, animals are usually protected from natural predators, food and water are readily accessible year-round and groups of animals are often confined to small spaces. Consequently, certain traits selected for in nature are no longer selected in captivity and other behavioral characteristics become more important. These changes in natural selection (in captivity) can result in genetic changes affecting behavior. One important evolutionary change accompanying the domestication process is a reduction in fearfulness of strange or novel objects ('neophobia'). This makes good sense since unfamiliar objects in nature are often dangerous (e.g. a predator, toxic food), whereas in captivity, strange objects usually do not jeopardize survival.

2. Ontogenetic or developmental processes

One of the more fascinating areas of study is how and when behavior develops during the course of an animal's lifetime. Some behaviors develop naturally on a relatively rigid time schedule and at a predictable time in life. Such traits are usually important for the animal's survival and (or) its ability to reproduce. Familiar examples are the development of suckling behavior in neonatal mammals, development of the locomotor abilities of young birds and development of certain social attachments (e.g. young bonding to mother and vice versa). At the other end of the spectrum are learned behaviors, which may have no timeline at all. Food preferences may change as the availability of food resources varies. Social relationships may change with the introduction of new individuals into a population. Prey species can learn the body language of predators signaling that they are hunting versus just moving about.

3. Proximate mechanisms

Some animal behaviorists spend their entire careers studying the proximate mechanisms that trigger or stimulate animals to behave in a certain way. Such stimuli may be internal (e.g. a physiological event), an external object or event, or some combination of the two. For example, domestic sheep (*Ovis aries*) have not lost the tendency of their wild ancestors to be seasonal breeders. Shortening day length (external event) triggers a change in the sheep's hormonal profile. In the male, testosterone levels increase (internal event), which results in an increase in the tendency to mate when encountering females (external stimulus) and increased aggressiveness when encountering other adult males (external stimulus).

4. Functional consequences

One cannot study animal behavior without becoming interested in the function served by certain behaviors. Studies of function address how behavior promotes survival, reproduction and animal well-being (i.e. how behavior facilitates adaptation to the animal's environment). For example, we could ask, 'What is the function of the social attachment of the young chicken (*Gallus domesticus*) to its mother?' The answer to that question may seem obvious; the young chick needs to follow its mother to take advantage of the protection the mother hen offers from inclement weather and predators. But further study reveals that the mother bird may teach her young other important lessons, such as what foods are good to eat, how to respond to other adult birds in the flock and how to navigate in their environment.

Relative to the chick, the young domestic kitten (*Felis domestica*) is even more dependent on its mother. At birth, the kitten is relatively immobile, its eyes have not opened and other sensory modalities are not fully functional. During this neonatal period, the mother cat licks her young to stimulate defecation and urination and she nurses it to provide important nourishment. By staying in the nest, the kitten obtains needed shelter and protection from the elements and predators. Later, the mother cat may take her kittens on hunting expeditions where they can learn how to capture prey. The mother cat's nest-building, grooming, nursing, protection from predators and prey-catching lessons are all subjects of interest, in themselves, but together form a complex of behaviors we call maternal care. Each of these behaviors has a function and collectively they serve to ensure the survival of the young by providing needed resources and by helping them learn how to live on their own.

Alternatively, each component of maternal care could be studied using all four approaches. Let us take, for example, the subject of mother cats (*F. domestica*) teaching their young how to hunt for food such as mice, small birds and other prey. From an evolutionary perspective, we could ask why domestic cats hunt alone rather than in groups? Does it relate to their basic social organization? We know that domestic cats typically spend much of their time alone rather than in groups, like lions (*Panthera leo*). Could this have something to do with the size of their prey? Domestic cats hunt small prey like rodents and small birds, which can be killed readily by a single animal, whereas lions typically prey on larger animals (e.g. hoofed ungulates), which often require several animals to capture. Thus, it is probably safe to say that in the ancestry of the domestic cat, prey size has provided little selection pressure for them to live in groups.

From a developmental perspective, we could study how the mother teaches her young to hunt. Does she show them by example where to hunt for prey, how to stalk prey once they are located and how to attack prey and execute the killing bite? That is, do young cats learn by watching their mother hunt or are some of the predatory behaviors inborn or innate? Do they acquire the taste for meat? Is the kill typically eaten whole or do they learn to dissect the prey before it is eaten? Do they learn that hunting at certain times of the day (or night) is more productive?

Of course, hunger is an important proximate mechanism or internal state which motivates the animal to set out on its hunting trips. What external cues might they use to locate good habitat for prey? What external stimuli are used to alert them to the presence of prey (e.g. sight, sound, smell)? What internal and external cues are important in initiating the capture of prey? What parts of the brain are involved in hunger and in prey-catching behavior? It appears there is no end to the questions we could ask.

We could also address this process from a functional perspective. Of course, the major function of predatory behavior is to obtain food. But we do not have to stop there. How has the physiology of the domestic cat been adapted to its carnivorous diet? Can rodents and small birds provide all the nourishment needed as adults or must they ingest other nutrients? We can even address predation from an ecological perspective. How does predation affect the population dynamics of various prey species? No matter what our perspective, there are plenty of topics to be researched. In the course of addressing one question, other questions typically arise. That is one of the reasons the study of animal behavior is so enjoyable.

Conducting Behavioral Research

Experimental approach

The example in the previous section on low-stress weaning in beef cattle gave some insight on how animal behavior can be studied experimentally. A question is raised (e.g. 'Is there a less stressful way to wean beef calves?') and a hypothesis is tested (e.g. fenceline weaning is less stressful than the traditional abrupt weaning method). You decide what behaviors you are going to measure that are indicative of stress (e.g. time spent feeding and pacing and number of vocalizations). These are called 'dependent variables'. Then you decide what behaviors, situations or treatments you are going to vary to test your hypothesis (e.g. fenceline separation versus abrupt separation). These are called 'independent variables'. The animals in the primary treatment(s) of interest (fenceline separation) constitute the experimental group. The animals in the standard or traditional treatment (abrupt separation) constitute the control group. The control group provides the 'baseline' data on which to compare the data generated by the experimental group. After you collect the data associated with each dependent variable, appropriate statistical tests are conducted to determine if the experimental and control groups differ significantly for each dependent variable. These statistical results help you determine whether your hypothesis is supported and what conclusions you can propose.

Descriptive approach

I mentioned earlier that it is important to know the basic behavioral biology of the animal before applied research can be conducted. Many textbooks on domestic animal behavior have chapters describing the behaviors exhibited by various species. In each chapter, there will be sections on behaviors associated with feeding, sexual behavior, parental care, communication and so on. It is important to know the basic behavioral repertoire of the species you are working with (often called an ethogram) before initiating your studies. Farmers, animal breeders, researchers, etc., typically have a working knowledge of the behavior of their animals by observing them in different situations over long periods of time. People not familiar with a species can quickly get 'up to speed' by reading about it. This knowledge is available because scientists and others interested in animal behavior have spent countless hours observing animal behavior both in the field and in captivity and have taken the time to put their findings in writing.

Describing and Interpreting Animal Behavior

Naming behavioral responses

When talking with others about animal behavior, we typically describe behaviors at three levels of complexity. As the level of complexity increases, the danger of misinterpretation increases. This is explained below.

First, when speaking of behavior at the motor or action-pattern level, we provide an objective description of what the animal is doing at a given point in time (e.g. running, climbing, pecking). Second, at the functional level, we speak of functional categories toward which action patterns are directed (e.g. maternal care, aggression). These functional categories typically involve more than one action pattern. For example, maternal care can include the action patterns of nursing, licking young, regurgitating food, etc. Aggression often includes the action patterns of baring teeth, biting, pecking, etc. Third, we sometimes use terms at an abstract level to describe mental states that cannot be defined objectively but may be described operationally by certain action patterns. For example, the terms 'emotional reactivity' or 'fearfulness' can be expressed through various behaviors, including vocalizations (e.g. alarm signals), locomotor behaviors (e.g. running away from something), or changing the position of appendages (e.g. ears, tails, wings) relative to the body.

At what level are the following named behaviors?

1. 'Courtship', as in 'The rooster courted the hen'.
2. 'Suckling', as in 'The calf suckled from the cow's udder within minutes after birth'.
3. 'Wary', as in 'Wild horses are extremely wary of humans'.

One should be cautious in naming behavioral acts, since personal biases and assumptions can influence the choice of names. It is usually safe to name behavioral responses at the action-pattern level simply because, at this level, you are merely stating what the animal is doing (e.g. running, pawing the ground) without offering subjective interpretations of the consequence or function of the behavior. It is sometimes wise to avoid functional titles or names such as 'aggressive', 'submissive' or 'sexual', since a given behavior (i.e. action pattern) may be observed in more than one context (i.e. it may have different functions). For example, the act of chasing (one individual pursuing another) may be observed in the context of play, aggression and courtship. Male pigeons (*Columba livia*) sometimes bring food to prospective mates as part of their courtship activities. It would be easy to mistake this as the male feeding its young. Some people believe that male dogs (*Canis familiaris*) mount other males to assert their social dominance, not to engage in sexual activity. The male house cat (*F. domestica*) often grabs the female's neck with his teeth when mating, a behavior frequently misinterpreted as aggression.

I once received an e-mail from a graduate student asking if I could help her in differentiating between aggressive and sexual behavior in Japanese quail (*Coturnix japonica*). The objective of her research was to describe the development of aggression in this species by observing heterosexual groups of young birds at successive ages. 'Aggressive behavior' was defined as a peck given or received in which the recipient either crouched in 'submission' or ran away. Once the birds began to reach the age of sexual maturity, she noticed chasing and pecking which appeared sexual in nature. Yet, sometimes, the male would fail to 'catch' the female and thus succeeded in only chasing and pecking her on the head, as when exhibiting aggression. In addition, she observed males mounting other males, which left her wondering whether these mounts were sexual or aggressive in nature. Can you sense her frustration?

Anthropomorphism

Anthropomorphism is a term given to interpreting animal behavior in terms of human experience. Anthropomorphic statements are used most commonly when discussing the mental states of animals. Some examples are provided below.

'The dominant ram held his head proudly.' Do animals possess any feelings akin to pride in humans?

'The pig was happy when it was fed.' We know that animals can appear to get excited or aroused. But to what extent can animals experience happiness as humans experience it?

'The dog was depressed when its owner went on vacation.' To what extent do dogs experience depression as experienced by humans?

Anthropomorphic statements usually are avoided in scientific gatherings because generally they involve unproven, subjective interpretations of behavior. The scientist that speaks of animal behavior in an anthropomorphic manner is risking his or her scientific credibility. However, I know scientists, including myself, who occasionally make anthropomorphic statements in more casual settings, particularly when the conversation involves our pets.

One benefit of being anthropomorphic is that it raises questions about the cognitive abilities of animals and the extent to which they have 'feelings' similar to humans. This topic has been especially 'hot' in recent years because of interest in the well-being of captive animals and their mental states under various circumstances.

Teleology

When describing innate or instinctive behavior patterns, it is sometimes tempting to ascribe to animals the capacity to foresee the end result of their behavior. We refer to this as teleology. However, we cannot assume that animals can anticipate the outcome or effect of their more innate behavior patterns. Here are some examples of teleological statements.

'The cow licked her newborn calf in order to stimulate it to stand and suckle.' Although licking the newborn serves this function in cattle (*Bos*), it is unlikely the cow understands the contribution of this behavior to the well-being of her calf. In the course of evolution, cows have been selected for those traits that favor the survival of their offspring. The cow in this instance responds to certain stimuli (internal and external) in a quite predictable way. A more preferred, non-teleological way to word this sentence would be 'By licking her newborn calf, the cow stimulated it to stand and suckle'.

'The cat licked its wound to prevent a serious infection.' Does the cat (*F. domestica*) really know that licking (and saliva) helps to prevent bacterial infections? A more preferred wording would be 'The cat prevented a serious infection by licking its wound'.

'Sheep stay in a flock to minimize the risk of predation.' Although the risk of predation may be greater for a sheep (*O. aries*) that is alone than for one that is part of a flock, it is quite likely that sheep have no foresight about this. They merely have inherited a predisposition to seek out and maintain close proximity to other members of their species (conspecifics). Flocking behavior has been selected for at some point in the species' ancestry. A non-teleological wording would be 'The flocking behavior of sheep minimizes the risk of predation'.

Just because certain responses are adaptive or beneficial to the animal does not mean the animal always has the foresight to recognize those benefits. The animal simply may be responding to stimuli in its internal and (or) external environment. Teleological errors are less likely when discussing learned behaviors because, in those instances, the animal often does understand the outcome or result of its behaviors. Teleology is all about the assumptions we make when communicating information regarding animal behavior. Like anthropomorphism, teleology is frowned on when misused in scientific gatherings.

Some Basic Principles of Animal Behavior

1. Behavior is a response to changes in the internal and (or) external environment of the animal.

Such changes in the external and internal environments are referred to as 'stimuli'. Stimuli may serve to arouse (prime) the animal, elicit a response and orient the animal with respect to the stimulus. A strange sound such as the bark of a dog (*C. familiaris*) may arouse a flock of sheep (*O. aries*), elicit an escape response (e.g. running) and orient the escape response away from the sound. An increase in the production of testosterone prior to the breeding season primes male sheep to exhibit aggressive behavior when encountering strange males. The sight of the strange male elicits approach and investigation, which may result in a well-oriented attack (butt).

The frequency with which individuals exhibit certain behavior patterns may be determined by changes in both the internal environment of the animal (e.g. hormonal changes) and the external environment (e.g. stimuli perceived through visual, auditory, olfactory or tactile modalities).

Most internal stimuli cannot be directly perceived and must be measured using relatively sophisticated laboratory equipment. For example, radioimmunoassay techniques are used to measure levels of androgens (e.g. testosterone) circulating in the blood of adult male mammals and birds. In contrast, many external stimuli which elicit behavioral responses (e.g. visual stimuli) are relatively easy for humans to perceive and study, while other external stimuli (e.g. olfactory stimuli) can be studied only with the help of complex instrumentation.

2. Not all stimuli impinging on an organism will elicit a response.

Von Uexküll (1934) was one of the first scientists to emphasize the point that an animal is capable of perceiving only a limited portion of its external environment, that is, its 'umwelt' ('world as perceived'). The sense organs of animals essentially 'filter out' a large amount of potential stimuli. For example, the olfactory capability of humans is very poor compared to most other mammals. We perceive only a fraction of the olfactory (chemical) stimuli perceived by a pet dog (*C. familiaris*), even though we may be exposed to the same chemical stimuli. Conversely, a pet dog has difficulty distinguishing between the colors green, yellow and orange and between red and orange, due to the fact that dogs lack green cones in the retina of their eyes (see also p. 171), one of the three photopigments needed to perceive the full color spectrum available to humans and other primates. This phenomenon is known as 'peripheral filtering' because the sense organs (filtering structures) are part of the peripheral nervous system.

In addition to limitations in stimulus processing imposed by the sense organs themselves, not all stimuli perceived by an animal take on functional significance. This occurs at the level of the central nervous system and is thus referred to as 'central filtering'. For example, even though a herbivorous animal can see or smell a piece of meat, the latter has little or no significance to the animal and essentially will be ignored.

Central filtering may occur as a consequence of learning. Animals may habituate to certain stimuli; that is, they may learn not to respond to certain stimuli as a result of repeated stimulation without a positive or negative consequence (see also p. 51). A stimulus that at first elicits fear responses eventually may elicit no response at all if the stimulus is repeatedly perceived as unimportant or neutral in significance to the individual. For example, motor vehicles passing an animal on the roadside initially may elicit fear or escape responses if the animal is unaccustomed to seeing or hearing them. But, with repeated exposure on a busy highway, vehicles eventually will elicit no response. However, if a vehicle stops on the road by the animal, the stimulus complex has now changed and the individual's fear responses are likely to be renewed. This example demonstrates that habituation can be very situation specific.

In other cases, animals initially may be unresponsive to a certain stimulus until it takes on significance through experience or learning. On our dairy farm, we kept the young calves (*B. taurus*) inside a barn from birth until they were several months old (often through the winter) and then turned them out in a pasture enclosed by a woven wire fence. On the first day in their new quarters, the calves would kick up their heels and start racing around in the pasture, only to crash headlong into the fence. Needless to say, it did not take long for the calves to learn that fences were significant. Rodents can be trained to press a bar or lever in their cage to get food or water. Initially, the bar in their cage is perceived but has little significance, or is ignored (centrally filtered). The bar takes on special significance only when it becomes associated with something to eat or drink.

3. Behavioral responses are not always predictable.

In a physics class many years ago, I learned that the laws governing relationships between certain stimuli and associated responses were very predictable. If a certain amount of force was applied to a lever, a predictable amount of weight could be lifted. If you hit a billiard ball precisely in the center with your cue stick, you can predict the direction the ball will travel. Needless to say, that is not the case in animal behavior. Animals often respond differently to stimuli at different times and under different circumstances. The propensity to exhibit sexual and aggressive behaviors changes with the seasons in seasonally breeding species. The likelihood that two individuals will engage in an act of aggression may depend on the presence or absence of other animals ('third-party effect'). Your dog (*C. familiaris*) may obey you, but not your friend. Any animal trainer will attest to the fact that animals sometimes will do just as they please. It is certainly a truism that animals do not behave like machines.

We can conclude that behavior is more than a simple response to a stimulus or set of stimuli. From the standpoint of predicting behavior, the proper question to ask is not 'What response is elicited by a given stimulus', but rather 'What is the probability that a certain response is elicited by a given stimulus under a given set of circumstances?' One of the goals of animal behavior research should be to determine such probabilities.

Behavioral Variability

The last principle cited emphasizes that animals are not robots that always behave in predictable ways. It would be useful at this point to discuss the subject of variability in greater detail. One way to approach this subject is to compare variation between individuals and variation in the behavior of a single individual over time (within-individual variation).

Between-individual variation

Not all individuals of a population respond in the same way to a given stimulus. This is not surprising, knowing that individuals differ in the behavioral-relevant genes they inherit. For example, the fruitless ('*fru*') gene in male fruit flies (*Drosophila melanogaster*) controls their libido and mating performance (Ryner *et al.*, 1996; Baker *et al.*, 2001). It is a high-level regulatory gene (i.e. it can turn other genes on or off) that equips specific centers in the brain to coordinate male courtship behavior. Males with severe mutations of this gene lose their motivation to follow other flies, play courtship songs with their wings or attempt mating (see also p. 22).

Animals may have similar genotypes but respond differently to stimuli based on different experiences. One of my graduate students, Dave Lyons, conducted a research project with twin dairy goats (*Capra hircus*) in which humans reared one individual of each twin set and the other twin was reared by its mother (Lyons *et al.*, 1988a). Once the goats had become adults, he compared the 'hand-reared' and mother-reared goats with respect to their relative degree of tameness toward humans and how they responded to being milked for the first time after giving birth to their own young. The hand-reared goats readily approached and physically contacted humans, whereas the mother-reared goats typically would keep their distance from humans (see also p. 230). When milked for the first time, the hand-reared goats exhibited less avoidance of the human milker, less kicking at the milking machine placed on their udder and more complete milk ejection.

Differences between individuals in both genotype and experience make it even more likely that different responses will be exhibited to certain stimuli. A strange dog (*C. familiaris*) is likely to respond differently toward you than your own dog, even though you may provide the same stimulus for both animals. Degree of familiarity (with you) gives one animal a different 'base' on which to respond. Your pet has an 'expectation' concerning you and your behavior, while an unfamiliar dog does not.

Within-individual variation

A single individual may respond differently to a given stimulus at different times. Three factors can have a major role in within-individual variation. Motivation often determines whether or not an individual responds to a stimulus in the first place and, if so, the intensity of that response. A hungry animal is likely to respond differently to a food item than the same animal that has just had a meal. The hunger drive is influenced by blood-sugar levels acting on the central nervous system, particularly the hypothalamus of the brain.

The context in which the stimulus is presented also may influence the response exhibited. Other competing stimuli may be vying for the animal's attention. Give a hungry dog (*C. familiaris*) a bone and it will probably lie down and chew on it. However, if another dog or a cat (*F. domestica*) appears, the dog may stop feeding and chase or attack the intruder. Or if the animal is sick or injured, it may refuse the bone entirely.

Learning can also affect how an animal responds to a given stimulus. If the dog in the example above learns that the intruding animal is not interested in its bone, it may ignore the intruder and continue eating. Two unfamiliar males of similar size meeting for the first time may engage in aggressive behavior. The loser of this battle likely will behave differently toward the winner the next time they meet, and vice versa. A dominant–subordinate relationship is established (i.e. learned), which changes how they subsequently respond to one another.

2   Interaction of Heredity and Environment

My aim in writing this book is to provide an understanding of the basic principles of animal behavior and how these principles can be applied in addressing practical issues of importance to persons responsible for the breeding and care of captive wild and domestic animals. Of all the principles discussed in the book, there is none more important than the fact that all behavior develops by an interaction of genes and environment. This chapter will give some background history on how this concept became so important in our study of animal behavior and some examples of how these interactions work in behavioral development and evolutionary time. The term 'ethology' is used here for the first time in this book. It is a term often used to describe the science of animal behavior that deals with the study of: (i) causation, including external (exogenous) stimulation and internal (endogenous) physiological mechanisms and states; (ii) development, including maturation (ontogeny) and experiential and genetic influences on behavior; (iii) ecology, including physiological and learned adaptations to the environment; and (iv) evolution, including the origins and modifications of behavior through evolutionary processes. Ethological studies are conducted both in the laboratory and in nature.

Innate (Instinctive) Behaviors

What do you think of when you hear the word 'instinct' in the context of animal behavior? Your first thought is probably a behavior that occurs naturally without being influenced by learning or experience. The concept of instinct originated before the time of Christ. It was assumed that animal behavior was governed by instincts and human behavior by reasoning. The 'instinct concept' gained a central position in scientific thought after the writings of Darwin. By this time, evolutionists had uncritically accepted the assumption that all behavior could be explained by instinct or reasoning. Still, the validity of the instinct concept had not been established scientifically. It was used because it filled a need in the theoretical system (i.e. it separated animals from humans). By the 1920s and 1930s, the term instinct was being badly abused and overworked and a critical debate ensued about what actually constituted an instinct.

The early ethologists and many casual observers of animal behavior preceding them attempted to define instinctive behaviors. Three of the criteria used were:

- Behavior that was stereotyped for the species and constant in form.
- Behavior that developed normally for animals reared in isolation.
- Behavior that developed despite a 'lack of practice'.

In essence, these criteria implied that instincts must arise independently of the animal's experience and must be distinct from acquired or learned behavior. In other words, instincts were genetically determined.

With time, these criteria became more and more subject to criticism. Some of the criticisms were as follows:

The first criterion that species-typical stereotyped behaviors were instinctive was criticized on the grounds that stereotypy, by itself, did not provide enough information about how a particular behavior developed. The early ethologists were awakened to the realization that stereotypy could be achieved not only if all the individuals in a population of a particular species (or sex within that species) inherited the same specific genes, but also if they learned the same things during early life or had the same experiences. To illustrate this point, I am sure you have noticed that livestock tend to spend their time in same-species groups. When sheep (*O. aries*) and goats (*C. hircus*) or cows (*Bos*) and horses (*Equus caballus*) are housed in the same field, typically it will be found that the species are segregated and not randomly mixed. This pattern of behavior is quite consistent and appears to satisfy this first criterion for an instinct very nicely. Can we then conclude that species affinity is an instinct?

One of my graduate students, Kim Tomlinson, and I put this question to the test by rearing young sheep (*O. aries*) and goats (*C. hircus*) together in isolated pairs from birth and subsequently testing them for their species preferences (Tomlinson and Price, 1980). We found that sheep reared with goats preferred the company of goats when given a choice and goats reared with sheep preferred the company of sheep (Fig. 2.1). Our control pairs (sheep reared with sheep and goats reared with goats) preferred to be in the company of members of their own species. We concluded that species affinities in these two species were learned early in life. This behavior is stereotyped for the species and consistent in form because, in the natural world in which these behaviors evolved, sheep were always reared by sheep mothers and goats were always reared by goat mothers. Because there is so little chance for error in the

Fig. 2.1. Male goat reared with sheep prefers the company of sheep. A pen of goats was approximately 25 m to the left in the photo (Tomlinson and Price, 1980).

development of species affinities, natural selection has never favored a gene-based developmental process.

The second criterion mentioned above, that a behavior is an instinct if it develops normally for animals reared in isolation from conspecifics (i.e. members of their own species), also seems logical on the surface but is beset with a number of practical problems that makes its use as a criterion for instinctive behavior highly questionable. One problem is that in designing an isolation or deprivation experiment, it is often difficult to determine what experiences might be important in the development of a given behavior. For example, if you were interested in determining whether the maternal behaviors of the domestic cat (*F. domestica*) were instinctive, what kind of deprivation experiment would you conduct? From what would you isolate your animal subjects? First, you would probably want to isolate your subjects from their mother, starting at birth because you do not want them to learn anything from being mothered. This, of course, would entail hand-rearing your subjects. Second, you would want to raise your subjects in isolation from their littermates and all other cats because the sensory experiences gained from being with other cats could have some effect on the subjects' subsequent ability to mother their own kittens. Third, you might want to isolate your subjects from self-stimulation, such as licking and other forms of body care. The stimulus feedback gained from these experiences could be related to the development of similar responses directed at their own offspring.

If you successfully deprive your cat subjects of all the above experiences and you observe no effect on their subsequent maternal care (relative to the normally-reared control subjects), you can conclude that the experiences withheld are not important in the development of maternal behaviors in this species. However, you still cannot be totally sure that experience is not important (i.e. that the behavior is instinctive), since other relevant critical experiences may *not* have been withheld. This again illustrates why knowing the biology of the species and its behaviors is so important!

What if a positive result is obtained in your deprivation experiment? What if maternal behavior *is* affected? Can you conclude that the experiences withheld are essential to the development of your subjects' behavior? What about the side effects of the deprivation? Could the general health of your animal subjects have been affected by all the experiences you withheld from your experimental subjects? Was your testing protocol appropriate to measure the behaviors in question and draw valid conclusions? Because of some of the difficulties associated with depriving animals of certain basic experiences, the isolation experiment is most useful for helping us determine if *specific* experiences are *not* important in the development of behavior.

Questions regarding the third criterion that instincts developed without the opportunity for practice centered on the topic of what constituted practice for a given behavior and when that practice occurred during development. It was thought originally that behaviors displayed soon after birth (or hatching) were instinctive because the animal would not have had the opportunity to practice these behaviors. Associated with this assumption was the idea that behavior did not develop until after birth. We now know that this assumption is incorrect. Many activities of the developing fetus during prenatal life prior to birth can be linked to behaviors exhibited by the young neonate after it is born. For example, mammalian fetuses display swallowing reflexes in which amniotic fluid is ingested. Human fetuses have been observed sucking their thumb prior to birth. Does it not seem reasonable to surmise that these prenatal events may constitute practice for sucking behaviors displayed at birth? Studies on the development of the pecking response in the young chick (*G. domesticus*) have shown that motor patterns associated with pecking (e.g. head lunging, bill opening) do not appear suddenly at the time of hatching. These, and many other motor patterns, are exhibited by the chick embryo as it develops in the egg (Kuo, 1932; Oppenheim, 1974). Furthermore, hatching would not occur if the chick could not peck its way out of the eggshell.

Chick (*G. domesticus*), mallard duckling (*Anas platyrhyncos*) and bobwhite quail (*Colinus virginianus*) fetuses are capable of perceiving sounds several days prior to hatching and can emit sounds starting at about 3 days prior. On the day before hatching, chicks and ducklings respond to their mother's calls by increasing rates of bill clapping and vocalizations (Gottlieb, 1968). These sensory experiences in the egg prepare them to recognize and respond to species-specific maternal calls immediately on hatching and thereafter. Neonatal responsiveness to maternal calls is important to the survival of chicks and ducklings since the mother leaves the nest as soon as her young have hatched and their ability to follow the mother in thick vegetation may depend, in part, on responding to her calls. Sound repetition rate appears to be a key component in the sound learning process (Gottlieb, 1982). Quail fetuses artificially exposed to maternal calls with sound repetition rates different from what is normal for the species subsequently preferred the altered maternal call (Lickliter and Stoumbos, 1992). These and other studies show that birds can transfer sensory information gleaned in the prenatal environment to events in the post-natal environment.

These examples with higher organisms point out that behavior does not necessarily begin *after* birth or hatching. Our current lack of knowledge of prenatal behavior and its subsequent effect on the developing organism is a handicap in understanding the various ways that 'practice' for certain post-natal behaviors can be attained, and how early in life these events occur.

As knowledge accumulated, it became apparent to the early ethologists that all behaviors developed through an interaction of both genes and environment. Behavior does not develop through the effect of genes or the environment acting alone or independently of one another. The notion that there were instinctive behaviors which developed through gene action alone became increasingly unpopular in scientific circles. Fuller and Thompson (1960, p. 2) emphasized this point when they wrote 'The dichotomy of inherited vs. learned behavior carried to its logical conclusion would define instinctive behavior as that which appeared in the absence of the environment and learned behavior as that which required no organism.'

In addition, people 'abused' the term 'instinct' by trying to use it to explain many behaviors that could not be explained by 'reasoning'. The idea that all behaviors could be explained by either 'instinct' or 'reasoning' gained disfavor and eventually 'instinct' was dropped from the vocabulary of ethologists in favor of the adjective 'innate' ('instinct' is a noun). 'Innate', as in 'innate behavior', is still used today to describe behaviors which are believed to have a relatively strong genetic component associated with their development. Unfortunately, usage of the term 'innate' can discourage studies on behavioral development without explaining how the behavior develops. In addition, 'innate' still carries the implication of non-intelligent, non-sophisticated behavior, in spite of our knowledge that inheritance provides the basis for cognitive development and the acquisition of many highly sophisticated behavior patterns (Gould and Marler, 1987). Recognizing these concerns, the terms 'innate' and 'acquired' in reference to certain behaviors can still serve a useful purpose in discussions of behavioral development if the evidence strongly supports their use. One could reasonably argue that the suckling response of the newborn mammal is *innate* and your dog *acquires* new behaviors when you take it to obedience training. If it were possible to measure the relative contributions of genes versus the environment in the development of various behaviours, there would certainly be many behaviors in which the contributions were nearly equal. It is clear that the right question to ask is 'how does a given behavior develop' rather than 'is it innate or learned'. By so doing, we can avoid errors associated with lumping behaviors into 'black and white' categories when, in reality, the development of all behaviors falls somewhere on a continuum.

These concepts were reinforced when scientific reports emerged suggesting that genes themselves could determine *what* can be learned and the extent to which experience can modify behavior. Wecker (1963) provided support for these ideas by studying habitat selection in wild and semi-domestic prairie deermice (*Peromyscus maniculatus bairdii*). He built a long, outdoor enclosure, half of which was in the woods and half in an open field. Habitat preference was determined by where each of his experimental subjects spent most of its time. Wecker found that no matter where his wild deermice were reared, either in a field, woods or in laboratory cages, they showed clear preferences for the field portion of his enclosure (Fig. 2.2). Mice randomly taken from a population of prairie deermice which had been bred and

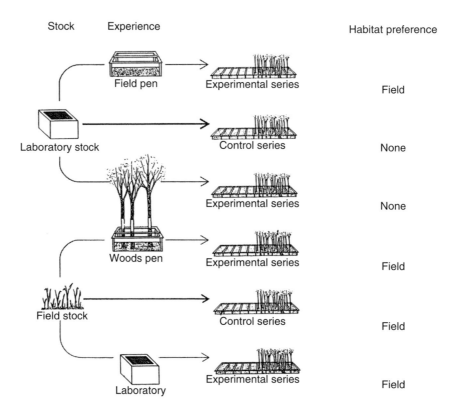

Fig. 2.2. Prairie deermice born to animals caught in the wild prefer a field habitat no matter where they are reared. In contrast, a laboratory stock of prairie deermice about 15 generations removed from the wild prefers the field environment only if given early experience in a field habitat (Wecker, 1963).

maintained in a laboratory environment (i.e. cages) for about 15 generations exhibited a different result. Laboratory-reared, semi-domestic mice showed no preference for either the field or woodland portion of Wecker's enclosure. When these semi-domesticated mice were reared in a separate small enclosure in the woods, the same result was obtained. Only semi-domesticated mice which were reared in a small field enclosure showed a preference for the field half of the test enclosure. Wecker concluded that after 15 generations of being housed in a laboratory environment, the semi-domesticated stock of mice had lost their 'natural' preference for field habitat. However, they still possessed a predisposition to learn from early experience in a field environment, but not the woodland environment for which the species had never been selected. The results point to the fact that preference for field environments in this species is normally under genetic control and that natural selection for habitat preference is relaxed in the laboratory environment to the point that, after 15 generations, early experience in the ancestral environment is necessary for the species' 'habitat' genes to be expressed. The finding by Wecker and others that genes can determine what can be learned and the extent to which experience can modify behavior put the final 'nail in the coffin' for usage of the term 'instinct' in scientific discussions.

3 Behavior Genetics

How Genes Influence Behavior

Qualitative nature of the response

Behaviors can be described in both qualitative and quantitative terms. The qualitative aspect refers to the motor patterns that constitute a behavior (e.g. running, climbing, pecking, barking, etc.). Genes heavily influence the structure (anatomy) of the organism and thus create structural biases for the kinds of motor patterns animals exhibit in response to specific stimuli. For example, domestic cattle (*Bos*) and sheep (*O. aries*) possess horns and exhibit aggression by butting with their heads. Horses (*E. caballus*) have no horns and show aggression by biting, kicking with their hind feet and rearing up and striking their opponent with their front feet.

Other evidence for genetic influences on the qualitative nature of behavioral responses can be illustrated by the behavior of interspecies hybrids. Closely related species will often interbreed in captivity. If two closely related species exhibit different behaviors when reared in the same environment, we can study the behavior of their hybrid offspring to obtain clues regarding the genetic influence on behaviors of interest. Dilger (1962) noted in his studies on lovebirds (*Agapornis*) that Fischer's lovebirds (*A. fischeri*) always carried nesting material in their bills, whereas peach-faced lovebirds (*A. roseicollis*) carried nesting material in their bills only 3% of the time. Instead, they typically used their bills to tuck nesting material under their lower back and rump feathers before flying off to the nest. Fortunately, the two species interbred in Dilger's laboratory, which permitted him to observe the behavior of their hybrid offspring. As adults, the hybrids initially carried nesting material in their bills only 6% of the time; 94% of the time they tucked the material under their feathers, like their peach-faced lovebird parent. However, the hybrids were very inefficient at tucking (compared to their peach-faced parent) and the nest material was frequently lost when they flew to their nest boxes. Interestingly, 2 months later, they were carrying nest material in their bills 41% of the time. After 3 years, they almost always used their bills for this purpose. Since the hybrids' first inclination was to carry nesting material tucked under their feathers, one could argue that the genes responsible for 'tucking' inherited from their peach-faced parent had a greater influence over the birds' initial behavior than the genes responsible for 'bill carrying' inherited from their Fischer's parent. Perhaps tucking is under greater genetic control than bill carrying because it is more difficult to learn. Could it be that the tucking response was poorly developed in the hybrids because they received tucking genes from only one parent?

Quantitative nature of the response

While the *qualitative* nature of the response refers to the kinds of motor patterns exhibited in response to a given stimulus, the *quantitative* aspect refers to how frequently a certain motor pattern is exhibited. Response frequency is one way to measure the motivational state of an animal. Differences in response frequency can be observed both between and within animal species. For example, domestic goats (*C. hircus*) are more predisposed to climbing steep inclines than domestic sheep (*O. aries*). Figure 3.1, which shows goats feeding on fruit in an argan tree in southwest Morocco, attests to the climbing skills of this species. Since both goats and sheep display the same basic motor patterns when climbing steep inclines, one can argue that the difference between goats and sheep is quantitative rather than qualitative. Likewise, domestic goats are more predisposed to rearing up on their hind legs in social interactions than domestic sheep (Fig. 3.2). The Rambouillet breed of domestic sheep exhibits tighter flocking behavior than the Suffolk and Hampshire breeds. Wild Norway rats (*Rattus norvegicus*) are more predisposed to dig burrows than domestic Norway rats. Retriever breeds of dogs (*C. familiaris*) are more predisposed to retrieve objects than individuals of most other breeds and response frequencies can vary between individuals. Response frequencies for a given individual are often different

Fig. 3.1. Domestic goats feeding on fruit in an argan tree (top and right side) near Agadir, Morocco (courtesy of Adel Kader).

Fig. 3.2. A young goat rears up in a display to a sheep penmate. Commercial breeds of sheep do not exhibit this behavior.

at different times of the day or year, or in different contexts (e.g. when hungry versus satiated).

One way the genes can mediate response frequency is through response thresholds. Response thresholds refer to the degree of stimulation necessary to elicit a given response. We say that goats (*C. hircus*) have a relatively *low* threshold for climbing because the intensity of stimulation needed to elicit a climbing response is relatively low. In contrast, domestic sheep (*O. aries*) have a *higher* threshold for climbing since a higher degree of stimulation is needed to initiate climbing in this species. If we speak of an animal with a low threshold for aggressive behavior, we are saying that it takes relatively little stimulation for this individual to initiate an attack on another animal. Such an animal will likely exhibit aggressive behavior more frequently than one with a relatively high threshold for aggressive behavior simply because it responds to both low and high levels of stimulation, whereas a conspecific with a high threshold responds to high-level stimulation only. An animal that is very hungry will likely have a low threshold of response to the sight or smell of appropriate food items. A satiated animal has a high threshold of response to food and may feed only on its most preferred items.

Neurophysiological Approaches to the Study of Behavior Genetics

Our current ability to study gene mutations or eliminate ('knockout') certain genes, or transfer genes from one organism to another will undoubtedly help us to understand the role of single genes in the development and expression of behavior. Genes can affect response thresholds by influencing the sensitivity of neural structures to both internal and external stimulation. When a female mammal is about to give birth, the posterior pituitary gland of the hypothalamus of her brain secretes greater than normal quantities of the hormone oxytocin. Oxytocin facilitates the process of labor (i.e. birth of young), milk ejection (i.e. milk release from the udder in response to suckling by the young neonate) and other maternal behaviors. Mutation of the *Peg*3 and *Mest* genes in domestic mice (*Mus musculus domesticus*, hereafter referred to as *M. musculus*) reduces the number of oxytocin-producing neurons in the hypothalamus to about two-thirds of those found in non-mutant mice, thus causing a reduction in the frequency with which important maternal behaviors are exhibited (Li *et al.*, 1999). Mutant mothers averaged eight times longer than non-mutant mothers to build a nest, 11 times longer to retrieve young displaced from the nest and they frequently failed to crouch over their newborn pups to keep them warm. In addition, the young of mutant mothers gained weight at a significantly slower rate.

A similar deficiency in maternal care was observed in domestic mice (*M. musculus*) whose *fos*B gene had been deactivated by a 'knockout' technique (Brown *et al.*, 1996). Affected mothers showed a dramatic neglect of their offspring but exhibited normal milk production, progesterone and oxytocin levels and normal olfactory capabilities. In one test of maternal behavior, the researchers removed pups from the nest and measured how long it took the mothers to retrieve their young back into the nest. All normal mothers retrieved all of their young within 4 min. Nine of ten knockout mother mice took more than 16 min to retrieve the first pup and only one mother retrieved all of her pups in 20 min. Which of the two groups exhibited the lowest response threshold for pup retrieval? The normal mothers were more sensitive to the absence of young from the nest, thus exhibiting the lower response threshold.

Further study revealed that when normal mothers were exposed to mouse pups, the production of the *fos*B protein was turned on in the preoptic area of the hypothalamus of the brain. This did not happen when the knockout mothers were exposed to pups. It was concluded that the preoptic area plays a central role in regulating maternal behavior in this species and that the *fos*B gene acts specifically on neurons in this structure.

A recent study by Stowers *et al.* (2002) has shown that a single gene in mice (*M. musculus*) determines their ability to identify the sex of strange mice using olfactory cues, and thus controls gender-specific behavioral responses (aggressive and sexual behaviors). The gene encodes the protein TRP2, which is expressed only in the nerve cells of the olfactory-sensitive vomeronasal organ (VNO) (see p. 178–179 for more information on the VNO). Sensory activation of the VNO neurons is abolished in the absence of TRP2. Knockout males (males with no copies of the gene and no TRP2) are indiscriminative in their mating behaviors; they attempt to mate with strange males, as well as females. Normal males with one or two copies of the critical gene attack males and attempt mating with females.

The fruit fly (*D. melanogaster*), used for decades in laboratory studies of genetics, has between 13,000 and 14,000 genes. Osborne *et al.* (1997) reports that one of those genes, '*for*' (or dg2), codes for a protein that helps to relay chemical messages inside cells, which determines how actively fruit flies engage in foraging. Their interest in this subject began when they noted that 70% of the fruit flies collected from local Toronto (Canada) orchards engaged in active foraging, while the remaining flies were relatively inactive. They called the former 'rovers' and the latter 'sitters'. It turns out that a mutation at the *for* locus results in lower than normal levels of one of the enzymes known as cyclic GMP-dependent protein kinases, or PKGs. Reduced PKG enzymatic activity is responsible for the sitter behavior in fruit flies. This effect was confirmed by inserting 'sitter' DNA into rover flies and then removing it. When the sitter DNA was present in the genome, the flies were sitters and when it was removed, the flies reverted back to being rovers. It is believed that PKG activity affects the excitability of nerve cells. Higher levels of the enzyme may cause nerve cells involved in sensing food or in controlling foraging movements to fire more readily, thus increasing foraging activity.

Neurogeneticists claim that we should not expect to find many unique genes for behaviors, considering that behaviors are likely to be shaped by genes that code for proteins with broad activities, affecting multiple traits. Generic messenger proteins such as PKGs and the cyclic AMP-dependent protein kinases are classic examples. As a case in point, the latter has a role in learning and memory in both fruit flies and mammals. Organisms may use the same gene in different cells at different times and in combination with other genes to generate their behavioral repertoire.

Scientists are especially interested when a single gene is discovered with major effects on a behavior or complex of behaviors. There are approximately 25 genes which control the major aspects of reproduction in the fruit fly, *D. melanogaster*. The 'double sex' gene oversees the building of the reproductive system and other essential structures. But it is the '*fru*' gene that controls the sexual motivation, courtship and mating behaviors of males (Ryner *et al.*, 1996; Baker *et al.*, 2001). It acts as a 'master regulator' gene. *Fru* does not affect the sexual behaviors of male fruit flies directly but controls a battery of genes that do. In some cases, *fru* merely organizes signals sent to other genes. I find that very interesting!

Animal Breeding Techniques Used in the Study of Behavior Genetics

Three animal breeding techniques used to study behavior genetics include: (i) breeding to identify major gene effects; (ii) hybridization experiments; and (iii) artificial selection experiments.

Breeding to identify major gene effects

One of the classic examples of major gene effects on behavior involves the behavior of domestic honeybees (*Apis mellifera*) when their hive becomes infected with American foul-brood (AFB). AFB is a disease (*Bacillus larvae*) affecting certain strains of domestic bees. Some bee strains are resistant to this disease because if a developing larva dies, worker bees uncap the cell containing the larva and remove the corpse from the hive. These resistant strains are referred to in the bee industry as 'hygienic'. Strains susceptible to the disease ('unhygienic') leave the diseased larvae in their cells. They neither uncap the cells promptly nor remove the dead corpses from the hive (Rothenbuhler, 1964a). Consequently, pathogens infect the corpses and the disease spreads throughout the hive.

Rothenbuhler (1964b) wondered if colony resistance to AFB was heritable, so he cross-mated hygienic and unhygienic strains. He found that all of the F1 hybrid colonies were unhygienic; they neither uncapped cells of diseased larvae nor removed them from the hive when the experimenter performed the uncapping. He then mated (backcrossed) F1 hybrid bees to hygienic bees and produced 29 colonies segregated by behavior into four classes.

- Eight unhygienic colonies. Did not uncap cells of diseased larvae. Would not remove larvae from cells if uncapped by experimenter.
- Six hygienic colonies. Uncapped cells of diseased larvae and removed corpses from hive.
- Nine unhygienic colonies. Uncapped cells but did not remove dead larvae.
- Six unhygienic colonies. Did not uncap cells but would remove diseased larvae if cells were uncapped by the experimenter.

Only one of the four classes was hygienic and statistical tests showed that the number of colonies in these four classes did not differ significantly from a 1:1:1:1 ratio. These results suggested two things to Rothenbuhler. First, uncapping and removing diseased larvae are controlled by separate genes at two different gene loci (points on a chromosome). If only one gene locus controlled both uncapping and removal, the colonies would have exhibited either both of these behaviors or neither and there would have been an equal number of hygienic and unhygienic colonies in the backcross generation, rather than the approximately 1:3 hygienic:unhygienic ratio that was obtained. Second, both uncapping and removing diseased larvae are expressed only when their respective recessive genes are paired in the homozygous condition. Otherwise some of the F1 hybrid colonies would have shown either uncapping or removal behaviors, or both, rather than the absence of both of these behaviors. More recent studies (Moritz, 1988; Lapidge *et al.*, 2002) have raised some questions regarding Rothenbuhler's conclusions. For example, Moritz suggests that a second gene for 'removal' may have been operating in Rothenbuhler's hygienic colonies and Lapidge *et al.* postulate that hygienic behavior is a quantitative trait with up to seven gene loci influencing its expression. However, there were some major differences in the methodology used in the latter study, which makes direct comparison with Rothenbuhler's data very difficult. One concern raised was that hygienic bees could have influenced the behavior of unhygienic bees. However, Arathi *et al.* (2006) observed infected bee colonies comprised of both hygienic and unhygienic individuals and found that the bees behaved true to their genetics. Specifically, the unhygienic bees did not learn from and were not stimulated by the activity of the hygienic bees, and vice versa.

Major gene effects on behavior are often found as pleiotropic effects of genes influencing anatomical or physiological characteristics. Pleiotropy refers to a gene's effect on the expression of multiple traits. Many years ago, I trapped some wild Norway rats (*R. norvegicus*) whose hair color was black instead of the normal brown (agouti) coloration. After taking them into my laboratory, I noticed that the black rats appeared to be less disturbed by my presence than their brown counterparts. I searched the literature for information on the black pelage color in Norway rats and found that black hair color was determined by a single recessive gene. I also found that Clyde Keeler previously had reported that black Norway rats were more docile than the normal agouti-colored rats (Keeler, 1942). Unfortunately, Keeler had not controlled for the 'genetic background' of the black and brown populations in his study (i.e. the black and agouti rats had not been drawn from related individuals). Since it was known that a single gene (allele) was responsible for black hair coloration in this species, and because it was recessive, an animal had to carry two alleles for this gene (i.e. homozygous condition) to be expressed. So, I mated black rats to agouti rats to get offspring with both a black gene and an agouti gene at this hair-color gene locus (point on a chromosome). When these animals matured, I mated these hybrids to one another to obtain both black and agouti offspring in the same litters (Fig. 3.3) in an attempt to control for the genetic background. Then, one of my graduate students, Celia Cottle, and I exposed the black and agouti offspring of the hybrid rats to four behavioral tests which, in my previous research, had clearly revealed behavioral differences between wild and domestic rats reared under identical laboratory conditions. The black and agouti rats did not differ in behaviors exhibited in three of these four tests (open-field test, platform jumping and response to a novel food item) but there were clear differences in the fourth test, which measured ease of handling (Cottle and Price, 1987). The black rats exhibited fewer escape and aggressive responses than the agouti subjects when touched, stroked and handled (Table 3.1). Interestingly, Hayssen (1997) has recently obtained the same result with black and agouti deermice (*P. maniculatus*). Biochemical studies have shown that the behavior differences between black and agouti mammals are related to the regulation and effects of melanocyte-stimulating hormone (MSH). The agouti allele at this coat-color locus interferes with MSH at its receptor on pigment cells and may interfere with melano-cortin receptors in neural tissues.

The black (non-agouti) coat-color allele in rats and mice has been called a major 'domestication' gene. Over 70% of approximately 140 inbred mouse strains and over 80% of approximately 50 inbred rat strains (which have been examined) are homozygous for the black coat-color allele (Festing, 1979, 1989; Staats, 1981). This could be one of the reasons domestic mice and rats are relatively easy to handle and manage compared to their wild counterparts.

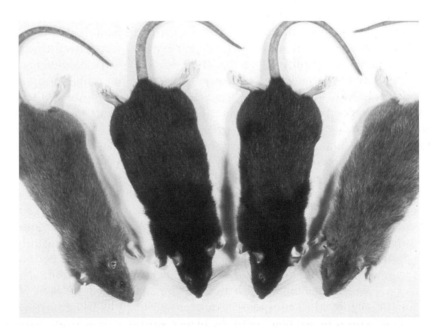

Fig. 3.3. Contrasting pelage color of black (center) and agouti (outside) Norway rats. Black coat color is conferred by a single gene in the homozygous recessive condition (Cottle and Price, 1987).

Table 3.1. Percentage of agouti (*N* = 62) and black (*N* = 28) Norway rats exhibiting the behaviors listed on the first day of handling tests (from Cottle and Price, 1987).

Behavior	Agouti (A_)	Black (aa)
Could be touched	64%	96%
Could be stroked	32%	89%
Attacked/attempted to bite	59%	25%
Jumped	91%	57%

Hybridization experiments

Cross-breeding individuals of different strains or breeds of a species is commonly practiced to improve the vigor and productivity of domestic animals. This process can provide us with certain clues regarding the inheritance of behavior, as we discussed in Dilger's experiments with lovebirds and their techniques for carrying nesting material (p. 18).

There are three possible outcomes to a hybridization experiment. One is for the F1 hybrids to be intermediate to their two parent strains with respect to some traits. For example, wild turkeys (*Meleagris gallopavo*) are typically more reactive to humans (i.e. exhibit stronger avoidance responses) than their domestic counterparts. Leopold (1944) found that hybrid turkeys obtained from cross-breeding wild and domestic stock were intermediate to the two parent strains in reactivity to humans. This result suggests that there are a number of genes controlling the reactivity of turkeys and they are expressed in an 'additive' or accumulative manner, rather than one or a very few genes dominating or masking the expression of other genes.

A second possible outcome is for the F1 hybrids to resemble one parent strain and not the other. This result is obtained when a gene(s) of one breed dominates or masks the corresponding allele(s) of the other breed. Genetic dominance can be expressed between alleles at a given gene locus (point on a chromosome) or between gene loci. In the latter case, the dominance is referred to as 'epistasis'. In Rothenbuhler's study with honeybees (p. 22–23), the alleles for failure to uncap and remove corpses from the hive were dominant over their corresponding hygienic alleles, which is why the hygienic alleles had to be present in homozygous recessive form for hygienic behavior to be expressed. It also explains why the F1 hybrids in Rothenbuhler's experiment were all unhygienic.

A third possible outcome is when the hybrids score higher than both parent strains on some measurable trait or behavioral performance scale. We refer to this result as demonstrating 'heterosis', also called 'hybrid vigor' or 'overdominance'. (Animal scientists sometimes claim heterosis when hybrids score significantly higher than the mean of the two parent strains averaged together.) Collins (1964) observed heterosis in a test of learning by domestic mice (*M. musculus*). Five inbred strains and the 20 possible hybrid combinations between them were tested for their ability to learn a conditioned avoidance response. Fourteen of the 20 hybrid groups exhibited better learning scores, on average, than the higher-scoring parent strain from which they were bred. The expression of heterosis in F1 hybrids indicates that the parent strains from which the hybrids were derived are highly inbred and are 'suffering' from 'inbreeding depression'.

Inbreeding depression is caused by the expression of deleterious recessive genes that would be masked by dominant alleles in an outbred population. Most deleterious genes are recessive and are expressed only in the homozygous condition. The interbreeding of close relatives increases the probability that, by chance, deleterious recessive alleles will be paired up at various gene loci. When different inbred strains of animals are cross-mated, many of those pairs of deleterious genes are broken up and *normal* vigor is restored. Inbreeding in some Holstein dairy cow (*B. taurus*) herds is estimated at 5–10%. One report claims that for every 1% of inbreeding in Holstein cows, the mortality rate will increase by 2% and lifetime milk production will decline by 27 kg. Breeders of thoroughbred race horses (*E. caballus*) over the past two centuries have selected for traits like speed, stamina and conformation and have largely ignored the genetic cost of inbreeding. Consequently, thoroughbreds are afflicted with a host of physical disorders. For example, more than 80% of yearlings show signs of cartilage degeneration at their joints and more than 95% have some degree of 'recurrent laryngeal neuropathy', a partial paralysis of the larynx that affects breathing. Ralls *et al.* (1979) examined the effects of inbreeding on juvenile mortality in captive wild ungulates. Inbred young experienced higher mortality than non-inbred young in 15 of 16 species studied. A larger percentage of young died when a female was mated to a related male than when she was mated with an unrelated male.

Artificial selection

In artificial selection, people determine which animals will be used as breeding stock for the next generation. Morphological, physiological and behavioral differences between the various strains and breeds of our common domestic animals attest to the effect of artificial selection. Consider the structure of our various breeds of dogs (*C. familiaris*) (Fig. 3.4). A strain of female mice (*M. musculus*) selected for large litter

Fig. 3.4. Size of various dog breeds (drawn to scale) subjected to artificial selection (courtesy of *Science*).

size for over 100 generations averaged over twice as many pups (at birth) than an unselected control line (21 versus 10, respectively) (Rauw, 2006).

There are several important distinctions between artificial and natural selection. First, artificial selection occurs prior to reproduction, whereas the influence of natural selection can be determined only after the animal's reproductive lifespan has ended and is measured by the number of reproducing offspring it has left to the next generation. Second, artificial selection is often applied to just a single trait, whereas natural selection acts simultaneously on all of the phenotypic characteristics (traits) of the animal.

Natural selection may determine the outcome of artificial selection. We can select our favored animals to provide progeny for the next generation but these individuals may fail to survive to breeding age, fail to mate or fail to produce viable offspring, due to factors beyond our control. We often artificially select captive animals for traits that are selected against by natural selection in the wild. For example, we have selected many of our production animal stocks for rapid growth rate and large body size. This selection has been at the expense of speed and agility. Speed and agility are relatively unimportant in captivity, where animals are assured food and water and are generally protected from predators. In captivity, phenotypic 'deviants' are often favored and propagated for their beauty (e.g. albinos, multi-colored animals) and novelty (e.g. dwarf strains of livestock, unusually feathered birds), whereas in nature, these same traits could confer greater susceptibility to predation.

Some of the best examples of artificial selection for behavior have been conducted on dogs, chickens, rats, mice, fruit flies and honeybees. For example, Page *et al.* (1995), working with domestic honeybees (*A. mellifera*), found a sixfold difference in pollen hoarding after only five generations of selection among colonies for high and low levels of stored pollen. Queen bees in the high line produced worker bees that specialized in pollen foraging, while low-line queen bees produced workers that specialized in collecting nectar (stored as honey). Among other studies, chickens (*G. domesticus*) have been selected for aggressiveness and mating ability. Siegel (1972) selected male chickens for high and low sexual performance (mating ability). At 7–8 months of age, males were individually exposed to a small female flock for 10min, during which time the number of completed matings were recorded. This was repeated eight times. In each generation, the ten males that obtained the most matings and the ten males that had the fewest matings were selected to serve as sires for the next generation. Males were chosen randomly in an unselected control population. Each selected male was mated to four females selected randomly from their own selection line and tested for eight 10min trials. Table 3.2 shows the wide divergence of the high and low sexual performance lines after 20 generations of selection. In fact, the sexual activity of the low performance line was so deficient in the later generations that artificial insemination had to be used to keep the line going.

Table 3.2. Mean (± SE) total number of mounts and completed matings in male chickens after 20 generations of artificial selection for sexual performance (from Bernon and Siegel, 1983).

Selection line	Mounts	Completed matings
High	31.2 ± 1.9	24.4 ± 1.4
Control	7.2 ± 0.8	5.2 ± 0.6
Low	0.5 ± 0.1	0.3 ± 0.1

These and many other selection experiments beg the question of what artificially selected populations can tell us about the relative roles of the genes and environment in the development of behavior? Six things come to mind; there may be others.

Heritability

From the response to artificial selection experiments, such as Siegel's work with chickens discussed above, we are able to obtain estimates of heritability, which very simply is the proportion of phenotypic variability due to genetic variation as opposed to variation from environmental factors. Traits with relatively high heritability estimates theoretically are easier to select for than traits with low heritability. Heritability estimates have been obtained for a host of behavioral traits, such as learning ability in rodents and sexual performance in livestock. For example, Pérez-Guisado et al. (2006) reported a heritability of 0.20 for dominant-aggressive behavior in 7-week-old English Cocker Spaniel puppies, indicating that 20% of the variability in dominant-aggressive behavior at that age can be attributed to genetic factors.

Mode of inheritance

Cross-breeding artificially selected populations (e.g. high and low lines of some trait) can help us understand whether the genes influencing the trait in question generally show intermediate inheritance (i.e. the progeny score intermediate to the two parent lines), dominance (i.e. the progeny resemble one of the two lines) or heterosis (i.e. the progeny perform higher than both parent lines), as discussed in the previous section on hybridization experiments (p. 25). In Siegel's artificial selection experiment with chickens just discussed, hybrid progeny of matings between high and low sexual performance lines were intermediate to the two parent stocks.

Maternal effects

Reciprocal cross-breeding (i.e. mating females of two breeds, strains or artificially selected lines of a species to males of the opposite breed, strain or line) will sometimes result in two F1 hybrid progeny groups which exhibit behavioral differences similar to the behavioral differences of the mothers (females) in the two populations. If both males and females in each hybrid group exhibit the behavior of their mother, we can suspect a maternal effect due to either the prenatal and post-natal environments, or both. No evidence for a maternal effect was obtained in Siegel's selection experiment with chickens; the progeny of reciprocal crosses scored the same, regardless of the line of the mother. In contrast, Scott and Fuller (1965) found a maternal effect when two dog breeds, Cocker Spaniels and Basenjis (*C. familiaris*), were reciprocally cross-bred. The F1 hybrids were compared on a number of behavioral traits. In most tests, the hybrids did not differ, but it was noted that the hybrids behaved the same as the breed of the mother in attractiveness toward humans (Table 3.3). Attraction behavior was scored as the number of times 13- to 15-week-old pups approached or followed the experimenter. The researchers theorized that the tendency to approach and follow

Table 3.3. Attraction to humans exhibited by reciprocal crosses of Cocker Spaniels and Basenjis in the F1 and F2 generations (from Scott and Fuller, 1965).

Parental cross	Attraction behavior score	
	F1 generation	F2 generation
Cocker ♀ × Basenji ♂	Similar to Cockers	Similar to Cockers
Basenji ♀ × Cocker ♂	Similar to Basenjis	Similar to Basenjis

humans was learned from the mother, but they noted that the behavior persisted only from weaning (about 8 weeks) until the pups were 13–15 weeks old.

If a maternal effect is found when two mammalian lines are reciprocally cross-mated, we do not know if the effect is due to the prenatal or post-natal environments, or both. Post-natal effects can be tested by fostering young immediately at birth to mothers of both the same (control for fostering, itself) and different stocks. If the behavioral differences between the two populations of fostered young are the same as the behavioral differences between the two reciprocal crosses, the post-natal maternal environment is likely responsible. If this is not the case, the prenatal environment may be involved to some degree.

Sex linkage

When the two F1 progeny groups produced by reciprocal cross-breeding of breeds, strains or artificially selected lines of a given species exhibit behavioral differences in one sex only, the difference can be attributable to sex linkage. This is in contrast to maternal effects in which the behavioral differences between the two hybrid groups are expressed in both sexes. Some behaviors are transmitted on the sex chromosomes and are thus inherited only from the male or female parent, depending on the species. Broodiness in chickens (*G. domesticus*) (i.e. the tendency to sit on a nest with eggs) is a sex-linked trait expressed only by females. The breed of the sire has a greater effect on the incidence of broodiness in hens than the breed of the mother. Hens of the White Leghorn breed of chickens have been artificially selected for non-broodiness because it can interfere with egg production. The Cornish breed of chickens is broody; it has not been selected for egg production and non-broodiness. If White Leghorn chickens are reciprocally cross-bred with Cornish chickens, the female F1 hybrids with WL sires are non-broody and the female hybrids with Cornish fathers are broody (Table 3.4).

Table 3.4. Phenotype (broodiness) of female offspring from reciprocal crosses of broody (Cornish) and non-broody (White Leghorn) strains of domestic chickens (from Craig, 1981).

Cross of parental stocks	Phenotype of F1 hens
White Leghorn ♂ × Cornish ♀	Non-broody
Cornish ♂ × White Leghorn ♀	Broody

In contrast to male mammals, male chickens possess the like pair of sex chromosomes (ZZ) and females possess the unlike pair (ZW). The Z chromosomes carry most of the genes for broodiness, while the W chromosomes are deficient in broodiness genes. Because F1 daughters always receive a Z chromosome from their fathers and a W chromosome from their mothers, their level of broodiness is determined by the breed of the father. F1 males do not show broodiness, in spite of receiving two Z chromosomes, because they lack the hormone, prolactin, necessary for the expression of broodiness. The lack of broodiness in male chickens illustrates the fact that genes affecting reproductive behaviors may not be expressed in the absence of an appropriate hormonal environment.

Correlated effects of selection

Artificial selection for one trait inadvertently may affect another trait through the process of pleiotropy, discussed on p. 23. Selection for large breast size in domestic turkeys (*M. gallopavo*) has resulted in a dramatic reduction in male mating ability. The US Department of Agriculture indicates that one can expect a 45–50% fertility rate in eggs when females are mated naturally; the fertility rate when females are artificially inseminated is 90–95%. For this reason, artificial insemination is the current standard in the turkey breeding industry. A relaxation of natural selection on mating performance has resulted in male turkeys that often exhibit low libido and, because of their large size, are very clumsy and inept in their attempts at natural mating.

Eysenck and Broadhurst (1964) artificially selected domestic Norway rats (*R. norvegicus*) for high and low defecation rates when placed in a novel environment and subsequently established the Maudsley 'reactive' and 'non-reactive' strains. Their selection resulted in two strains of rats differing not only in novelty-induced defecation rates but also in 24 of 32 different behavioral tests and 19 of 24 different physiological measures. It is likely that some of these correlated effects were influenced either directly or indirectly by strain differences in general emotional reactivity produced by the selection process and manifested in the rats' reactivity to the various testing and sampling techniques.

A study of over 13,000 dogs (*C. familiaris*) representing 31 dog breeds in Sweden revealed that artificial selection of dogs for use in dog exhibitions (shows) generally resulted in greater fearfulness and reduced playfulness, aggressiveness and curiosity in potentially threatening situations (Svartberg, 2006). Selection of dogs for work has resulted in greater playfulness and aggressiveness. Considering that the most popular dog breeds are those which are more playful and sociable toward humans, these relationships suggest that selection for use in dog shows generally produces animals which are often less desirable as pets.

Genotype–environment interactions

A genotype–environment interaction occurs when strains or breeds of animals perform differently, relative to each other, in different environments. Artificially selected populations that differ in behavior can be used to demonstrate such interactions. Figure 3.5 illustrates genotype–environment interactions one might observe when

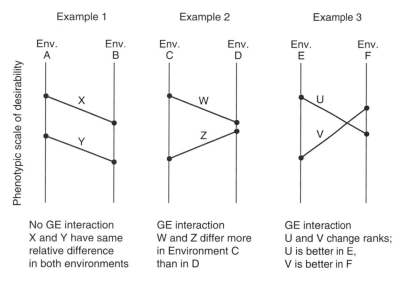

Fig. 3.5. Models illustrating the presence or absence of genotype–environment (GE) interactions. Within each example, mean performances of each stock are shown on scales of phenotypic merit within each of two environments. GE interactions are present when stocks perform differently, relative to each other, in different environments, as in Examples 2 and 3 (Craig, 1981).

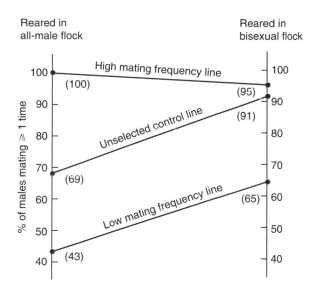

Fig. 3.6. Percentage of male chickens mating when three genetic stocks were reared with and without females. A genotype–environment interaction was found as differences between stocks were greater when males were reared in all-male flocks than with females (Craig, 1981, based on data from Cook and Siegel, 1974).

two behaviourally distinct populations are both reared in two different environments. In the first example, there is no genotype–environment interaction because the relative difference between the two populations remains the same when individuals are reared in the two environments, 'X' and 'Y'. In the second example, the two populations differ in environment 'X' but score the same in environment 'Y'. In the third example, the first population scores higher than the second in environment 'X' but lower in environment 'Y'. In the second and third examples, the difference between the two selected populations varies in the two environments, thus demonstrating genotype–environment interactions. Cook and Siegel (1974) compared the difference in number of completed matings for his high- and low-line male chickens (*G. domesticus*) reared both with females and without females. Figure 3.6 shows that the difference in mating performance scores was greater when the males were reared without females than with females. This genotype–environment interaction suggests that male sexual performance in this species can be influenced by their early social environment and that selection for low levels of performance has reduced the sensitivity of males to female stimuli. In contrast, the sexual performance of the high-line males did not change; they did not require early exposure to females to exhibit their characteristic high levels of sexual performance.

4 Early Experience and Behavioral Development

Introduction

Early behavioral experiences can have lasting effects on the developing organism. For example, early handling by humans has been shown to accelerate growth and physical development in a number of domestic animal species. Hemsworth *et al.* (1981a) found that domestic pigs (*Sus scrofa*) given 'pleasant' handling experiences with humans (mostly tactile stimulation) gained weight at a faster rate than pigs that were exposed to 'unpleasant' handling experiences. Dogs (*C. familiaris*) subjected to early handling as puppies exhibited reduced emotional reactivity (Gazzano *et al.*, 2008). Early experience with different food items can shape adult food preferences. Lynch *et al.* (1992) found that Australian sheep (*O. aries*) were reluctant to eat certain grains unless they had exposure to them early in life, preferably in the presence of their mothers. Lyons *et al.* (1988a) confirmed that hand rearing domestic goats (*C. hircus*) resulted in greater tameness toward humans than normal mother rearing. Animals reared in physical isolation tend to be more hyperactive and aggressive toward species members than individuals reared in social groups. Valenstein and Young (1955) demonstrated that male guinea pigs (*Cavia porcellus*) require early social experience with conspecifics to show normal levels of sexual performance as adults. The aim of this chapter is to discuss some of the major concepts and principles that govern the role of the environment in behavioral development and then to illustrate how experience affects the behavior of animals in several selected areas of interest.

Phenotypic Plasticity

West-Eberhard (2003) stated, 'Phenotypic plasticity is the ability of a single genotype to produce more than one alternative form of morphology, physiological state or behavior in response to environmental conditions'. Inherent in this statement is the fact that phenotypic plasticity confers adaptability. Yet, we know that this does not apply to all behaviors under all conditions. In reality, it is adaptive for some behaviors to be flexible and others to be more rigidly fixed or 'hard wired'. Waddington (1966) introduced the term 'canalization' to describe this phenomenon. Canalization as applied to behavioral development concerns the degree to which the development of a behavioral trait is resistant to modification by experience and the environment. Take, for example, the development of food preferences. Some animals are 'specialists'; their diet is very limited in scope. The Australian koala (*Phascolarctos cinereus*) feeds almost exclusively on eucalyptus leaves, which is adaptive because in Australia eucalyptus trees are widespread and leaves are abundant year round. Each koala inherits the morphology to harvest eucalyptus leaves and the physiology to digest them and have its nutritional needs met by this one plant material. The development of food preferences in this species is highly canalized. Other animals are 'generalists' and eat a wide variety of food items. Bears (*Ursus*), pigs (*S. scrofa*) and chickens (*G. domesticus*), for example, are omnivorous and eat both plant and animal matter. Whereas specialists usually live in environments where a necessary or preferred food item is always present, generalists tend to live in environments where the abundance of specific food resources may change with the seasons, or from one year to the next. Their diet may depend on what is available at the time.

Nest building by birds reflects a highly canalized set of behaviors. Some species nest in trees; some build their nests on the ground. Some nests are made of mud, others of fibrous materials or sticks. Each species is highly predictable in its nest-building behaviors.

The motor patterns used by each species in their aggressive behaviors are highly canalized. Cattle (*Bos*) butt head-to-head and engage in pushing and shoving matches. Horses (*E. caballus*) kick and bite. Domestic cats (*F. domestica*) bite and claw. The development of the motor patterns used in aggressive behavior in each species is highly canalized. It is adaptive for each individual to recognize the opposing party's intentions immediately and to engage each other under the same set of 'rules'. On the other hand, it is also adaptive for individuals to exhibit flexibility, or plasticity, in the use of these aggressive motor patterns. Knowing when to initiate an aggressive inter-action and when to back off can have immediate survival value. This knowledge is typically acquired through experience or, as we will see in Chapter 12, by observing other animals. In Wecker's research (1963) on habitat selection in prairie deermice (*P. m. bairdii*) (p. 16–17), the wild stock exhibited a more highly canalized habitat response than the semi-domestic stock; they chose the field habitat, even when reared in the woods or in laboratory cages. In contrast, the semi-domestic stock required early experience in a field to show a preference for the field environment.

Sensitive Periods in Behavioral Development

The relative effect of experiences may be a matter of timing (i.e. the stage of the animal's development). There are periods during development when certain experiences will have an optimal or maximal effect on behavioral development. These periods are often referred to as 'sensitive periods'. Sensitive periods have been proposed for the socialization of domestic cats (*F. domestica*) and dogs (*C. familiaris*). By socialization I mean the development of an affinity for the company of a certain class of social objects, usually members of one's own species. For example, the sensitive period for socialization in domestic cats is believed to be between 3 and about 9 weeks of age. That is not to say that cats cannot be socialized later than 9 weeks; it is simply more difficult. Social experiences during this stage of development will have a greater effect on the animal's subsequent social behavior than at any other equivalent time period.

Imprinting

'Imprinting' is a term used to describe learning that is restricted to a relatively short period of time, usually early in the life of an organism. It has been used to describe early learning associated with the development of social and sensory-based preferences.

There are two aspects of social imprinting, 'filial imprinting' and 'species imprinting'. Filial imprinting constitutes the learning process by which very young animals become socially bonded to their parent(s). In precocial species, filial imprinting occurs very soon after birth or hatching and is expressed by the tendency of the young to follow their parent. Hess (1959) studied the sensitive period for the development of the following response in mallard ducklings (*A. platyrhynchos*) by exposing newly hatched birds to a moving model of a male mallard duck at various ages following hatching (Fig. 4.1). In addition to the visual stimulus, the model duck emitted a human vocal imitation of a female mallard duck. About 24 h later, each duckling was observed for its preference in a choice test between the previously used male model and a female model (very different visual characteristics) under various conditions of movement and vocalizations. Ducklings were scored for the number of approaches and following responses they made to each of the two models. Hess determined that the choice of the male model was maximal when the birds had been exposed to it initially between 13 and 16 h after hatching (Fig. 4.2). Interestingly, this is also the period when the motor abilities of the young duckling become fully developed and their fear responses to strange objects become evident (Fig. 4.3). Further research revealed that moving, vocal models stimulated stronger following responses than stationary, quiet models.

There is evidence that young birds of precocious species can learn the specific vocalizations of their mothers prior to hatching. This prenatal learning facilitates the following response of the young bird to its mother, and subsequent social bonding. Social bonding to the parent is further facilitated by positive stimuli that become associated with the parent, such as food, warmth and 'security' when exposed to fear-provoking and other aversive stimuli. Negative experiences associated with other animals, including rejection behaviors exhibited by unfamiliar adults, also facilitate the bonding of a young animal to its parent. Following the parent is

Fig. 4.1. Young duckling following imprinting stimulus (Hess, 1959).

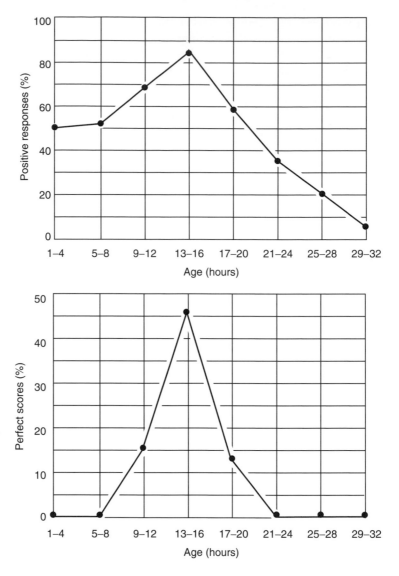

Fig. 4.2. Top: Percentage of positive responses (responses made to the male model) by young ducklings based on the number of hours following hatching when they were exposed to the male imprinting stimulus. Bottom: Percentage of perfect scores (positive responses to the male model only) made by mallard ducklings based on age when they were exposed to the male model. The period of maximal sensitivity for development of the following response appears to be 13–16 h after hatching (Hess, 1959).

adaptive in that it keeps the young in close contact with its parent, and thus the care and protection the parent provides.

Species imprinting refers to the learning process by which young become social-ized (i.e. develop an affinity) to their species. Species imprinting is to a class of social stimuli rather than a specific individual, as in filial imprinting. Whereas filial imprint-

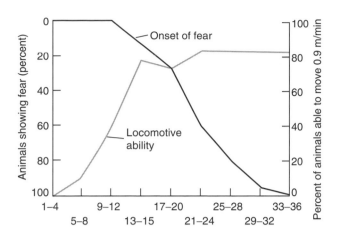

Fig. 4.3. Percentage of chicks showing fear and percentage of chicks able to move 0.9 m/min based on number of hours after hatching. The onset of fear curve and the curve for the development of locomotor ability together outline the period of maximal imprinting sensitivity for the following response (Hess, 1959).

ing occurs in the hours and days following birth or hatching, species imprinting occurs somewhat later (Vidal, 1980). Scott (1962) shows that the sensitive period for the development of species socialization in domestic dogs (*C. familiaris*) is from 3 to 12 weeks following birth, with the period of maximal sensitivity between 6 and 8 weeks. If a pup is taken from the litter prior to the period of socialization and reared by humans in isolation from other dogs, it will likely develop a strong attachment to people and exhibit undesirable social behaviors (e.g. excessive fear or aggression) as an adult when encountering other dogs. Conversely, lack of exposure to humans during the sensitive period will likely result in relatively weak ties to people.

Species imprinting is adaptive in that it facilitates integration into the social group. This is especially important in learning the skills necessary for survival and reproduction. Species imprinting is the basis for reproductive isolation; that is, the preference of an animal to mate with members of its own species (conspecifics) rather than alien species. In sexually dimorphic species such as mallard ducks (*A. platyrhynchos*), the brightly colored male looks very different from the mottled-brown female. By imprinting on their mothers, male mallards will naturally seek out females of their species for courtship and mating once they reach sexual maturity. But this creates a dilemma for the female mallard; imprinting on its mother could direct it to the wrong sex (or species) during the breeding season. This dilemma is solved by the fact that female mallards tend to prefer males that give them the most attention, namely males of their species. Female mallards show a preference for males of their own species, even when reared by females of a different species, providing males of their own species are most demonstrative (Kruijt, 1985). Of course, in nature, young female mallards are exposed to male siblings as they are growing up, so both experience and an innate attraction to attention-giving males greatly reduce the chances of mating errors.

Although most of the work on species imprinting has been conducted on birds, there is evidence that a similar phenomenon is operating in many social mammals. In addition to the study by Tomlinson and Price (1980) already discussed (see p. 13),

Kendrick *et al*. (1998) cross-fostered lambs (*O. aries*) on goat mothers and goat kids (*C. hircus*) on sheep mothers at birth and monitored their sexual preferences over a 4-year period. Cross-fostered males strongly preferred to socialize and mate with females of their adopted species. Cross-fostering effects on females were weaker. All normally-reared control animals preferred the company of their own species. When the cross-fostered subjects were presented with facial cues from sheep and goats, they again showed preferences for their adopted species. As in the live species preference tests, the response was stronger in males than females. Thus, the face appears to be an important species recognition/preference stimulus in sheep and goats. Like the studies with precocious species of birds, species imprinting in sheep and goats appears to be stronger in males than females. If males consistently seek out members of the species represented by their mothers and are demonstrative toward females of that species, the likelihood that females will attempt to mate with males of a different species is much reduced.

Many of the early ethologists believed that species imprinting was irreversible (i.e. species affinities could not be changed). Research has shown this to be not entirely true. Forced cohabitation with a different species can result in the formation of new preferences. In our study with male sheep and goats (Tomlinson and Price, 1980), animals reared with the opposite species came to prefer members of their own species (i.e. reversed their preferences) if housed exclusively with them for about 60–90 days. Kendrick *et al*. (1998) reported that the species/sexual preferences of their male sheep and goats were not affected by living exclusively with their genetic species for a 3-year period. In contrast, preferences of the females in their study were reversible after 1–2 years. Of course, in nature, the likelihood is next to nil that animals will be forced to live exclusively with animals other than their genetic species. Species affinities are continually being reinforced by living with conspecifics in social groups.

Large birds (e.g. geese, ducks and cranes) reared by humans can be trained to follow (in flight) their caretakers traveling on motorcycles, automobiles and even aircraft. Ellis *et al*. (2003) describe how cranes, geese and swans have been led on their first southward migration with either ultralight aircraft or vehicles on the ground in order to teach them new migration paths. Is this an example of filial or species imprinting?

Olfactory imprinting

Exposure to certain odors early in life can result in subsequent olfactory preferences. Olfactory imprinting has been demonstrated in all vertebrate groups, including mammals, birds, reptiles, amphibians and fish. The widespread occurrence of chemo-sensory learning suggests that it has an important function in the lives of animals. Blass and his associates (Pedersen *et al*., 1982; Fillion and Blass, 1986) found that newborn rats (*R. norvegicus*) (blind and deaf at birth) locate and attach to their mother's nipples through searching for their mother's familiar odor. If the mother rat's nipples are washed, they will not attach to them. In one of their experiments, Blass injected citral, a tasteless, lemon-scented compound, into the mother's amniotic fluid before the rat pups were born. After birth, the rat pups attached to her nipples only if the lemon-like scent of citral was in the air. In another experiment, Blass discovered that rats would remember odors associated with pleasurable experiences in the days after they were born. Pups were exposed to citral either while being stroked by the experimenter (to mimic grooming by the mother) or while being given amphetamine (a stimulant drug). In both cases, the newborns would attach to their mother's citral-coated nipples. However, these treatments had to be administered prior to the 8th day following birth. After day 8, citral did not modify the pups' nipple preferences. A subsequent experiment demonstrated that rats exposed to citral either *in utero* or as newborns later preferred to mate with citral-scented sexual partners.

My first paper on animal behavior (independent study paper in college) was a review of the sensory cues used by salmon in their migration from the ocean back to their natal stream to spawn. Salmon hatch from eggs in freshwater rivers and streams and soon migrate to the ocean, where they spend from 2 to 5 years (depending on the species), only to return to the same stream where they were hatched. Hasler *et al.* (1987) showed that each stream has its own unique odor and that each salmon uses its odor memory to find its way back to its natal stream. The odor of this stream becomes imprinted on the young hatchlings during a sensitive period for learning soon after hatching and directs their return migration several years later. The fish locate the home stream by seeking the ever more concentrated home-stream odor (i.e. chemical gradient) as they swim upstream.

Auditory imprinting

A kind of auditory imprinting has been demonstrated in the development of song in the males of certain passerine species of birds. Marler, Baptista and Petrinovich and colleagues have provided an excellent example of early auditory learning in the song development of the white-crowned sparrow (*Zonotrichia leucophrys*) (see Marler and Slabbckoorn, 2004). This species has a very wide range and songs of male sparrows from different regions have their own unique local dialects. Young males taken from the nest at hatching and reared alone in soundproofed chambers will eventually sing a simplified version of the 'normal' species song. This so-called 'innate song' develops regardless of the region from which the birds were hatched. Since isolate-reared birds never learned the dialect of their home region, it was hypothesized that the local dialect was learned by listening to adult birds and modifying their own innate species song accordingly. Males captured in the field during their first autumn and reared alone in the laboratory were found to sing the normal song for their region the following spring, when they were first capable of singing. This result suggested that young males learned their local dialect well before they could sing themselves, a hypothesis subsequently confirmed by playing tape-recorded songs to males that had been isolated from the time of hatching. It was found that the young male could learn its own or a different dialect if it heard the appropriate tape recording between approximately 30–50 days after hatching, even though it could not practice singing the song until many months afterwards. Tape recordings played after 50 days had little effect on the song that developed, suggesting a sensitive period for song learning in this species. Interestingly, song learning can be extended to 100 days if a live bird tutor is used.

It is also important to note that sparrow genes may limit which songs can be learned early in life. Juvenile male white-crowned sparrows will not learn to sing tape-recorded songs of just any species played to it during the sensitive period. This makes good sense because, in the field, young juveniles are often exposed to the songs of alien species. However, in one study, juvenile males learned the song of the male strawberry finch (*Amandava amandava*) when they were allowed to interact with a live bird tutor between 50 and 100 days after hatching. Likewise, a laboratory study (Mundinger, 1995) with the Roller and Border inbred strains of canaries (*Serinus canaria*), which sing distinctly different songs, revealed a genetic predisposition to sing like adults of their own strain. When isolate-reared birds were given a choice of a Roller and a Border canary tutor song, the Rollers selectively learned from the Roller song and the Borders learned only from the Border song. Interestingly, hybrids of the two strains learned from both tutor songs, suggesting multifactor inheritance in regard to the kinds of syllables and song patterns they acquired.

While the genes may limit what songs can be learned by juvenile males of this species, Konishi (1965) showed that song development was contingent on the birds being able to hear themselves sing (i.e. auditory feedback). Konishi deafened birds at various stages of development by removing the cochlea of the inner ear. Some birds were deafened shortly after hatching, some after exposure to the normal species song for their region but before they could sing themselves, and some after they had successfully sung the normal species song. The first two groups never developed even the innate species song characteristics of intact isolate-reared birds. Rather, they sang a series of disconnected notes that only faintly resembled the innate white-crowned sparrow song. The group deafened after they had successfully sung the normal species song continued to sing this song accurately. Konishi concluded that the white-crowned sparrow must hear itself sing in order to produce the innate species song and to modify this song in such a manner to match the song stored in their memories. Thus, the basically inherited potential for singing the white-crowned sparrow song can only be realized by hearing other males sing and by the experience of obtaining auditory feedback from hearing itself sing.

Not all birds produce species-typical vocalizations which develop in the same manner as the song of the white-crowned sparrow. For example, learning is not involved at all in the development of vocalizations in domestic chickens (*G. domesticus*). Chickens produce a variety of calls, including alarm calls, food calls, distress calls, aggressive calls, laying calls emitted by hens just before laying an egg and crowing exhibited only by males. Schjelderup-Ebbe (1923, cited in Konishi, 1963) found no difference in the calls produced by group-reared chickens and chickens reared in isolation. Konishi (1963) compared the calls of intact chickens with birds deafened at the time of hatching (cochlea removed). As adults, the calls of deafened chickens could not be distinguished from the calls of intact birds. It was concluded that chickens did not need to hear the vocalizations of other birds nor receive auditory feedback from hearing themselves to produce the sounds characteristic of their species.

Selected Examples of Early Experience Effects on Adult Behavior

Maternal effects on growth and physical development

Neonatal rat pups (*Rattus*) and human babies need their mother's touch (i.e. physical contact) to show normal rates of development (Kuhn and Schanberg, 1998). Tactile stimulation from mother rats (*R. norvegicus*) promotes growth protein synthesis and weight gain in their offspring (Kuhn *et al.*, 1978; Schanberg and Field, 1987). Without maternal contact or an artificial substitute, rat pups fail to synthesize growth proteins, even though they continue to synthesize other proteins. Studies have shown that prematurely born human babies thrive better when given extensive skin-to-skin contact. Babies given back massages, neck rubbing and forced moving of arms and legs showed more rapid weight gains (50% increase over 12 days) and improved neurological development.

The key to this response appears to be levels of the brain peptide, beta-endorphin (Greer *et al.*, 1991). Maternally deprived rat pups have higher than normal levels of beta-endorphin in their brains. By injecting this peptide directly into the brains of normally mothered rat pups up to 3 weeks of age (weaning age), one can mimic the growth-stunting effects of maternal deprivation. Beta-endorphin affects the way the entire body responds to two early regulators of growth, insulin and growth hormone. Grooming of rat pups by the mother keeps brain beta-endorphin levels in check, an effect that can be duplicated by stroking rat pups with a wet paintbrush. Maternal licking at an early age also stimulates reflexive urination in rat pups.

Maternal effects on emotional reactivity of rats

Fear is one of the best-understood emotions in terms of the neural circuitry and brain chemistry involved. The amygdala is an almond-shaped structure in the center of the brain that is one of the most important brain sites for processing fear responses by mammals. The amygdala becomes active and blood flow to this area increases when animals are exposed to fear-provoking stimuli. Researchers have shown that a cluster of neurons in the amygdala, called the lateral nucleus, receives fear messages from the sense organs and another cluster, the central nucleus, sends these messages out to other brain structures.

Scientists have also investigated the biochemical events taking place in the amygdala and other parts of the brain when animals experience fear. They have found that intense emotions experienced early in life can trigger not only behavioral changes at the time but changes in brain biochemistry, which persist long after these emotions have passed, and even into adulthood. Individuals who typically are fearful or calm not only exhibit characteristic behavior patterns but also have distinct patterns of brain activity as well. To illustrate this point, Hall *et al.* (1998, 1999) and Matthews *et al.* (2001) subjected young domestic rats (*R. norvegicus*) to two different types of stressors early in their lives. One set of rats were stressed by being reared in social isolation after weaning. A second group was stressed by repeated separation from their mothers prior to weaning. When tested as adults, isolation-reared rats appeared 'overexcited' when exposed to food cues or a novel environment. It was subsequently found that the rats in this group had higher than normal levels of the neurotransmitter, dopamine, in areas of the brain known to be involved in addiction and motor control. Maternally deprived rats appeared 'sluggish' and 'dull' in their reactions to changes in their environment. They had lower than normal levels of the 'mood-mediating' neurotransmitter, serotonin, in parts of the brain that process emotions and memory. In other research, Liu *et al.* (1997) found that rats (*R. norvegicus*) born to especially attentive mothers (e.g. excessive grooming and nursing) grew up to be more 'resilient' and less 'anxious' in potentially stressful situations than rats given normal amounts of maternal care. Subsequent studies (Caldji *et al.*, 1998) showed they had more receptors for neurotransmitters that inhibit the activity of the amygdala and fewer receptors for the so-called 'stress hormone', corticotropin-releasing hormone. These and other studies have demonstrated the lasting effect that early experiences can have on an animal's physiology and behavior.

Development of food preferences

Inherited morphology and physiology set limits on basic food selection (e.g. carnivore, herbivore, etc.). Within those parameters, early experience with different food stuffs can determine what will be selected or rejected as adults. These experiences may begin prior to birth or hatching. In mammals, the mother's diet during gestation can influence the flavor and olfactory preferences of her young, a phenomenon demonstrated in humans, domestic sheep, rabbits, dogs and rats (see Wells and Hepper, 2006). For example, Hill and Przekop (1988) discovered a prenatal sensitive period for the development of sensitivity to sodium chloride in the domestic rat (*R. norvegicus*). Development of neural structures associated with perception of sodium in the diet is dependent on exposure of the developing fetus to sodium in its mother's diet. Restriction of maternal dietary sodium on or before embryonic day 8 reduced the taste responses of the chorda tympani nerve (to sodium chloride) to 40% that of control rats born to mothers on an unrestricted sodium diet. Instituting the low sodium diet on the 10th day of gestation did not produce an effect. Thus, the sensitive period for development of sensitivity to sodium chloride extinguished quickly during a 2-day period (between the 8th and 10th day of gestation). Interestingly, these events occur before the initial formation of taste buds in this species.

Hepper (1988) demonstrated fetal learning of the flavor of garlic in domestic rats (*R. norvegicus*). Experimental rats were born to pregnant females who were fed garlic, starting on day 15 of their 21-day gestation period. Control rats were born to mothers denied garlic. Twelve days after birth, the experimental and control rats were tested for their preference for the odor of garlic and onions. Offspring of mothers given garlic showed a significant preference for garlic, while the control rats showed no preference for either garlic or onions. To test for the possibility that the experimental rats could have smelled garlic on their mother's body after birth, a second trial was run in which the young were fostered on 'clean' mothers immediately at birth. These young also showed a preference for the odor of garlic at 12 days of age, confirming that the odor preference is acquired *in utero*, presumably by the flavors infiltrating the amniotic fluid bathing the fetuses.

Following birth, interactions with parents and conspecifics can contribute to the development of food preferences. Galef, Clark and their colleagues (see Galef *et al.*, 2006), also working with domestic rats (*R. norvegicus*), found that young rat pups became familiar with the flavors of foods ingested by their mother by suckling and ingesting her milk. Flavors present in mother's milk cause young rats to prefer foods with similar flavors. Also, when seeking solid food for the first time, young rats are attracted to adults engaged in eating and will sample some of the food items they see being consumed. Young rats also show a preference for foods with odors detected on the breath of conspecifics. Residual odors deposited on trails leading to feeding sites and on and around foods while feeding also attract hungry rats. Rats prefer foods scent-marked by other rats and scent-marked feeding sites. While these experiences help the young rat to learn which foods are nutritious and safe, mother rats and adult conspecifics do not provide cues which directly assist them in avoiding toxic foods.

Similarly, studies with sheep (*O. aries*) have shown that food preferences can be modified by early experience, especially experience with foods in the presence of their mothers. A series of Australian studies (reviewed in Lynch *et al.*, 1992) sought to determine why range sheep would not eat unfamiliar grains made available to them during times of drought when their normal diet of grasses and forbs was in short supply. The researchers systematically exposed young lambs to wheat grains, either in the presence or absence of their mothers. Lambs whose mothers ate wheat in their presence readily accepted wheat as adults. Lambs exposed to wheat in the absence of their mothers were reluctant to sample the grain and eat it as adults (Fig. 4.4). The studies showed that the lambs did not learn solely by watching their mothers eat. Acceptance of wheat was enhanced if the lambs sampled the wheat themselves, following the model of their mothers.

Herbivorous animals select their diet based on visual cues, taste, odor, texture, availability and variety. Not all food preferences are learned; some are based on relatively strong innate predispositions. Villalba and Provenza (1999) found that ruminants (ungulates with a four-chambered stomach) could readily associate certain flavors with foods high in protein or energy. When offered an appropriate choice of foods, lambs (*O. aries*) maintain a constant ratio of protein to energy in

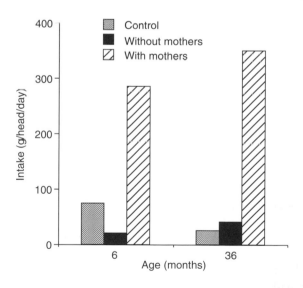

Fig. 4.4. Mean wheat intake (g/sheep/day) of groups of sheep offered wheat at 6 and 36 months of age after having been exposed to wheat before weaning with or without their mothers present. Controls had no experience with wheat prior to the week of testing (Lynch *et al.*, 1992).

their diet. Their preference for energy foods (in grams of starch) is naturally much greater than their preference for protein (in grams of crude protein), presumably because their energy requirement on a daily basis is roughly five times greater than their need for protein. In addition, sheep are very efficient at storing energy and recycling nitrogen. Villalba and Provenza reported that their lambs' preference for foods rich in energy decreased after a high-energy meal and increased after a high-protein meal. The ratio of protein to energy in their chosen foods was thus continually balanced to maintain the proper protein–energy ratio. In addition, they noted that the food (flavor) preferences of free-living ruminants typically changed over time in response to seasonal changes in the protein–energy ratio within available food items.

It is clear that animals do not always naturally prefer food items that rectify deficiencies in their diet. Much about diet selection in animals is learned by experiences with elders and individual experiences with different food items. In general, animals adjust their learned food preferences based on availability, taste (i.e. flavor) and post-ingestive physiological reactions. Flavorful foods that in the past have *not* become associated with gastrointestinal upset (e.g. feelings of nausea) are preferred over food items which prove to be toxic. In addition, because toxicity exists on a continuum, avoidance of toxic food items is dose-dependent; that is, the more toxic the item, the more it is avoided. This presents a problem for herbivores, since a meal may consist of both nutrient-rich and toxin-rich plants and many plants, especially browse, have both nutrients and toxins in one food type. Fortunately, the higher the basic nutritional quality of the animal's diet, the less likely it will react negatively to toxins in that diet (Baraza *et al.*, 2005). In reality, herbivores choose food items based on the integration of the animals' dose-dependent learned responses to both the positive and negative post-ingestive effects of available foods (Duncan *et al.*, 2006).

Development of temperament in male cattle

Prior to domestication, the ancestors of our common domestic farm animals lived in social groups. Young of the species almost always grew up in the company of conspecifics. On farms today, we sometimes house animals individually to: (i) prevent them from injuring one another; (ii) as part of disease prevention programs; (iii) to ensure each animal get its fair share of feed; or (iv) to limit breeding to an optimal time of life or season. One often-overlooked consequence of individual housing is its effect on the temperament of animals, particularly in human–animal interactions. This is especially important when we are dealing with large, potentially dangerous (to humans) animals like bulls (*Bos*) and stallions (*E. caballus*). This fact was dramatically displayed to me in a study on the sexual behavior of Hereford beef bulls (*B. taurus*) (Price and Wallach, 1990a). All of our subjects were removed from their mothers at birth and hand reared either in individual pens or in group pens. The animals reared in individual pens could see, hear and smell other subjects, but physical interactions were prevented. Between 3 and 6 months of age, we tested each male for its sexual interest (i.e. mounting behavior) in sexually receptive females. We also observed the behavior of the animals in the group pens and noted frequencies of aggressive interactions (e.g. butting and head-to-head sparring). Even at an early age, when one group-reared animal butted another, the latter would often retaliate with a butt of its own. Starting at about 12 months of age, my assistant and I noticed that the individually reared males were becoming more difficult to handle. At 19 months of age, we initiated tests to compare the sexual performance of the two treatment groups. We immediately noted that the individually reared bulls resisted being driven to the test pen and, once there with the stimulus (estrous) females, were reluctant to leave the pen. On a couple of subsequent occasions, individually reared bulls attacked my assistant and I. In contrast, the group-reared animals exhibited a respectable amount of fear when we approached them and would go to and from the test arena without hesitation or incident. Literally fearing for our lives, I terminated the study at that point.

Reflecting back on our experience with the individually reared bulls, I realized that we reared our subjects in the same manner that dairy bulls are traditionally reared. Dairy calves are taken from their mothers at birth and reared in individual pens (hutches) for a period of months to control the amount of milk and feed each individual receives and to prevent disease transmission. Males that are not sold for veal are often placed in individual stalls until used for breeding. As many of you know, dairy bulls are notoriously dangerous to humans. In fact, one of my relatives was killed by a Holstein bull.

So, why are individually reared bulls so dangerous? I believe there is a relatively simple answer; they never learn the give-and-take associated with aggressive behavior (i.e. they lack social inhibitions). Group-reared bulls learn from an early age that if they initiate a butt or a sparring match, their opponent is likely to reciprocate. Males reared in physical isolation from other males never learn these lessons and become socially uninhibited, not only toward conspecifics but also toward humans. When testosterone levels elevate at sexual maturity, humans can be the recipient of their uninhibited aggressive behavior. Many human fatalities from bulls are first-time events (attacks) because the handler thought the bull could be trusted. The tame, tractable individual, hand reared from an early age, suddenly becomes an 'ugly monster' and catches the handler off guard. Hand-reared male livestock and horses should always be handled with the utmost of caution.

Development of sexual behavior in male pigs

Social isolation is unnatural for social animals and can lead to undesirable temperament, as discussed in the section above. Social isolation can have other effects for male pigs (*S. scrofa*), namely sexual incompetence. Hemsworth *et al.* (1977) reared male pigs from 20 days to 7 months of age using three treatments, either socially restricted (visual and physical isolation in individual pens), in groups of six males, or in groups of six males and three females. Males were subsequently tested for sexual performance every 2 weeks between 7 and 13 months of age by exposure to sexually receptive females for 20 min. The socially restricted males showed a marked impairment in sexual performance (Fig. 4.5). Latency for first mount (i.e. time from start of test to first mount), the total number of mounts and the total number of successful

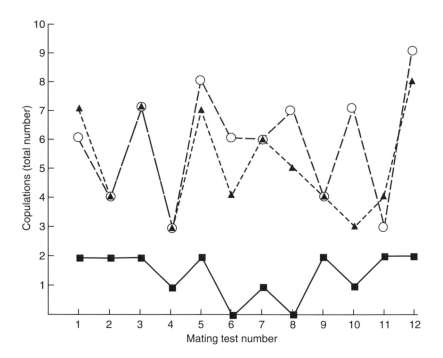

Fig. 4.5. Total number of copulations achieved by boars reared in individual pens (■ _ _ ■), in all-male groups (○–––○) and in male–female groups (▲–––▲) during each of 12 sexual performance tests administered between 7 and 13 months of age (Hemsworth *et al.*, 1977).

copulations were significantly lower for the socially restricted males than for males in the other two treatments, which did not differ. Furthermore, sexual performance did not improve with age or sexual experience during the 6-month test period. Interestingly, rearing *female* pigs in socially restricted conditions had no effect on their sexual performance relative to females reared in heterosexual groups. It should also be mentioned that the sexual performance of the individually reared beef bulls discussed in the section above was indistinguishable from their group-reared counterparts. One should not generalize findings from one species to another.

5 Learning

Definition of Learning

There are many definitions of learning. The one I prefer to use is that learning is a relatively permanent change in response over time as a result of practice or experience. 'Permanent' in this definition implies that the change in response is not due to physical or mental fatigue, or sensory adaptation in which the sense organs temporarily lose their sensitivity to a stimulus due to prolonged exposure. The 'practice or experience' phrase in the definition excludes changes in response due to maturation or the effects of injury. Because we cannot directly observe the changes in the central nervous system when learning occurs, we must measure learning operationally, that is, by changes in behavior or performance.

Some Factors Which Can Affect Learning Ability

Learning ability varies with species, breed and individual, and individuals typically exhibit changes in learning ability over time. Some factors which can affect performance, and thus our assessment of learning ability, are provided below.

Temperament or emotional reactivity

Animals which are hyperactive or emotionally reactive (i.e. fearful) typically learn more slowly. A number of behavioral scientists (e.g. Barnett, 1963, p. 152; Robinson, 1965, p. 514) have hypothesized that the laboratory rat (*R. norvegicus*) would prove to be inferior in learning ability to their wild counterparts, based on the assumption that natural selection for learning ability has been relaxed in laboratories, where learning is relatively unimportant for survival and reproduction. Interestingly, behavioral research has generally failed to support this hypothesis. Laboratory rats have been found superior to wild rats in maze learning (Stone, 1932), appetitive conditioning (Boice, 1968), avoidance learning (Boice, 1970) and escape learning (Price, 1972). The inferior performance of the wild subjects in each case was explained by their relatively high level of emotional reactivity. Experimenters commented that after entering the test apparatus, wild rats would crouch ('freeze') in a corner or make little effort to investigate the test environment. In contrast, domestic rats typically would explore the test apparatus and stimuli associated with the learning task. Some of the wild rats' reactivity may have been generated by being transported from their cages to the test apparatus and (or) the novelty of the test environment itself. In any case, highly reactive or fearful animals are more reluctant to attend to stimuli associated with learning tasks than animals with 'calmer' temperaments.

Motivation to explore

Animals with similar temperaments may differ in their motivation to explore and investigate new objects and situations. Inquisitive animals theoretically should learn more quickly than less responsive animals and 'stay-at-home' types. Much learning is trial and error. Investigating one's environment is conducive to learning by trial and error. In our study on the development of species affinities in sheep (*O. aries*) and goats (*C. hircus*) (p. 13), we had to wire shut the doors of all the pens housing our goats. The goats were very good at learning how to unlock the doors by manipulating the sliding bolt locks with their tongue and nose. In contrast, the sheep never learned how to unlock their doors. Anyone who has worked with domestic sheep and goats can attest to the fact that goats are more exploratory than sheep.

Motivation for rewards

Learning is more rapid when animals are highly motivated to receive the rewards offered. Food, tactile stimulation and verbal praise are often used when training animals. Some animals are more motivated than others to receive such rewards. Foods are more rewarding when the animal is hungry. Some breeds of dogs (*C. familiaris*) are more attracted to people than others (e.g. Labrador retriever versus Basenji) and thus are more motivated to receive a pat on the head or verbal praise for a correct response in a training task. While working for my college degree, I took a course in experimental psychology. The laboratory part of the course involved training a domestic rat (*R. norvegicus*) to perform various tasks (e.g. press a bar, pull a chain, etc.) to obtain water. On each test day, I removed the water bottle from the rat's cage several hours before testing to ensure that the subject was sufficiently thirsty to be motivated to receive a water reward. Each reward was limited to a drop of water to prevent the animal from quenching its thirst too quickly. By limiting the quantity of water in each reward, we could run more trials on any given test day.

Structural and sensory biases

Species, breeds and individuals often differ in sensory capabilities (acuity) or locomotor abilities, which can affect their performance in various learning situations. In some cases, these differences are based on the animals' basic structure or morphology. We have all seen pictures of chimpanzees (*Pan troglodytes*) which have learned to use blades of grass or sticks to extract termites and ants from tiny holes. Primates can learn to use such tools because the digits on their hands offer the manual dexterity to hold and manipulate small objects. This kind of tool use would not be possible in hoofed ungulates (e.g. pigs, cattle, horses), for example, because they lack the anatomical features to support this behavior.

In nature, pigs (*S. scrofa*) are extremely adept at using their snouts to gain access to buried food objects such as plant tubers and invertebrates. Grazing animals such as horses (*E. caballus*) and cattle (*Bos*) seldom feed on items beneath the ground surface and, when snow covers the landscape, use their hooves to gain access to vegetation. A pig's snout is designed for exploring and uprooting vegetation, whereas a horse's hooves are designed for supporting weight and facilitating locomotion. Thus, a learning task requiring individuals to uncover buried food items could easily favor pigs.

The manual dexterity of primates should also facilitate locating and gaining access to buried food items. However, the olfactory capabilities of pigs (*S. scrofa*) are notably superior to primates and, in a learning task requiring animals to locate buried food items, the pigs have an advantage based on their superior sense of smell.

Sensory biases for learning certain things are ultimately based on structural differences in the sense organs and the various sensory processing centers of the central nervous system. For example, color vision provides an important sensory bias in many learning situations. Species differences in color vision can influence the rate at which animals learn to discriminate between objects in their environment (e.g. foods, mates, landscape features, etc.). Free-living primates learn at an early age to use color to locate certain food items such as fruits.

Psychological biases

It is somewhat arbitrary to distinguish structural and sensory biases from psychological biases because the latter are often based on differences in the structure of the nervous system and associated biochemistry. I have done so merely to emphasize the fact that differences in learning ability can be based on inherited tendencies that bear little relationship to external anatomical features or perceived sensory abilities. Among dog (*C. familiaris*) breeds, the Labrador retriever is not as adept as the Border Collie in learning to gather and move groups of sheep, even though there is no perceivable advantage of collies based on structure or sensory abilities. The Border Collie has been artificially selected for the tasks required of sheepdogs and consequently possesses a psychological bias for learning these tasks.

Tryon (1940) artificially selected domestic rats (*R. norvegicus*) for their ability to learn to traverse a complex maze to obtain a food reward. After just a few generations of selection, his 'maze-bright' rats clearly outperformed his 'maze-dull' rats in learning the maze. Interestingly, Tryon's maze-bright rats were no better than randomly selected rats in learning tasks that did not involve a maze (see McClearn, 1963). Searle (1949) found that the maze-bright rats were inferior to the maze-dull rats in a water escape test and no better in other tests. He also found that the maze-bright rats were more motivated to work for food, which could explain their superior performance in the land-based maze. The learning ability of the maze-bright rats did not generalize to all learning situations but was specific to their performance in land-based mazes where hunger was the primary motivating factor and food was the goal.

Prior experience in learning certain tasks can transfer to new but similar tasks and facilitate performance. 'Learning to learn', or 'latent learning' as it is sometimes called, can also be considered a type of psychological bias. Experience in learning mazes can facilitate learning a new maze. Experience in learning to locate one kind of food item can transfer to learning how to find a different type of food (i.e. development of 'search images'). We experience this phenomenon in our own lives when we transfer information learned in one situation to another.

Conclusions

When we consider all of the factors that can affect performance in learning tasks, we quickly realize that superior performance in one task does *not necessarily* predict superior performance in a different task. Therefore, it is risky, if not unprofessional, to compare the intelligence or general learning ability of different breeds or species of animals. We can easily design learning tasks that make animals 'look good' or 'look bad'. If our goal is to compare the learning ability of different groups of animals (breeds, species), we are obliged to test our comparison groups in a variety of learning tasks, which together minimize the biases that can affect performance. That can be a formidable task. Perhaps, some day, we will be able to measure directly what goes on in the brain when animals learn. That may give us our best estimate of learning potential.

Reinforcement and Punishment

Before we discuss the various kinds of learning which influence the behavior of animals, we must have a clear understanding of the terms 'reinforcement' and 'punishment'. Reinforcement refers to factors, events or experiences which *increase* the likelihood that a behavior will occur. Positive and negative reinforcement refer to adding or removing something, respectively. Positive reinforcement refers to *adding something the animal wants* to increase the likelihood that a behavior will be exhibited in the future. Let us say you command your dog (*C. familiaris*) to 'sit'. It sits and you immediately reinforce (positively) your pet with a tasty food item. That is positive reinforcement. Negative reinforcement refers to *removing something aversive* from the situation to increase the likelihood that a behavior will occur in the future. As an example, you invite your cat (*F. domestica*) to come in out of the rain. Coming in the house is reinforced by cessation of being sprinkled with water.

In contrast, punishment refers to factors, events or experiences which *decrease* the likelihood that a behavior will occur. Positive punishment refers to *adding something aversive* to the situation to decrease the likelihood that a behavior will occur. Your dog attacks a skunk (*Mephitis*) and gets squirted in the face. Hopefully, this experience is unpleasant enough that it will stay away from skunks in the future. (Sadly, it took four encounters for my daughter's Labrador retriever to learn to avoid skunks.) Negative punishment refers to *removing something the animal wants* to decrease the likelihood that a behavior will be exhibited in the future. You come home and your dog immediately jumps up on you. If you give it attention, which is what it wants, you are positively reinforcing this unwanted behavior. But, if you turn away from your dog and ignore it, you are removing or withholding the attention that it wants, thus providing negative punishment.

The lines between reinforcement and punishment in learning are not always clear. For example, it may be difficult to determine whether an animal learned to refrain from exhibiting a behavior because of being *punished* for the behavior or because it received *reinforcement* for the cessation of the behavior, or both. One could also argue that negative reinforcement cannot be effective without the animal first experiencing positive punishment. For example, if dog A initiates an unprovoked attack on dog B and gets 'beaten up' before it has a chance to assume a submissive posture or retreat, does dog A learn to behave submissively toward dog B because it was beaten up by dog B (positive punishment) or because the punishment stopped when it behaved submissively or retreated (negative reinforcement)? Perhaps both are involved in different degrees, depending on the circumstances. For one thing, being beaten up was probably painful for dog A and the painful bites ceased when dog A behaved submissively or retreated. If I had to choose between the two, I think the punishment received by dog A when it attacked dog B would have the greater influence on its subsequent behavior.

Kinds of Learning

Habituation

Habituation can be defined as the persistent decrease in frequency and (or) intensity of a response due to repeated stimulation in the absence of reinforcement and punishment. 'Persistence' in this definition distinguishes habituation from the effects of fatigue and sensory adaptation. We know that very young animals typically show fear and avoidance responses toward a wide variety of novel stimuli. Some of these stimuli will eventually become ignored because neither reinforcement nor punishments become associated with them. Through central filtering (see p. 9), which accompanies habituation, animals gradually cease to respond to certain neutral stimuli (Fig. 5.1).

Applications of habituation are innumerable. Horses (*E. caballus*) trained for traffic and crowd control in large cities are systematically exposed to loud noises before taking them out on the streets. Cattle (*B. taurus*) will enter and traverse

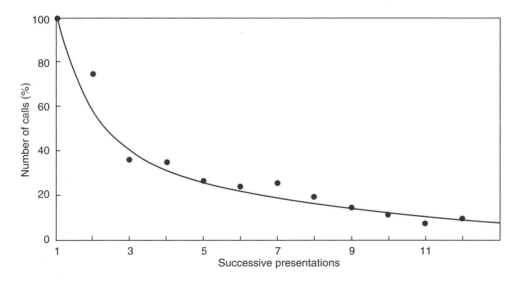

Fig. 5.1. The habituation curve of alarm calls given by chaffinches in response to a live owl placed in their aviary. Each dot represents the average number of alarm notes per bird for that day expressed as a percentage of the number recorded on day 1 (Hinde, 1954).

working chutes and alleyways more readily if they are allowed to walk calmly through the chutes the first few times without being restrained. Young dairy heifers should be introduced to the milking parlor before giving birth to habituate them to the milking environment they will subsequently access twice a day. Animals should be exposed to test arenas prior to conducting research so that the response to the test environment itself does not become a confounding factor (i.e. independent variable) for the results obtained.

Habituation is context or situation specific. Stimuli habituated in one context may not be habituated in a different situation. You have probably driven along roadways where wild animals such as deer appear to pay little attention to your moving automobile. Try stopping sometime and see how quickly they become alerted or disappear. By stopping your vehicle, you have changed the context in which the stimulus (your automobile) is perceived.

Habituation is perhaps the most widespread form of learning in the animal world. It has been demonstrated in many of the most primitive and simply constructed animal species.

Classical conditioning

Both autonomic (involuntary) and non-autonomic (voluntary) responses are subject to conditioning. Autonomic, or involuntary, responses are subject to classical conditioning. Classical conditioning is sometimes referred to as Pavlovian conditioning, since the most commonly cited example of classical conditioning was provided by Pavlov, the famous Russian scientist who classically conditioned dogs (*C. familiaris*) to salivate on the sound of a bell. This is the scenario. When exposure to a neutral stimulus (bell) becomes closely associated in time with presentation of a stimulus (food) which elicits an *unconditioned autonomic response* (salivation), the neutral stimulus (bell) will, in time, elicit that same autonomic response (salivation) in the absence of the *unconditioned stimulus* (food). Salivation in response to the bell alone is referred to as the conditioned response and the bell becomes the *conditioned stimulus*. In classical conditioning, learning is based on the 'principle of association' or the 'contiguity principle' because all that is needed is for a previously neutral stimulus to become associated with a stimulus which triggers an involuntary response.

Milk release (ejection) in dairy cows (*B. taurus*) is another example of an involuntary response subject to classical conditioning. Massaging the cow's udder by a human or calf (unconditioned stimulus) will stimulate milk release (unconditioned response). Placing a cow in a milking stall can become associated in time with udder massage, so that eventually the former, alone, will elicit milk release. We had a few cows on our dairy farm that would start ejecting milk as soon as they were brought into the milking parlor. We always milked them first to avoid losing a portion of their milk to the floor (and cats).

Luteinizing hormone (LH) and testosterone are autonomically released by male rats (*R. norvegicus*) when exposed to sexually receptive females. Graham and Desjardins (1980) demonstrated that if methyl salicylate (oil of wintergreen), a highly fragrant odor foreign to rats, is paired with female rats, the odor alone will eventually trigger the release of LH and testosterone (in males). Methyl salicylate becomes the conditioned stimulus. A similar classically conditioned response was demonstrated with female sheep (*O. aries*) (Gelez *et al.*, 2004). Abrupt exposure of ewes to a ram or its odor toward the end of the non-breeding season activates LH secretion and estrous cyclicity. When oil of lavender was placed on rams, ewes would eventually show an increase in LH to just the odor alone. In contrast, LH secretion did not change in control ewes which had not been exposed to oil of lavender. These experiments demonstrate how environmental cues, which animals learn to associate with sexual stimuli, can alter physiological responses to these stimuli.

Operant conditioning

Voluntary responses which the animal controls, itself, are subject to operant or instrumental conditioning. Animals engage in a variety of spontaneous behaviors (e.g. exploration, digging, climbing, etc.) during the course of their day. The frequency with which each of these behaviors is exhibited is known as the operant response rate. Each operant rate of responding can be increased or decreased by reinforcement or punishment presented during or immediately following the response. Suckling by the newborn ungulate (e.g. calf, lamb, kid goat) has a relatively high operant response rate. These little eating/drinking 'machines' initially will try to suck on almost anything that touches their muzzles. This response is adaptive, considering that sucking the mother's teat is required to obtain nourishment starting in the hours following birth. Suckling behavior is positively reinforced by the milk obtained so that the newborn quickly becomes conditioned to go to its mother's udder (conditioned stimulus) when it gets hungry or thirsty.

Your new 10-week-old puppy is not housebroken and you want to train it to let you know when it has to relieve itself. Your pup occasionally whimpers, whines or goes to the door of your apartment but the operant response rate of each of these behaviors is relatively low (i.e. occurs infrequently). That is unfortunate because these are cues your pup can use to signal that an event is about to happen. Fortunately, you can use operant conditioning to increase the operant rate of responding as you seek the goal of housebreaking your new little friend. This topic will be discussed further on p. 57.

I explained earlier that, in my experimental psychology laboratory, I was required to train a rat (*R. norvegicus*) to press a lever to obtain water. The instructor gave me a small enclosure (Fig. 5.2) with a lever projecting from one wall so that, when the rat pressed the lever, I could direct a drop of water into a small cup near the lever. At first, my thirsty rat explored its new enclosure but showed little interest in the lever. Eventually, it pushed the lever (a voluntary response) and I positively reinforced it with a drop of water. After this experience, it concentrated its exploration in the vicinity of the lever so that the second lever press occurred in a much shorter period of time. Several lever presses later, the rat connected its lever pressing with obtaining water, and operant conditioning was achieved. The stimulus (lever) became associated with a voluntary response (lever pressing) and the response was instrumental in providing the rat with positive reinforcement (drop of water). Figure 5.3 illustrates a typical learning curve for a rat conditioned to perform a simple task to obtain a food or water reward.

How do classical and operant conditioning differ from a conceptual standpoint? Classical conditioning is based on establishing an association between two stimuli (e.g. a bell and food). The animal's conditioned response (e.g. salivation in response

Fig. 5.2. An operant conditioning apparatus requiring the animal (e.g. rat) to press a lever to obtain food or water (courtesy B.F. Skinner Foundation).

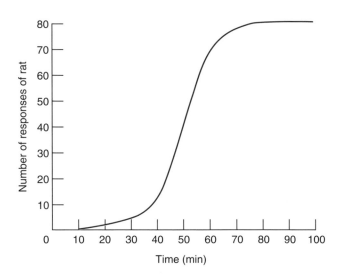

Fig. 5.3. Cumulative number of responses made by a hypothetical rat in an operant conditioning apparatus as it learned to press a bar to obtain a food reward. The response rate (frequency of bar presses) was initially low but accelerated rapidly once the rat learned that bar pressing was rewarded with food. The rat stopped responding when it became satiated.

Chapter 5

to hearing the bell) is involuntary, and thus beyond its control. Operant conditioning is based on establishing an association between a stimulus and a voluntary response, as well as the direct effect of the response (reinforcement or punishment). For example, a dog can be trained to bark (response) when it sees a person holding a treat (stimulus) in order to receive the treat (positive reinforcement) (Fig. 5.4).

Experimental psychologists distinguish between primary and secondary reinforcers when training animals with operant conditioning techniques. Primary reinforcers typically have biological reinforcing properties (e.g. food, water, shelter, warmth, tactile stimulation such as scratching and petting, etc.). Secondary reinforcers are stimuli (e.g. other animals and objects) associated with primary reinforcers in time and (or) space. Milk is a primary reinforcer for the suckling response in young mammals. The mother providing the milk serves as a secondary reinforcer. The fact that young mammals are usually rejected when they approach alien mothers for milk strengthens the association with their own mother. People frequently serve as secondary reinforcers in the lives of domestic animals by being the agent through which primary reinforcers are delivered.

The timing of reinforcement or punishment relative to performance of the response is extremely important in operant conditioning. If reinforcement or punishment is delayed (relative to the response), the animal may not associate the reinforcement or punishment with its actions. Animals tend to associate reinforcement or punishment with what they are doing at the time or immediately before, not something they were doing several minutes, or even seconds, before that. I read about a woman who wanted to train her horse (*E. caballus*) to hold its foot up so she could examine it. She decided to use food as a reward. Each time she examined the horse's foot she would let go of it and then reach for the food. By the time she gave the horse its reward, it had put its foot down. The horse thought it was being rewarded for putting its foot down and eventually started stomping its foot on the ground, expecting a reward. Perhaps you can also envision the next scenario. A dog (*C. familiaris*) runs away from its owner and the owner calls for it to come back. The dog ignores the owner for a while and then returns. The owner, in his anger, physically or verbally punishes the dog when it finally reaches his side. The next time the dog strayed, it ignored the owner's commands to 'come' even longer. In this scenario, the owner thinks he is punishing his dog for running away when actually he is punishing the animal for coming back, the exact behavior the owner is trying to encourage. Offering positive reinforcement when the animal returns may be the best way to encourage the animal to return promptly

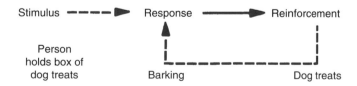

Fig. 5.4. Diagrammatic representation of an operant conditioned response. A dog is taught to bark to receive a food reward. The presence of someone with a food treat (conditioned stimulus) and having learned that barking (conditioned response) results in delivery of the treat (positive reinforcement), the dog is repeatedly motivated to bark. The lines leading to the conditioned response (barking) are broken because barking is voluntary (modified from Hart, 1985).

the next time it runs away. Electric fences used to contain livestock and, more recently, cats and dogs in their owner's yards ('invisible fences') are effective for the same reason; the punishment is immediate.

'Clicker training' was developed to bridge the time between a response and delivery of a positive reinforcer such as food. It is particularly useful when the distance between the trainer and animal prohibits immediate reinforcement. First, train the animal to associate the clicking sound with food. Do this by sounding the clicker, followed by immediately giving the animal a preferred food item. Once the animal has learned that the clicking sound means food, you can use the clicker to tell the animal immediately (and remotely) it has done something right and food is forthcoming. The woman with the horse in the example above would have profited from using a bridging stimulus such as a clicker. If she had activated a clicker before she let go of the horse's foot, the animal might have known it was being rewarded for holding its foot up, not putting it down. Once the animal is trained and knows what it is being rewarded for, the bridging stimulus can be used more sparingly, or eliminated.

Immediacy in reinforcement is also important in classical conditioning. However, there is one exception to the rule. This is seen in the development of conditioned taste aversions. Conditioned taste aversions occur when an animal associates the taste of a food item with illness (nausea) and subsequently avoids eating that food item. You know from your own experience that food-induced nausea usually does not occur immediately after eating tainted food. Likewise, herbivores which eat toxic plants typically experience a gastrointestinal upset some time after ingesting them. In spite of this delay, they will usually show an observable subsequent aversion to these same plant items. Garcia and colleagues (Garcia and Koelling, 1966; Garcia et al., 1966, 1976) experimentally demonstrated this phenomenon in domestic rats (R. norvegicus) and other species. When rats were given food or water with a distinct taste and containing a toxin (e.g. lithium chloride), or the rats were injected with a toxic substance some time later (e.g. up to an hour) to induce gastrointestinal disturbances and nausea, the animals were reluctant to ingest this same food or water again. The delay in reinforcement (nausea) associated with the development of conditioned taste aversions is markedly different from other types of conditioning in which a delay in reinforcement of even several seconds can impair learning. Interestingly, Garcia found that visual and auditory stimuli could not be classically conditioned to nausea. This makes sense because nausea usually follows eating something recognizable by taste and odor and is not as readily associated with experiences and places identified by visual or auditory cues. On the other hand, animals readily associate visual and auditory cues with externally induced painful experiences and show related 'conditioned fear' responses.

Conditioned taste aversions can present a problem for persons given the task of poisoning harmful pest species if the animal initially does not ingest a lethal dose of the poison. Wild animal pests such as rats and mice are hesitant to eat unfamiliar food items and often ingest small amounts the first time a novel food is sampled. Habituating the animal to the 'base diet' minus the poison helps to ensure that a lethal dose is consumed when the poison is first added.

When there is risk of poisoning non-targeted species, a different approach can be used. Non-lethal doses of illness-producing chemical compounds may be used to create aversions to a specific protected species or food item. For example, placing an illness-inducing substance like lithium chloride in sheep (O. aries) carcasses can condition some coyotes (*C. latrans*) and other predators from harassing and killing sheep once the tainted meat is consumed. Placing non-lethal doses of poison in bird eggs has been used to discourage animals (e.g. rodents, skunks) from pillaging bird nests. One practical problem with the non-lethal approach is in sustaining the aversion. Use of the illness-producing chemical must be continued on a regular basis to prevent the target animals from sampling untreated food and thus renewing their interest in the protected species or food item. In addition, some species are more difficult to condition than others. For example, Norbury *et al.* (2005) obtained limited success in training captive-reared ferrets (*Mustela furo*) to avoid eating bird eggs using conditioned taste aversion techniques. In general, the non-lethal approach has not proved to be an effective and practical way to condition predators from taking specific prey species and food items.

Sometimes, we humans want to train an animal to perform a response which is not part of its normal behavioral repertoire, or one which is exhibited so infrequently that the opportunity to reinforce the animal is seldom possible. Fortunately, there is a process psychologists call 'shaping', which can be used to encourage an animal to perform spontaneously a behavior it would otherwise seldom exhibit, so that conditioning can occur. Returning to my laboratory exercise on conditioning a rat to press a bar to obtain water, I found on the first day of testing that my rat probably was not going to press the bar (to be reinforced) in the time allotted to me. So, I initially reinforced my rat with a drop of water whenever it moved close to the lever. This resulted in the rat concentrating its activity in the region of the lever. I next reinforced the rat when it touched the lever. After several reinforcements for touching the lever, I waited until the rat put sufficient weight on the lever to depress it. From that point on, the rat pressed the lever often enough for me to complete the conditioning process.

As stated earlier, housebreaking a young puppy can be difficult because the impending elimination cues you have to work with occur so infrequently. The process can be facilitated by shaping so that you can reinforce more effectively the signals you want your pup to provide (e.g. whines, going to the door, etc.) when it has to relieve itself. First, your pup should become familiar with eliminating outdoors. This can be achieved by taking it outside on a regular basis, if possible, and giving it verbal and tactile praise when it relieves itself. Next, when inside, watch for any cues that elimination might be forthcoming (e.g. restlessness, circling, squatting, etc.). When these are seen, quickly pick up your pup, take it outside and then reinforce it for a successful elimination. Do the same thing even if you catch your pup in the act of eliminating. Hopefully, your pup will soon learn that outside is the place to go. That is where it is rewarded for eliminating (positive reinforcement) and that is where the aversive treatment of being picked up quickly and carried outside is terminated (negative reinforcement). You could also argue that if being picked up and carried outside is aversive in any way, you are punishing (positive punishment) your dog for trying to relieve itself inside your apartment. Once your pup learns that outside is the place to relieve itself, it is more likely to whine or run to the door when it has to eliminate. Encourage these signals by letting your pup outside (negative reinforcement) and rewarding it for eliminating (positive reinforcement). In time, your pup will consistently let you know when it has to go out.

Experimental psychologists sometimes use the terms 'avoidance conditioning' or 'escape conditioning' to refer to two related forms of operant learning. In nature, prey species learn to associate certain visual, auditory or olfactory cues with the presence of a predator. In fact, you have probably seen on TV how large African ungulates can tell whether or not predators are hunting by observing the body language (i.e. visual cues) they exhibit. If a prey animal *consistently* runs into its burrow or flees from the area whenever it sees a certain type of predator (i.e. before it is stalked or attacked), we can argue that avoidance conditioning has occurred. The sensory cues provided by the predator aids the animal in avoiding the predator. Attention to the predator's body language and successful avoidance of the predator are negatively reinforcing (i.e. the possibility of an aversive experience is removed and fear is reduced). If the prey animal ignores or does not perceive the sensory cues provided by the predator and responds to the predator (e.g. runs into its burrow, outruns the predator) only when it attacks, the animal is said to exhibit escape conditioning. Behaviors which allow the prey to escape the predator are also negatively reinforcing because of fear reduction. These examples illustrate that reinforcement does not have to be a tangible entity like food, water or shelter.

Wild and domestic animals typically exhibit a fear of unfamiliar or novel places and objects, a phenomenon often referred to as 'neophobia'. Fear of the unfamiliar has adaptive value for animals living in nature and these tendencies are not entirely lost during the process of domestication. First experiences in novel places or with novel objects have a particularly important effect on the animal's subsequent responses to these stimuli. In the late 1960s, I spent a period of time live-trapping wild Norway rats (*R. norvegicus*) in the Syracuse, New York, landfill and on nearby farms for my research. I quickly learned that trapping rats on farms was much more difficult than trapping in landfills. On the farms, my live-traps represented a novel object in an otherwise stable environment, something to be avoided or approached cautiously. To overcome this fear, I initially baited the traps with the spring-loaded doors wired open. After a couple of nights, my bait started disappearing and I knew it was time to begin trapping. The rats had become habituated to my traps and were not particularly afraid to enter them. Trapping in the landfill was a different scenario. Each day, tons of garbage and refuse were dumped, creating a constantly changing environment. The rats living in the landfill were used to unfamiliar objects being added on a regular basis and did not hesitate to enter my traps on the first night. The farm rats readily recognized the novelty of my unfamiliar traps, whereas in the land-fill, the novelty of my traps appeared to be ignored. Klemann and Pelz (2006) noted a similar dichotomy in the behavior of wild Norway rats on a farmstead in Germany. Rats visiting a bait station in the piggery where the food supply was limited and disturbance frequent were active by day and night, whereas the rats in the granary where grain was stored (stable and abundant food supply) and where there was little disturbance were strictly nocturnal. In addition, the rats accessing the piggery visited the bait station more frequently and were more aggressive than rats using the granary. Through habituation and conditioning, animals learn appropriate (i.e. adaptive) responses to stimuli in the environment in which they live. They also learn to discriminate between stimuli which are rewarding and those to be avoided.

Discrimination learning

Animals learn to discriminate from a very early age. Mothers learn to discriminate their young from the young of other females and young learn the characteristics of their mother. Animals learn to discriminate between aggressive and non-aggressive conspecifics. Males learn to discriminate between sexually receptive and non-receptive females. Ruminant species prefer hay cut in the afternoon to hay cut in the morning, because of its higher non-structural carbohydrate content (Burritt *et al.*, 2005). Discrimination learning occurs over and over again as animals adapt to the environment in which they are found and respond to changes in that environment.

Both classical and operant conditioning can be involved in discrimination learning. Cues used in discrimination learning may elicit both involuntary (autonomic) and voluntary responses. For example, prey species use visual, auditory and olfactory cues to discriminate between predators and non-predators. Cues associated with predators can elicit fear (involuntary response) and (or) flight (voluntary response). Fear provides the motivation to flee. Flight, if successful, helps to reduce fear, as noted above.

Discrimination learning has been used to study the sensory capabilities of animals. We know a great deal about the color vision of mammals based on their performance in discrimination learning tasks. For example, Jacobs (1993) trained dogs (*C. familiaris*) to respond to colored lights (disks) on a display panel (Fig. 5.5). In each test, two lights (disks) were the same color, while a third was different. The dogs were trained to use their noses to push on the disk with the color that was

Fig. 5.5. Behavioral tests for color discrimination in the dog. Two of the lighted discs on the display board are the same color and light intensity. The color of the remaining disc is different and is illuminated to eliminate differences in light intensity between the three discs. The dog must press the odd-colored disc with its nose to receive a food reward. The difference between the two colors is progressively reduced until the dog fails to discriminate. Tests are repeated using colors of various wavelengths, thus providing information on the dog's ability to discriminate between the various segments of the color spectrum (courtesy of Gerald Jacobs).

different. If the dogs pushed on the disk with the correct (different) light, they were given a snack. They received nothing if they pushed on either of the two similarly colored lights. Jacobs systematically varied the color of the lights on the disks across the entire color spectrum. He also varied light intensity on the disks so the dogs could not distinguish between colors based on intensity or shades of gray. Jacobs found that his dogs could discriminate readily between red and blue, colors at opposite ends of the color spectrum. They could even distinguish between the similar hues of blue and violet at the blue end of the spectrum. However, they could not discriminate between greenish-yellow, yellow, orange and red because dogs, like many non-primate mammals and color-blind humans, lack 'green' cone-cells in the retina of their eyes. Performance was random when these latter colors were presented on the disks. This topic will be revisited in the chapter on communication (p. 171–172).

Likewise, the ability of dogs to detect certain chemical substances can be determined using discriminative techniques. Walker *et al.* (2006) trained dogs to indicate to their handler which of five Teflon boxes contained *n*-amyl acetate (*n*AA). Precisely controlled concentrations of *n*AA were systematically reduced during tests over several weeks, until the dogs responded to the boxes randomly. The threshold for perception of *n*AA was the lowest concentration that each dog could successfully detect.

Discrimination learning has also been used to compare the learning ability of different breeds and species of animals. For example, one of my former students, Doug Mader, compared thoroughbred and quarter horses (*E. caballus*) in a discrimination learning task requiring the animals to distinguish between different visual stimuli (Mader and Price, 1980). Quarter horses learned the task quicker than thoroughbreds and younger horses learned at a faster rate than older ones.

Extinction

Extinction can be defined as the waning of a learned response due to a lack of reinforcement (or punishment). If reinforcement is withheld from an animal which has been trained to perform a given task, such as pressing a lever to get food, the magnitude or frequency of the conditioned response will decline over time (Fig. 5.6). Changes in response rate during extinction resemble the waning of response rate during habituation.

If an animal is given the opportunity to perform a learned response some time after a previous extinction experience, it will often show a temporary recovery in response magnitude or frequency (Fig. 5.6), a phenomenon referred to as 'spontaneous recovery'. However, the temporary initial increase in response rate will not be as great as in previously reinforced trials. If the animal's performance is extinguished again and again, consecutive recoveries of response rate will be smaller and smaller.

Interestingly, the rate at which extinction occurs to an operant conditioned response can be related to the effort (i.e. demand) required in making the response. Capehart *et al.* (1958) trained domestic rats to press a lever to receive a reward. The force (weight) required to depress the lever varied with different groups of rats. When reinforcement was withheld, rats with the 'heaviest' levers extinguished the fastest (Fig. 5.7).

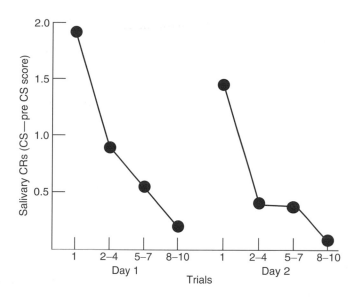

Fig. 5.6. Extinction and spontaneous recovery of salivary (classical) conditioning in dogs. On each day, animals received ten trials in which the conditioned stimulus (CS) was presented alone, without food, at an average interval of 3–5 min between trials. At the end of Day 1, the animals had stopped salivating to the CS, but on the first trial of Day 2, the salivary response (CR) was nearly as strong as it was at the beginning of Day 1 (Wagner *et al.*, 1964).

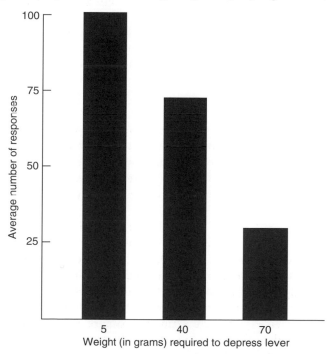

Fig. 5.7. Rate of extinction is related to the effort required to perform an operant conditioned response. When rats were required to use more force to depress a lever, fewer responses were made during extinction (Capehart *et al.*, 1958).

Schedules of reinforcement

The rate and pattern of responding during extinction can be influenced by the rate or schedule of reinforcement prior to extinction. Continuous reinforcement refers to reinforcing the animal each time it performs a learned response. Response frequency declines rapidly when reinforcement is withdrawn for animals trained under a continuous reinforcement schedule. Using a human analogy, you might say the animal 'expects' to be rewarded each time it performs the learned response and quickly gets 'discouraged' when reinforcement is not forthcoming. Under a fixed ratio schedule, the animal is reinforced after a certain number of responses (i.e. every Nth response). Some animals reinforced in this manner will anticipate being rewarded after so many responses and will respond more quickly to reach the Nth response in a shorter period of time. Animals trained under this schedule are more resistant to extinction; response frequency declines more slowly when reinforcement is withdrawn. You might say they do not anticipate a reward after every response and are more persistent in responding during the extinction process. The variable ratio schedule provides reinforcement after a variable number of responses have been made by the animal. The variable reinforcement schedule eliminates the ability of the animal to anticipate reinforcement based on response frequency. It, too, results in a slower rate of extinction because the animal has learned not to expect a reward for every response. In the fixed interval schedule, reinforcement is delivered for the first response made after a fixed time period has elapsed following the previous reinforcement. Under this schedule, some animals will stop responding for a while after being reinforced and then resume responding when the time interval is almost up. This suggests that they have the capacity to judge time intervals, a topic discussed in the next chapter (p. 78–79). In the variable interval schedule of reinforcement, the animal is rewarded for the first response it makes following passage of variable periods of time. Animals trained under this schedule do not know when the next response will be rewarded and thus continue responding at a more uniform rate than with the fixed interval schedule.

In general, the decline in response frequency with extinction is slower when animals have been trained on reinforcement schedules with relatively large ratios of reinforcement (responses per reward) and relatively long time intervals between reinforcements. Most trainers use continuous reinforcement during the acquisition of a learned response and then slowly shift to a ratio or interval schedule of reinforcement to maintain the learned response.

Other forms of learning

The scientific literature abounds with examples of adaptive behaviors that appear to involve higher cognitive abilities. I have divided these examples somewhat arbitrarily into three categories, namely tool use, observational learning and abstract thinking.

Tool use

There are many examples of tool use in the animal kingdom and most of these examples involve obtaining food and the processing of food items for consumption. Chimpanzees

(*P. troglodytes*) use sticks and twigs to pull clinging termites from holes in their mounds, blades of grass to extract clinging ants from their nests and rocks to crack open nuts. Sea otters (*Enhydra lutris*) hold a rock on their chest and crack shellfish, such as mussels and other mollusks, against it. Egyptian vultures (*Neophron percnopterus*) use stones to break open eggs. Herring gulls (*Larus argentatus*) drop clams, whelks and other shellfish on to hard surfaces such as a seawall, roadway or parking lot to break open the shells. New Caledonian crows (*Corvus moneduloides*) not only use probing tools to procure food in hard-to-reach places but they are capable of making tools out of objects in their environment, such as twigs and leaves (Kenward *et al.*, 2006).

Tool use in animals typically develops through an interaction of inherited species-typical anatomical characteristics and behavior (Beck, 1980). Social factors may accelerate its development when animals learn aspects of tool use from observing conspecifics. Tool use does not necessarily require a high level of cognition. Animals from many different genera use tools to gain access to food. Successful food retrieval (tool use) in naive New Caledonian crows is preceded by an inherited predisposition to manipulate objects found in their environment. For example, young naive crows have been observed picking up twigs with their beaks and moving them back and forth, as if they were using them to probe into a crevice or hole. There is no evidence that these precursor movements are reinforced in any way (Kenward *et al.*, 2006). Manipulation of twigs and other objects is refined through reinforced individual experience (i.e. trial-and-error learning). The fact that the efficiency of twig manipulation improves gradually rather than abruptly argues that the animal does not experience rational problem solving of an insightful nature. Interestingly, young crows showed a preference to handle objects they had seen being manipulated by their human caretakers, and juveniles that received demonstrations of twig tool use from their caretaker learned to insert twigs in holes more quickly than birds that had not seen such demonstrations.

Studies with primates show that tool use develops in a similar way to crows. There is a long period of object manipulation and exploration, with reinforced behaviors becoming more and more complex before tool use is mastered.

Observational (social) learning

There is evidence that tool use is passed on from generation to generation, with younger animals learning from older animals by observing their behavior. Group-living animals use body language to understand the most basic intentions of others (i.e. what they are trying to do). It only follows that animals possibly *could* learn to perform certain tasks by watching and imitating others. Behaviorists who have studied non-human primates in the wild have noticed repeatedly that certain behaviors (e.g. specific tool use, eating habits) practiced in one geographical area are not necessarily practiced in another area. The closer the groups live to one another, the more likely they are to share certain behaviors, presumably by the mixing of animals between different populations. Physically isolated populations are less likely to share unique behavioral patterns. Orangutan (*Pongo pigmaeus*) researchers working at six different sites in Borneo and Sumatra found 24 examples of behaviors practiced by at least one of the wild groups of orangutans studied but not the others (van Schaik *et al.*, 2003). Ten of these behaviors involve specialized feeding techniques, including tool use, and six are alternative forms of social signals. There is strong evidence that the young orangutans in each population learn these behaviors from their elders in each generation and that the unique behaviors are not due to habitat differences at the various sites.

Observational learning, or copying, greatly reduces the time and energy requirements associated with trial-and-error learning and, in some cases, costly mistakes. It can provide information on the best places to live, the best foods to eat, how to gain access to foods, which foods to avoid, mate selection, the meaning of communicatory signals, social status, recognition of predators and techniques to avoid them, and so on. The question one must ask, though, is whether an animal is truly imitating (copying) others or whether it is simply more likely to learn certain things (by simple conditioning or trial-and-error) merely by being attracted to the presence of conspecifics in certain places or near certain objects, or by the behavior of others that draw attention to certain key stimuli. Interestingly, true cognitive observational learning, that is, observational learning that does not involve a motor task, has only been documented in primates. In a recent study (Subiaul *et al.*, 2004), rhesus monkeys (*Macaca mulatta*) were trained to respond in a prescribed order to different photographs displayed on a touch-sensitive monitor. Because the position of the photographs on the monitor varied randomly from trial to trial, sequences could not be learned by motor imitation (i.e. choosing one location followed by the next in a specific order). The monkeys learned new sequences more rapidly after observing an 'expert' conspecific execute the sequences than when they had to learn new sequences by themselves using trial and error.

In reality, investigations which claim to study observational learning in domestic animals may have as much to do with drawing attention to stimuli associated with the task as with learning the task itself. Trial-and-error learning is facilitated when animals focus their attention on specific salient stimuli. Veissier (1993) gave cattle (*B. taurus*) the opportunity to observe trained demonstrator cattle push a panel with their noses to get a food reward before being tested themselves. Subjects which had observed the demonstrators performed no better than naive subjects when subsequently given the opportunity to learn the operant task. However, the observer cattle spent more time investigating the stimuli involved in the task during the limited time allotted.

Animals living in large social groups may be more observant of others, and thus less dependent on themselves for finding needed resources and avoiding predators. Studies on observational spatial memory illustrate this point. For example, both Mexican jays (*Aphelocoma ultramarina*) and Clark's nutcrackers (*Nucifraga columbiana*) cache (store) seeds and other food items, which allows them to survive periods of food shortages. They also remember where they have observed other birds caching food and will pilfer these food stores when given the opportunity. Mexican jays are as accurate at retrieving food from caches they have observed being created as they are at retrieving their own. In contrast, Clark's nutcrackers are much better at remembering the location of their own caches than the caches of conspecifics. Bednekoff and Balda (1996) explain this species difference (remembering the location of observed caching) based on differences in sociality. Nutcrackers live alone or in small family groups, while Mexican jays are more social.

Abstract thinking

Animals can acquire and use a wide range of abstract concepts involving geometric relationships, number, color and tool use. Chimpanzees (*P. troglodytes*) have been observed purposefully stacking boxes on top of one another to reach an overhead food reward (Fig. 5.8) or using a stick to retrieve an object beyond their reach. How do such problem-solving behaviors develop? Do animals use information previously gained by accident or through trial and error (i.e. standing on a box makes me taller) and then apply this information to solve a related problem (i.e. stacking boxes on top of one another makes me taller still)? To what extent can they apply rational or abstract thinking to foresee the end result of a series of actions? It seems that the jury is still out on that question.

Chimpanzees (*P. troglodytes*) and gorillas (*Gorilla gorilla*) can communicate with humans by visual sign language and object manipulation. We also know that they understand the meaning of gestures and other 'words' in their 'vocabulary' (Sevcik and Savage-Rumbaugh, 1994). Bottlenose dolphins (*Tursiops truncatus*) can distinguish the different meanings of certain words used in different orders or sequences (Herman, 2006). For example, dolphins can be trained to learn the meaning of words such as 'fetch', 'ball' and 'hoop', transmitted as artificial sounds to the animals using underwater speakers, and then perform various tasks using different combinations and sequences of the words they have learned. (The level of understanding was measured by the accuracy and reliability with which the animals carried out the instructions.) One report indicated that dolphins responded correctly about 80% of the time to more than 600 different two-word 'sentence' instructions. With experience, dolphins have been taught to respond to sentences up to five words long. Their ability to master sentence form and use may be related to the fact that dolphins are very social and their social relationships may be governed by complex social rules and conventions.

Animals can be trained to understand the meaning of number words or Arabic numeral symbols. African grey parrots (*Psittacus erithacus*) can determine the number of objects in a given class, up to six (Fig. 5.9; Pepperberg, 2006). If a chimp sees a person put two oranges in one box and two in another box, it will correctly choose the sum of four out of a line-up of three cards, each with a different Arabic numeral. To what extent do such behaviors indicate intelligence of a higher order? The performance of chimpanzees with numbers might suggest that they process them like we do. However, it should be noted that to acquire the arithmetic ability just described, chimps can require thousands of training trials over several years just to reach an integer list of nine. A human child who has acquired the numbers one, two and three will grasp the idea that the integer list is based on a succession of numbers and will eventually learn the numbers four and up quite quickly. In contrast, chimps require the same amount of time to learn *each* successive number on the integer list. Before conclusions are made, it is important to point out that the learning process of children versus chimps in a laboratory can be quite different. Children typically learn the order of the numbers before a meaning is attached to them, whereas chimps in a laboratory environment typically are taught the meaning of each number, one by one, without learning the numerical order.

Psychologists have found that children begin to recognize themselves in a mirror at 18–24 months of age. Mirror self-recognition is believed to mark the beginning of self-awareness, introspection and the ability to perceive the mental states of others. Gallup (see review by Gallup *et al.*, 1995) devised a 'mark test' to determine whether animals are capable of recognizing themselves in a mirror. The mark test consists of marking an animal with a spot of dye on its face while anesthetized and then later

Fig. 5.8. One of Wolfgang Köhler's chimpanzees using boxes and a stick to reach food (modified from Sparks, 1982). Do you think this technique was acquired through 'insight' learning ('a picture in the mind') or a series of trial-and-error experiences?

Fig. 5.9. Alex, an African grey parrot, has been used in a series of cognition studies involving vocal communication with his trainers (courtesy Irene Pepperberg).

giving it a mirror to see if it examines (i.e. touches) the otherwise unseen mark. Both chimpanzees (*P. troglodytes*) and orangutans (*P. pygmaeus*) have been observed engaged in facial self-examination while looking in a mirror. Monkeys and African grey parrots (*P. erithacus*) fail to use a mirror for body inspection.

Interestingly, monkeys and baboons have been observed reaching behind mirrors as if they expect another animal to be hiding there. There is some evidence that dolphins (*T. truncatus*) possess self-recognition (Reiss and Marino, 2001). In a series of tests, dolphins were marked with either circles and triangles or 'sham marked', in which the marker was filled with water and left no mark on the animals. The marks were placed on various parts of their bodies that could not be seen without the use of a mirror. After being marked, each dolphin swam directly to the mirror to investigate ('gaze at') the place where it had been marked, often twisting and turning its body to expose the correct spot.

Primates are remarkably sophisticated in their social relationships. They can form alliances, cooperate, deceive and bear grudges. Some of the great apes appear to show empathy for the feelings of others. For example, chimpanzees (*P. troglodytes*) have been observed rushing up to a member of its group who has just been defeated in a fight and placing its arm around the defeated animal, a behavior not seen in monkeys. It is still not clear if this behavior is a true form of consolation since it does not appear to alleviate stress in the victim (Koski and Sterck, 2007). Chimpanzees also display a sense of reciprocity over food sharing. Individuals that readily share food with others are more likely to become the recipient of future food sharing, or other favors such as grooming or sex. Chimps with 'stingy' reputations are more likely to be denied food items controlled by others. To what extent do these behaviors reflect a higher level of consciousness, a sense of what is 'right' or 'wrong', or a primitive form of morality? Some scientists believe that social animals tend to be more intelligent than non-social species. This makes sense, considering that social animals may be selected for their ability to communicate in rather sophisticated ways.

In spite of decades of testing, no one has found indisputable evidence that chimpanzees or other non-human primates can infer what others are thinking (i.e. that they can 'read' the minds of conspecifics). To illustrate, suppose animal A observes animal B hiding a piece of food under bush X. Soon after, animal B leaves the area and animal C comes along and accidentally finds animal B's food. Animal C takes the food and hides it under bush Y. Later, animal A sees animal B returning. Where does animal A expect animal B to look for its food? We humans would naturally expect animal B to look under bush X where it originally hid the food. Non-human primates and human children up to the age of about 5 expect animal B to look under bush Y, where the food is now located. They cannot correctly predict animal B's actions because they cannot infer what animal B is thinking. Psychologists believe that understanding the mental states of others is a specific adaptation separate from general-purpose intelligence, such as that used in simple conditioning.

6 Biological Rhythms

Introduction

The behavior and physiology of organisms often exhibit periodic changes that are cyclical in nature. The study of these biological rhythms is relatively new. At first, most of the work in this area was descriptive but, more recently, scientists have made great strides in identifying their molecular and physiological bases.

The primary questions being asked by scientists working in this area are: (i) What is the role of external environmental events in maintaining these rhythms of behavior and physiology?; (ii) What is the role of internal endogenous rhythms of molecular and physiological function in the organism that are independent of external environmental influences?; and (iii) Where in the organism do these endogenous physiological rhythms originate? While all three of these questions may arouse our intellectual curiosity, the role of external environmental events in maintaining behavioral and physiological rhythms is of particular interest to those working with domestic and captive wild animals.

Periodic External Environmental Events

The early literature on this subject often used the German term 'zeitgeber' (meaning 'time-giver') to refer to the periodic environmental events which control biological rhythms. The external 'time-givers' of particular importance to domesticated animals are ultimately related to the changing relationships of the earth and sun. Two cycles are of particular importance:

- Yearly or seasonal cycles associated with the earth's rotation around the sun (i.e. annual rhythms).
- Day–night periodicity associated with the earth turning on its axis (i.e. 24 h rhythms).

Annual Biological Rhythms

Perhaps the best-known examples of annual rhythms in behavior concern the seasonally based reproductive, migratory and hibernation behaviors of many bird and mammal species. When certain rodent species living in relatively cold climates are exposed to normal seasonal changes in day length and temperature, they exhibit an annual periodicity of wakefulness and hibernation of very close to 365 days. This annual cycle can be readily duplicated in a laboratory setting, even under constant environmental conditions. For example, when golden-mantled ground squirrels (*Citellus lateralis*) were placed in a room under a *constant* light:dark cycle (i.e. no change in the amount of light or darkness per day) and a constant cold temperature, they exhibited cycles of wakefulness and hibernation of *about* a year in length corresponding to changes in body weight and food consumption (Pengelley and Asmundson, 1974). These 'free-running' rhythms of wakefulness and hibernation are almost always less than 365 days in length, with many animals exhibiting cycles in the 300-day range. Some squirrels continued to show their free-running annual rhythm in the laboratory for 4 years (Pengelley *et al.*, 1976). These investigations demonstrated that the squirrels possessed a cyclical endogenous 'annual clock' mechanism of about a year that, in nature, is typically brought into phase with normal seasonal changes in day length and temperature corresponding to 365 days. These endogenous, 'free-running' annual rhythms of behavior and physiology in animals have been referred to as 'circannual' because they are *about* a year in length. The Latin word 'circa' means 'about'.

Seasonal cycles of mating activity have great practical significance to breeders of captive animals. For example, Hulet *et al.* (1974) described the seasonal changes in estrus and ovulation in female sheep (*O. aries*) taken from a single population but maintained in the states of Idaho and Texas. The incidence of estrous behavior and ovulation was relatively high during the fall and winter months and low between April and August at both locations (Fig. 6.1). However, the seasonal changes in behavior and physiology were not as great in Texas as in Idaho, perhaps because seasonal changes in day length are not as great in the southern USA as in more northern states. Other researchers have also noted that breeding seasons for livestock are less seasonal the closer one gets to the equator.

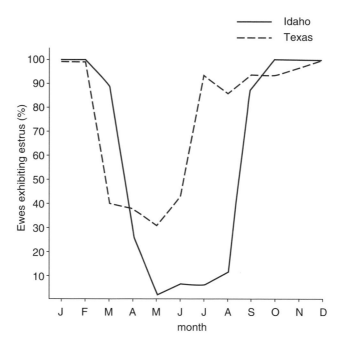

Fig. 6.1. Seasonal variation in the incidence of estrus for Rambouillet ewes drawn from the same populations and located at two different latitudes (Idaho and Texas) in the USA (from Hulet *et al.*, 1974.)

There are circumstances when it is desirable to breed captive animals during the 'off-season'. For example, horse (*E. caballus*) breeders sometimes use artificial light to advance the normal spring breeding season (Sharp *et al.*, 1975; Palmer and Guillaume, 1998). To do this, mares are typically exposed to 'long days' in winter by subjecting the horses to 14.5–16.0 h of light per day, starting in December until the the natural days are longer than 14.5 h. This same result can be obtained by exposing mares to an hour of artificial light about 10 h after sunset (Scraba and Ginther, 1985). In California, there is a year-round demand for goat milk and cheese. Because goats (*C. hircus*) are seasonal breeders (mate in the fall and give birth in the spring), there is a surplus of goat milk in the spring and summer months and not enough milk supply in the fall and winter. One large goat dairy decided to put some of their goats on an artificial light regimen to stimulate breeding in the spring of the year so their herd would produce a more consistent year-round supply of milk (BonDurant *et al.*, 1981). Starting on 1 February, the goats were placed in a barn with artificial fluorescent lighting 20 h/day (Fig. 6.2). The artificial lighting was terminated 70 days later, thus exposing the goats to a shorter day than normal for that time of year, which simulated the decreasing day length of the fall breeding season. Both males and females exhibited significantly greater breeding activity than the non-treated control goats (Table 6.1). This relatively simple approach to off-season breeding may be limited only by the facilities available to expose animals to supplemental artificial lighting.

As a graduate student at Michigan State University, I was hired to be the caretaker of my major professor's colony of deermice (*Peromyscus*). One species, *Peromyscus maniculatus gracilis*, proved to be very difficult to breed in laboratory cages. Knowing that in nature *P. m. gracilis* breed best in the spring of the year when day length is increasing, and considering the fact that our windowless laboratory was kept at a constant 12 h of light per day year round, we surmised that the constant

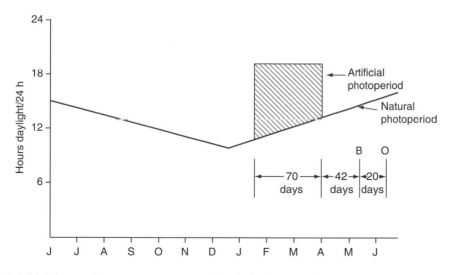

Fig. 6.2. Lighting and management protocol for inducing out-of-season breeding of domestic goats. Day length was artificially prolonged for 70 days starting in January. Male goats (B) were introduced 42 days after the termination of the artificial photoperiod and first female estrous (O) was observed 20 days later (BonDurant *et al.*, 1981).

Table 6.1. Photoperiod induction of estrus and ovulation in California dairy goats (from BonDurant *et al.*, 1981).

Treatment	No. of subjects	Number of females that		
		Were mounted by male	Ovulated	Conceived
Artificial lighting	19	16 (84%)	15 (79%)	12 (63%)
Natural lighting	6	1 (17%)	0	0

light:dark cycle might be part of the problem. Hence, we decided to increase the daily amount of light in the laboratory by 3 h, in 1 h increments, over a 2-week period. An increase in litters born was noted by the 5th and 6th weeks after initiating the light increase and, by the 7th week, the number of females giving birth was seven times greater than the average of the 13 weeks prior to the light increase and six times greater than the weekly average for the previous year (Price, 1966). For some reason, photoperiod changes proved to be more important for reproduction in our *P. m. gracilis* mice than for the other species of deermice in the mouse colony.

Daily (24 h) Rhythms

Twenty-four-hour rhythms have been described for a wide variety of behaviors and physiological parameters. Perhaps the best-known behavioral variables are the daily rhythms of activity and ingestive behaviors. Body temperature, heart rate and circulating hormone levels also cycle on a 24 h basis.

Animals are classed as diurnal if they are predominantly active by day, nocturnal if mostly active by night and crepuscular if predictively active at dawn and (or) dusk. Most of our common domestic animals are basically diurnal or crepuscular, although they may be somewhat active at night. Domestic cats and rodents tend to be nocturnal.

A variety of external environmental factors have been postulated to maintain these 24 h rhythms, including light:dark cycles, daily changes in temperature, food availability, social cues and such geophysical phenomena as geomagnetism, polarized light and cosmic radiation. The two factors most thoroughly studied are daily light: dark cycles and temperature fluctuations.

Somewhat regular changes in temperature occur over a 24 h period. However, daily temperature fluctuations are not as consistent or predictable as changes in light: dark (L:D) cycles. Studies with animals subjected systematically to regular daily temperature changes have shown that 24 h rhythms are relatively independent of changes in temperature. On the other hand, animals subjected to a 24 h cycle of light and darkness exhibit an accurate 24 h rhythm of activity and inactivity (Fig. 6.3).

Twenty-four-hour rhythms not only influence daily cycles of behavior and physiology but also enable species using the sun or moon as a compass for navigation to compensate for the daily movement of these celestial bodies across the sky. Without such a time sense, the sun or moon would be an unreliable indicator of compass direction. Froy *et al.* (2003) showed that monarch butterflies (*Danaus plexippus*) in eastern North America migrate to Mexico for the winter using a time-compensated

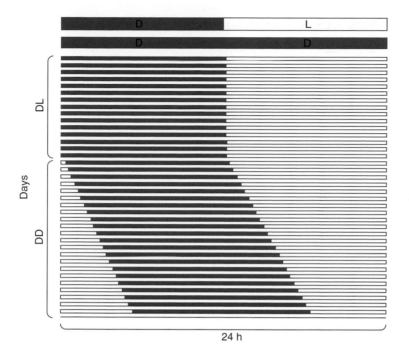

Fig. 6.3. Schematic illustration of hamster wheel-running activity in 12 h of light and 12 h of darkness (DL) and then in constant darkness (DD). Each horizontal strip represents 24 h and the solid dark bars represent periods of wheel running. Being nocturnal, hamsters are active during the hours of darkness. In constant darkness, the hamster's circadian rhythm of activity is expressed (i.e. 'free runs') with the animal initiating activity a little later each consecutive day (Silver, 1990).

sun compass to navigate in a south-westerly direction. Monarchs maintained in the laboratory on an L:D cycle that simulated natural conditions in eastern USA showed a normal south-west orientation, no matter what time of day they were released in the wild. They could compensate for the different positions of the sun along its arc at different times of the day. However, if the monarchs were maintained under constant light for a number of days, they would fly directly toward the sun when released, no matter where it was in the sky. The butterflies needed L:D entrainment to use the sun as a compass guide.

The activity rhythms of animals subjected to continuous light or continuous darkness in a laboratory setting are typically *about* 24 h in length (23 h, 30 min to 24 h, 30 min) but rarely exactly 24 h. These so-called 'free-running' rhythms are endogenously controlled and are referred to as 'circadian', from the Latin words 'circa' meaning 'about' and 'diem' meaning 'day'. The circadian rhythms of diurnal animals tend to be longer than 24 h and nocturnal animals tend to have circadian rhythms shorter than 24 h. The average for humans is close to 25 h and domestic mice (*M. musculus*) have circadian rhythms averaging about 23.7 h. In either case, these endogenous circadian rhythms are subject to entrainment by external 24 h light:dark cycles. The daily light:dark cycle is a very effective 'zeitgeber' or entraining agent.

Shifting the Clock Relative to Local Time

If an abrupt shift is made in the L:D cycle, as when we take a plane from San Francisco to Paris, our endogenous clock shifts but the change occurs gradually over days until it has fully entrained to the L:D cycle in Paris. Domestic rats (*R. norvegicus*) require 10+ days to reset their circadian rhythms to a L:D cycle 12 h out of phase with local time. Domestic hamsters (*Mesocricetus auratus*) can require up to 3 weeks to reset their activity rhythms to a 12 h shift in the L:D cycle. Figure 6.4 illustrates the gradual shifting of a hamster's running-wheel activity in response to abruptly advancing the L:D cycle by 4 h.

If the L:D cycle of migration-ready monarch butterflies (see previous section) is advanced by 6 h in the laboratory, the average compass direction of their subsequent flight is south-east (rather than south-west), maintaining the same angle to the sun they normally would assume if the sun was 6 h farther along its arc from east to west. The insects were dependent on their time sense to use the sun as a compass, and a L:D shift for their geographical location resulted in a predictable deviation in compass direction traveled.

The circadian rhythms of animals will not entrain to L:D cycles considerably different from 24 h. In mammals, the range is from 20 to 28 h. Tribukait (1956) subjected domestic mice (*M. musculus*) to a 24 h L:D cycle and then gradually increased and decreased the period length to 16 and 29 h. (The hours of light and darkness were equal in all cases.) He found that his mice would entrain to L:D cycles from 21 to 27 h in length. Schedules shorter or longer than this would cause the mice to revert to their normal circadian periodicity. He noted that entrainment to these limits could be achieved only if the L:D cycle was moved gradually. If the cycle length was shortened abruptly from 24 to 22 h, entrainment failed.

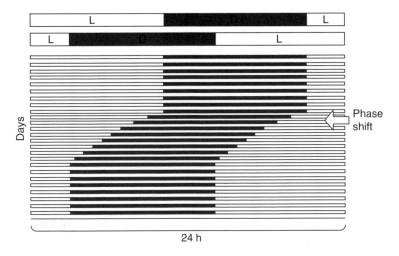

Fig. 6.4. Schematic illustration of hamster wheel-running activity during a 12:12 L:D cycle. When the onset of darkness is advanced by 4 h (see arrow), the activity rhythm *gradually* shifts over days to become entrained to the new photoperiod. As previously, each horizontal strip represents 24 h and dark bars represent wheel-running activity (Silver, 1990).

Endogenous Basis of 24 h Rhythms

Circadian rhythmicity in mammals is dependent on the normal functioning of the suprachiasmatic nucleus (SCN) in the anterior hypothalamus of the brain (Fig. 6.5). The SCN serves as a 'pacemaker' by coordinating oscillating components in other organs of the animal's body (e.g. heart, liver, kidney). It also secretes at least one 'locomotor activating factor' at one phase of the L:D cycle and at least one 'locomotor inhibitory factor' at another phase. The receptors for the secreted SCN factors are believed to be located in the subparaventricular zone near the third ventricle of the hypothalamus. Recent research (Cheng *et al.*, 2002) revealed that the factors referred to above (i.e. the SCN's 'time message') were embodied in a protein called prokineticin 2 (PK2). Levels of PK2 secreted by the SCN oscillate during a 24 h period and control the circadian locomotor activity rhythm.

When the SCN of a mammalian brain is removed, the animal is still capable of entraining to a light:dark (L:D) cycle. However, circadian rhythmicity is lost under constant lighting conditions (i.e. no external entraining agent) (Fig. 6.6). Circadian rhythmicity can be restored in the SCN-deficient animal by neural grafts of SCN cells from another animal and the rhythmicity expressed is that of the donor (Ralph *et al.*, 1990).

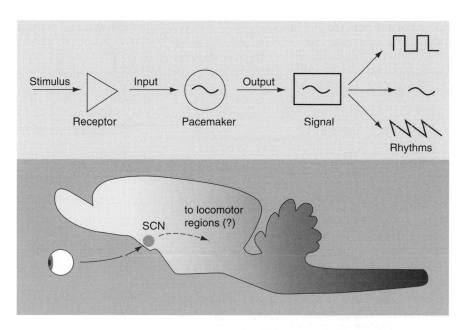

Fig. 6.5. Top: Circadian systems contain three basic elements. Rhythms are set in synchrony with information from environmental cues (e.g. light:dark), which is transmitted along an input pathway to a pacemaker or clock that lies at the heart of the system. The pacemaker relays an output signal, which drives rhythms of organismal activity such as locomotion and reproductive, cellular or metabolic activity. Bottom: In the rat, information about light enters the eyes and is relayed to the suprachiasmatic nucleus (SCN). The rat's SCN contains a master circadian clock, which sends signals to other parts of the rat's brain that control various activities such as periods of physical activity and rest (Takahashi and Hoffman, 1995).

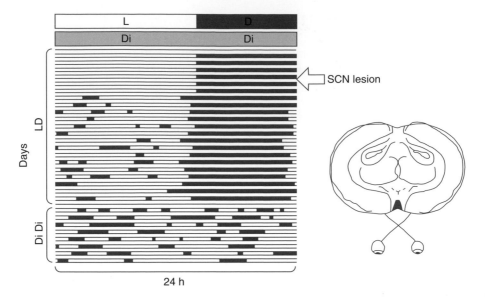

Fig. 6.6. Schematic illustration of the consequences of removal of the suprachiasmatic nucleus (SCN) in a hamster. Before the SCN lesion, wheel-running activity occurred only during hours of darkness. After lesioning (arrow), the hamster remained active during the dark hours but some activity also occurred during the light phase. It was still subject to entrainment to the L:D cycle. When the hamster was placed in constant dim light, it did not show a synchronized free-running rhythm, suggesting that the SCN was responsible for the animal's circadian activity cycle (Rusack, 1977; Zucker, 1980; figure from Silver, 1990).

Rodents have functioning circadian clocks in the fetal stage and these clocks are entrained by signals from the mother. Marsupials such as the opossum (*Didelphis virginiana*) are born at an earlier stage of development and most SCN neurogenesis occurs post-natally.

The retinohypothalamic tract connects the retina to the SCN and thus delivers messages regarding L:D cycles to the SCN. These messages entrain the activities of the SCN to an exact 24 h rhythm. Retinal rod and cone cells, which are responsible for vision in most vertebrates, are not required for photoentrainment. Rather, in mammals the entraining photoreceptors are believed to be a small subset of retinal ganglion cells. The light-absorbing pigments (proteins) for these cells in the retina are melanopsin and probably cryptochrome, which appear to be capable of transmitting light information to the SCN in the absence of rods and cones.

In mammals, L:D signals important to rhythmicity originate only in the retina of the eye, whereas in non-mammalian vertebrates (fish, amphibians, reptiles and birds), light receptors other than the retina (extraocular photoreceptors) are found in several places in the brain. These cells respond to light that penetrates the skin, skull and overlying brain tissue and their outputs act directly on the clock centers in the brain. Eyes (retinas) are not needed for these non-mammalian vertebrates to synchronize their activity to light:dark cycles but, when present, they increase the sensitivity of synchronization.

Circadian rhythmicity in birds and many other non-mammalian vertebrates is dependent on both the SCN and the pineal gland. The pineal gland (see Fig. 6.7 for

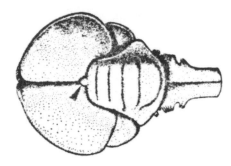

Fig. 6.7. Location of the bird pineal gland (arrowhead) in a dorsal view of the brain.

location in the brain of birds) is more important to endogenous rhythmicity in some species (e.g. sparrows) than others (e.g. starlings and chickens). As in mammals, ablation of the SCN and pineal gland eliminates endogenous rhythmicity.

Melatonin, a hormone secreted by the pineal gland, also has an important role in biological rhythms (Gorman *et al.*, 2001). Melatonin secretion affects seasonal rhythms of reproduction and molting in many species by controlling their physiological responses to photoperiod (i.e. changes in day length) via hormone secretion by the pituitary gland. Melatonin is secreted primarily during the hours of darkness. The *duration* of night-time melatonin secretion determines the species' response. Sheep (*O. aries*) show an increase in gonadal activity in response to relatively long durations of night-time melatonin secretion as day length becomes shorter. Lincoln (1992) has shown that a 16 h-long infusion of melatonin in pinealectomized sheep induces a short-day response (e.g. breeding activity), while an 8 h infusion does not. The opposite is true in the hamster (several species), which typically breeds in the spring when day length is increasing. Relatively long-duration infusions of melatonin in pinealectomized Djungarian hamsters (*Phodopus sungorus*) inhibit gonadal activity and short-duration infusions increase reproductive activity (Carter and Goldman, 1983).

Cyclical melatonin release can also affect 24 h rhythms. In some species, particularly non-mammalian vertebrates, the pineal gland serves as a 'pacemaker', controlling the functioning of circadian clocks located elsewhere in the body. Removal of the pineal gland in these species usually results in the animals becoming arrhythmic in their motor activity under constant light (or dark) conditions. Daily injections or presentations of melatonin in drinking water can entrain the circadian activity rhythms of various species of pinealectomized reptiles, amphibians and birds (Underwood, 2001).

Genetic Basis of Circadian Rhythms

Recent research has shown that the length of the endogenous circadian rhythms of animals can be influenced by genetic mutations. In the early 1970s, Konopka and Benzer (1971) discovered the *per* ('period') gene in fruit flies (*D. melanogaster*), which plays a critical role in maintaining their circadian rhythms of activity and inactivity. Mutations in the *per* gene can alter a fly's circadian rhythms in any one of several ways, abolishing them altogether, for example, or making them longer or shorter. Later, a gene called *tim* ('timeless') was found in fruit flies, which also makes a key protein component of their clock mechanism. *Per* and *tim* proteins work together to generate an oscillating cycle of activity which establishes the fly's daily rhythms of physiology and activity. In addition, some five other genes are believed to play a role in the functioning of the clock in this species. Mammalian homologs (i.e. similar functioning genes) have been found to all of these fruit fly clock genes. Takahashi and co-workers (reviewed by Takahashi and Hoffman, 1995) found a *clock* gene in domestic mice (*M. musculus*), which produces the normal 23.7 h circadian cycle of resting and running-wheel activity when the mice are kept in total darkness. A mutation of the *clock* gene results in a circadian rhythm close to 25 h, on average, in the heterozygous condition (of alleles) and a 28 h endogenous clock in the homozygous condition, which is completely lost after about 2 weeks in total darkness. Ralph and Menaker (1988) discovered a mutation of the *tau* gene in the domestic hamster (*M. auratus*), which produces either a 22 h (heterozygous gene condition) or 20 h (homozygous condition) circadian rhythm in constant darkness. Animals that carry the mutant alleles exhibit abnormal entrainment to 24 h L:D cycles, or are unable to entrain.

Interval Timing Clock

Some animals are capable of judging the passage of time. We mentioned in the chapter on learning how animals being conditioned on a fixed time-interval reinforcement schedule will refrain from responding for a while after reinforcement and then show a sharp increase in response rate as the time-based interval ends. In many bird species, the male and female trade off incubation duties at predictable intervals. For example, the female ring dove (*Streptopelia risoria*) incubates her eggs for all but several hours of the day,

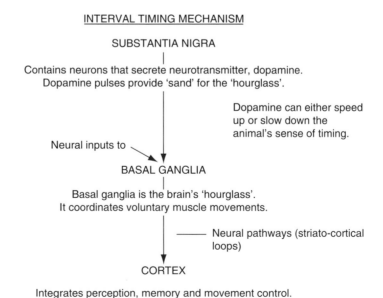

Fig. 6.8. Schematic of the interval timing mechanism (Hinton and Meck, 1997).

when she is relieved by the male. The female uses her *circadian* clock to begin incubating at the same time each day and the male uses his *interval* clock to determine the length of his stay on the nest (Gibbon *et al.*, 1984). Laboratory experiments have shown that if the male's arrival is delayed for a few hours in the morning, the male does not want to leave the nest when the female arrives in the afternoon (i.e. his time on the nest is not up) and the pair may actually fight over whose turn it is to tend the nest.

Fortunately, much of the physiological basis for the interval timing mechanism in animals has been determined (see review by Hinton and Meck, 1997) (Fig. 6.8). The basal ganglia of the brain coordinates voluntary muscle movements and sends messages to the thalamus and then on to the cortex of the brain via the striato-cortical loops. The cerebral cortex integrates perception, memory and movement control. The basal ganglia serve as a metaphorical 'stopwatch' or 'hourglass' for interval timing. The hourglass needs a reservoir of 'sand' to register the passage of time. The substantia nigra contains neurons, which secrete the neurotransmitter dopamine. Pulses of dopamine provide 'sand' for the 'hourglass', which 'fall' into an accumulator, the caudate-putamen, a major part of the basal ganglia. Dopamine secretion from the substantia nigra can either speed up or slow down the animal's sense of timing. Studies have shown that lesions in the caudate-putamen or substantia nigra stop the interval-timing clock. Humans with Parkinson's disease typically have difficulty judging the passage of time and, interestingly, are deficient in dopamine.

Introduction to Chapters on Social Behavior

Many definitions of social behavior have been proposed. Scott (1958) defined social behavior as 'that behavior which is either stimulated by or has an effect on another member of the same species'. I like Scott's definition because it implies a central role for reciprocal communication in social interactions. The topic of social behavior encompasses a major part of the rest of this book. This is appropriate since most of our domesticated animals are highly social. In the next few chapters, we will discuss reproductive behaviors, including mating systems and information important to understanding sexual and parental behaviors. We then move to the topic of communication, with particular emphasis on the role of olfaction in the lives of domestic animals. The next chapter deals with conflict behaviors (agonistic behavior) and is followed by the subject of social organization, how domestic and captive wild animals achieve a degree of order in their social lives. Lastly, we tackle the topic of personal space and social dynamics, exploring some of the finer points of animal sociality.

Before we start the next chapter, I would like to raise a question that my students often asked. Are interspecies interactions (e.g. dog–human interactions) social or must we confine social behavior to interactions between members of the same species? I must admit that I have had trouble answering this question in a precise manner. Scott's definition of social behavior limits it to members of the same species. This makes good sense when speaking of free-living wild animals, which typically only form social bonds with conspecifics (species members). However, interspecies bonding is not uncommon for captive animals. Almost anyone will agree that a certain degree of bonding (with us) takes place when a dog or cat is taken into our home. We become attached to our animals and, in most cases, they appear, through body language and behavior, to become bonded to us. It would not be too far-fetched to assume that our pet dog (*C. familiaris*) relates to us like its wolf (*C. lupus*) ancestors relate to members of their 'pack'. In other words, if captive animals become socially bonded to humans, their interactions with humans should be considered social in nature. Interactions between two species (e.g. rat and cat) are not social, unless a social bond has been formed, possibly through filial or species imprinting. Using this criterion, an action whereby an individual of one species alerts an individual of another species to impending danger (e.g. approach of a predator) would not be considered social unless the two individuals live together and are socially bonded to some degree. Do you agree or disagree?

A related point is that animals found in groups are not necessarily social or interacting socially. Flies can be independently attracted to dung, or different species of ungulates can be attracted to a waterhole without interacting socially. We simply refer to such groups as 'aggregations'.

7 Mating Systems and Reproduction

Mating systems determine the relative roles of males and females in providing progeny for the next generation. There are two basic types of mating systems exhibited by vertebrates, monogamy and polygamy (Emlen and Oring, 1977).

Monogamy

The term monogamy is derived from the Greek words 'monos' meaning 'single' and 'gamos' meaning 'marriage'. In monogamy, neither sex monopolizes additional members of the opposite sex. Reproductive activities typically involve one male and one female. Pairs may associate for a season (e.g. many songbird species) or for as long as their mate lives (e.g. geese, gibbons). Reproductive fitness is often maximized through shared parental care of the offspring. Both male and female have significant roles in the nurturing of young. For males, monogamy reduces the risks associated with prolonged competition for females (e.g. injury, energy depletion).

Some investigators insist that monogamy should include the sharing of parental care. In the monogamous California mouse (*Peromyscus californicus*), bonded females ignore other males and bonded males shun other females, even those in estrus. Tests conducted on 28 families of California mice in the wild showed that all young were sired by their mother's pair-mate. Biparental care is a requirement in this species (Gubernick and Alberts, 1987). The female cannot rear the litter of one to three pups by herself. The young are rather helpless (altricial) when born and cannot control their body temperature. Mother and father take turns huddling over the young to keep them warm, a critical requirement for successful nursing and pup survival. If the male leaves the area or is taken away, the female will abandon or kill her pups.

Insights into the sexual and physiological basis of monogamy have been obtained by comparing two closely related rodents, the monogamous prairie vole (*Microtus ochrogaster*) and the polygamous montane vole (*M. montanus*). Male–female pairs of prairie voles in nature share a common nest, home range, mate defense and a low incidence of re-mating if one of the pair dies. Shapiro and Dewsbury (1990) found that when trios of captive prairie voles consisting of a single male and two females were placed in enclosures, they spent 59% of their time in side-by-side contact. An identical laboratory study with montane voles showed that they spent only 7% of their time in contact. Male prairie voles preferentially paired and nested with one of the females in their cage, whereas male montane voles did not show a preference for either female, even after mating.

Carter and Getz (1993) examined the hormonal changes occurring in prairie and montane voles during their respective reproductive cycles. Copulation in the monogamous prairie voles releases oxytocin in females and vasopressin in males, both powerful brain neuropeptides (hormones) needed for pair bonding. Oxytocin is associated with maternal behavior and lactation, while vasopressin is associated with male aggression and paternal behavior. Pair bonding did not occur when the action of these hormones was experimentally blocked. These same two hormones are released by copulation in the more polygamous montane voles. However, vasopressin affects brain circuits differently in males of the two species (Fig. 7.1). Male prairie voles had high numbers of vasopressin receptors in the thalamus and olfactory regions of their brains, while male montane voles showed more vasopressin activity in the lateral septum, which has been implicated in aggression. Experimental trials have shown that pair bonding can be induced in male prairie voles by injections of vasopressin. Within a day, the experimental prairie voles showed a preference for their mates and defended them against rivals. In contrast, vasopressin injections did not facilitate pair bonding in male montane voles.

Scientists have distinguished between social monogamy and sexual monogamy. Social monogamy refers to monogamous pair associations involving one male and one female. Sexual monogamy refers to pairs that confine sexual relations to one another. Social monogamy does not guarantee sexual monogamy. Social monogamy is not very common in the animal kingdom. It is estimated that only 3–10% of

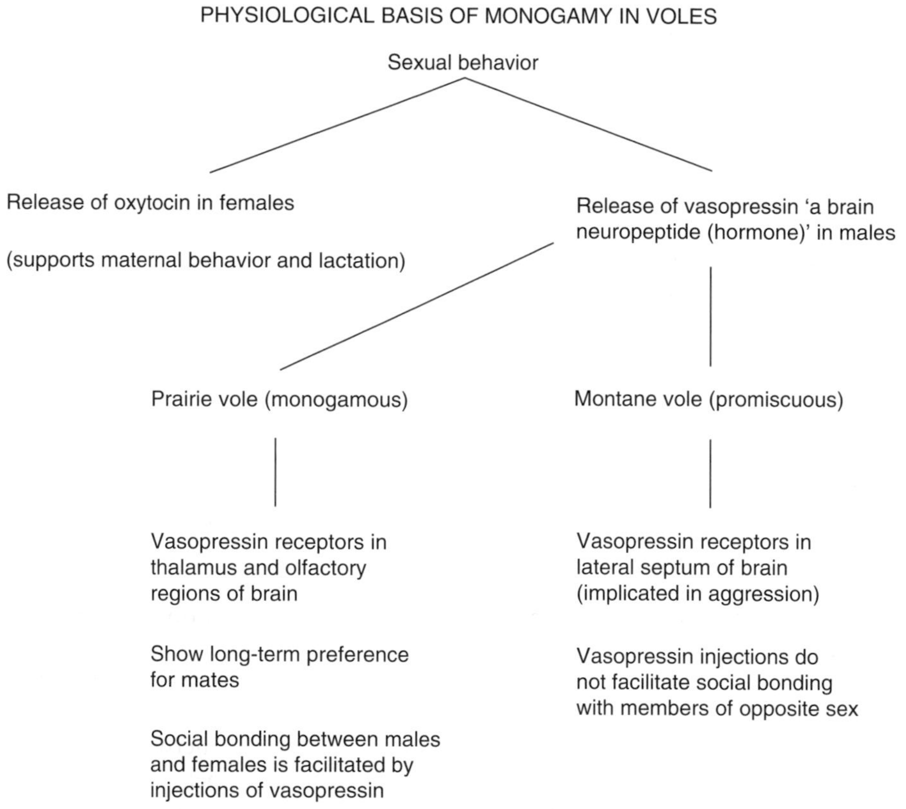

PHYSIOLOGICAL BASIS OF MONOGAMY IN VOLES

Sexual behavior

Release of oxytocin in females

(supports maternal behavior and lactation)

Release of vasopressin 'a brain neuropeptide (hormone)' in males

Prairie vole (monogamous)

Montane vole (promiscuous)

Vasopressin receptors in thalamus and olfactory regions of brain

Show long-term preference for mates

Social bonding between males and females is facilitated by injections of vasopressin

Vasopressin receptors in lateral septum of brain (implicated in aggression)

Vasopressin injections do not facilitate social bonding with members of opposite sex

Fig. 7.1. Schematic of the physiological basis of monogamy in voles. Vasopressin, a brain neuropeptide, facilititaes social bonding between male and female prairie voles (monogamous) but not montane voles (promiscuous) (Carter and Getz, 1993).

mammals are socially monogamous. A smaller percentage is sexually monogamous. Studies on the DNA of chicks of some 180 socially monogamous species of songbirds indicate that only about 10% are sexually monogamous. Gowaty and Karlin (1984) conducted a long-term study of the 'monogamous' Eastern bluebird (*Sialia sialis*) and reported that 5% of the adult males and 15% of the adult females in their study population were caring for at least one offspring not their own. A similar study with red-winged blackbirds (*Agelaius phoeniceus*) revealed that nearly half of the nests examined contained at least one illegitimate chick fathered by a male holding a nearby territory (Gibbs *et al.*, 1990). Females of several avian species actively seek extra-pair copulations. Female hooded warblers (*Wilsonia citrina*) use a special song to solicit such matings.

In captive populations of monogamous animals, care must be taken if one wishes to breed males to more than one female. Enders (1945) reported that fox breeders were careful not to allow males to remain with females for more than a few hours. Longer cohabitation frequently results in the development of female preferences, which can interfere with the male breeding successfully with other females.

In nature, we assume that socially monogamous pairs consist of individuals who are mutually compatible; that is, they display relatively low levels of aggression toward one another and relatively high levels of affiliative behaviors, such as maintaining close physical proximity and frequent engagement in mutual grooming and sexual behaviors. Often, however, when humans pair captive animals, little concern is given to whether pairs are behaviorally compatible. As a result, pairings often fail to produce offspring and sometimes individuals are injured, or even killed. Spoon *et al.* (2006) explored the relationship between behavioral compatibility and reproductive success in cockatiels (*Nymphicus hollandicus*), a socially monogamous species with biparental care and variability in compatibility among randomly chosen pairings. As hypothesized, cockatiel pairs with significantly higher compatibility ratings laid more eggs, exhibited greater coordination of incubation duties, had greater hatching success of fertile eggs and raised more chicks to independence.

Polygamy

In polygamy, individuals frequently control or gain access to multiple mating partners. This may occur either simultaneously or sequentially. Pair bonds are either temporary or non-existent (i.e. the relationship is promiscuous). Parental care is not shared. There are two forms of polygamy, namely polygyny and polyandry. In polygyny, males have multiple mating partners in each breeding season. In polyandry, females mate with more than one male. Most of our common domestic animals are polygynous.

Polygamous species are favored for domestication since only a few males are needed to impregnate many females, thus reducing the total number of animals needing to be accommodated and needed to maximize the reproductive success of the population. The lack of pair bonding and promiscuous mating behavior permit males to impregnate a large number of females in a relatively short period of time. Nearly all mammal and bird species which have been domesticated for meat (and egg) production are polygamous and mate promiscuously. For example, broiler chickens (*G. domesticus*) are typically bred by placing eight or nine males with 100 females at

20–30 weeks of age (Hazary *et al.*, 2001) to maintain optimum fertility. One exception is the rock dove or pigeon (*C. livia*), which must pair bond to reproduce successfully.

Emlen and Oring (1977) discuss three forms of polygyny (Table 7.1). In female (or harem) defense polygyny, males control access to females directly, usually by virtue of female gregariousness. Male elk (*Cervus canadensis*) in North America defend groups of females (from other males) during the fall breeding season (Fig. 7.2). Control of females is aided by their gregarious behavior. In resource defense polygyny, males control access to females indirectly by monopolizing critical resources (e.g. territories). Individual male ring-necked pheasants (*Phasianus colchicus*) (Fig. 7.3a) defend territories during the breeding season. Female pheasants are attracted to the resources (e.g. food, shelter, etc.) in those territories, giving the male territory holder preferential access to these females. Male dominance polygyny occurs when it is not feasible for males to monopolize mates or critical resources. Males aggregate during the breeding season and females select mates from these aggregations. The more dominant males mate with multiple females. Male and female sage grouse (*Centrocercus urophasianus*) gather in groups on traditional 'strutting grounds', or 'leks', during the breeding season in North America (Fig. 7.4). Dominant males are most successful in attracting females and intimidating rival males, and thus have multiple mating opportunities.

Polyandry is much less common than polygyny. In some polyandrous bird species, females lay multiple clutches of eggs and males do all the incubating. Female American jacanas (*Jacana spinosa*), large wading birds found in Central and South America, control access to multiple males indirectly by monopolizing critical resources (resource defense polyandry). Jenni (1974) found that female breeding sites were divided into relatively small male-held territories. These female 'super-territories' may encompass the nesting areas of several males. Females lay clutches of eggs for males on their respective territories and provide replacement clutches if nests are lost

Table 7.1. Polygamous mating systems in vertebrates (from Emlen and Oring, 1977).

Polygyny (males)	
Female (or harem) defense polygyny	Males control access to females directly, usually by virtue of female gregariousness.
Resource defense polygyny	Males control access to females indirectly by monopolizing critical resources (e.g. territories).
Male dominance polygyny	Males aggregate during the breeding season and females select mates from these aggregations (e.g. leks).
Polyandry (females)	
Resource defense polyandry	Females control access to males indirectly by monopolizing critical resources (e.g. territories).
Female dominance polyandry	Females do not defend resources essential to males but through dominance gain access to males.

Fig. 7.2. Bull elk with his harem during the fall breeding season (courtesy of Agustin Orihuela).

through predation. Breeding females are much larger than males, dominate males and provide little parental care. Females specialize only in egg production. In female dominance polyandry, females do not defend resources essential to males but, through dominance, gain access to males. Honeybee (*A. mellifera*) queens do not defend critical resources themselves but, by their status in the hive, mate with multiple males (drones).

The Greater Rheas (*Rhea americana*) of Argentina are particularly interesting because the males are *simultaneously* polygynous and the females are *sequentially* polyandrous (Codenotti and Alvarez, 2001). As the mating season approaches, adult males associate with and defend a harem of 2–12 females. Together, they participate in nest building, mating and egg laying over a period of 30–45 days. The females then depart, leaving the male with incubation duties and the care of young. The temporarily unattached females subsequently become integrated into another male's harem, lay eggs and the cycle is repeated.

Mating systems for a species can differ with the environment. African lions (*P. leo*) are typically polygamous in the Serengeti and monogamous in the Kalahari Desert. Biparental care of young may be a critical requirement in the harsh Kalahari environment.

Mating systems can have profound effects on species integrity. There are some 19 recognized populations or races of Canada geese (*Branta canadensis*) living in the USA. Differences are based primarily on variation in plumage coloration and size. On the other hand, most duck species found in the USA look the same externally, whether they are hatched on the east or west coast. Biologists have determined that the

Fig. 7.3. Male ring-necked pheasant in full breeding plumage (a) and characteristics of the head of male pheasants in non-breeding and breeding condition (b). Note the engorgement of the wattles and length of the ear tufts in the breeding male (right) (Hill and Robertson, 1988).

differences between Canada geese and ducks in morphological integrity are based on their social behavior and mating systems. Canada geese are monogamous and mate for the life of their partner. The male participates in the care of the young and both mother and father stay with their young through the first year. Adult geese lead their young on their first migration and then back to the breeding grounds. Consequently,

Fig. 7.4. Male sage grouse displaying to female on a lek (mating area) (courtesy of Tom Tietz).

populations of Canada geese do not intermix freely with one another and, as a result, different races have developed in different parts of the country. In contrast, pair bonds in most duck species last only until the female begins to incubate her eggs. The male then leaves and seeks new mating opportunities. The female abandons her young late in the brood-raising period and offspring are unlikely to associate with their parents for the rest of their lives. Males and females disperse during migration and join birds from other areas. As a result, pair bonds are formed between ducks hatched in different parts of the country, precluding the development of races or subspecies.

Sexual Selection, Sexual Dimorphism and Mate Choice

Sexual selection can be defined as natural selection between members of the same sex (intrasexual selection), which results in the evolution of traits that enhance an individual's ability to acquire mates. Sexual selection is a form of natural selection in which individual males or females compete with conspecifics of the same sex to gain the favor of the opposite sex.

In the polygynous mating system, one male can impregnate many females. Consequently, there is male–male competition for the opportunity to mate with females, and the breeding success of males depends, in part, on their ability to prevent other males from gaining access to females. The largest, most attractive and most dominant males, by virtue of their competitive abilities, will be most successful in gaining access to females and will leave the most offspring for the next generation. Competition between males can be costly in terms of energy expenditure, injury and survival.

Male–male competition and the resulting intrasexual selection among males is significant in species exhibiting a polygynous mating system because males are expendable. One male can inseminate many females. In addition, males do not participate in parental care. Their primary reproductive function is to impregnate females. In most cases, male–male competition can take place without seriously interfering with the reproductive potential of the population as a whole.

Breeding success among polygynous species is not only dependent on the ability of males to compete with one another for access to females but on their relative attractiveness to females once contact has been made. In most polygynous species, females choose their mating partners and males compete to be chosen. It is believed that female choice in mating is commonplace in the polygynous mating system, presumably since females invest so much more than males in each offspring (Trivers, 1972). Attractiveness to females may depend on such secondary sexual characteristics as body size, physical adornments (e.g. brightly colored feathers and body parts, antlers, horns, tusks, etc.), elaborate vocalizations or control of resources (e.g. food supply, territory, etc.), as well as the relative vigor of the male as it interacts with a female. Many secondary sexual characteristics are controlled by androgens (e.g. testosterone). Thus, it is not surprising that females often choose as mating partners the dominant males with relatively high levels of androgen.

Males better at gaining access to females, and who are most attractive to females, will be favored in the sexual selection process. Larger males will be favored over smaller males and males which vigorously attend to females will more likely attract their interest. More impressive adornments or weapons will be favored over less impressive ones (Fig. 7.5). In the ring-necked pheasant (*P. colchicus*), tail length,

Fig. 7.5. Male peacocks have been selected for large, colorful tail feathers, which they display when courting females.

length of the ear tufts and the presence of black points in the wattle (Fig. 7.3b) positively influence female choice, but not wattle size or color or brightness of the plumage (Mateos and Carranza, 1995). The result is sexual dimorphism between the sexes, in which males are larger than females (except in species exhibiting polyandry) and possess more impressive adornments, weapons or vocalizations (Fig. 7.6a,b). For example, mature male bighorn sheep (*O. canadensis*) weigh 75% more than adult females (Pelletier and Festa-Bianchet, 2006) and possess much larger horns for their age. Alexander *et al.* (1979) provide a long list of sexually dimorphic mammalian species. Interestingly, the polyandrous female American jacana (*J. spinosa*) is 50% larger than the male.

It seems reasonable to assume that secondary sexual characteristics are expressed most extravagantly by the more vigorous adult males in the population. The size and attractiveness of male secondary sexual characteristics and their ability to use them to gain a mating advantage should be positively related to their overall quality in terms of ability to produce large numbers of offspring (i.e. their fitness). In recent years, there has been interest in whether females can 'assess' the relative health and vigor of males by their secondary sexual characteristics. Evidence supporting this hypothesis is increasing steadily. For example, male blackbirds (*Turdus merula*) with a brighter bill color (orange) are more attractive to females than males with dull colored bills. Faivre *et al.* (2003) reported that bill color in blackbirds reflected the strength of their immune system. When the immune system of male birds was challenged experimentally, bill coloration became duller. A similar study by Blount *et al.* (2003) showed that females preferred male blackbirds that had been fed carotenoids. Carotenoid supplements not only boosted the birds' immune system, but brightened their bills. Of course, female preference for males with brightly colored bills does not mean that females 'understand' the significance of their choices for themselves and their offspring.

There is also evidence that females can use chemosensory cues to assess mate quality. Fisher and Rosenthal (2006) found that female swordtail fish (*Xiphophorus birchmanni*) preferred the chemical cues of males in good nutritional health to males that were food deprived. Preference was measured by their relative attraction to water used in housing the two treatment groups of males. Females did not discriminate between the odors of well-fed and under-fed females, suggesting that their response to males was based on sex-specific chemical cues, not the odors of dietary excreta.

Females are believed to obtain information about mate quality by observing male–male interactions. Doutrelant and McGregor (2000) reported that their female Siamese fighting fish (*Betta splendens*) mated preferentially with the winners of the male–male aggressive interactions they had observed, whereas females that had not observed male–male contests behaved the same toward winners and losers.

Another line of evidence shows that females prefer males with genetically based indicators of health and vigor ('good genes hypothesis'). For example, male gray tree frogs (*Hyla versicolor*) attract females with calls that last from 0.5 s to 2 s. Females avoid males producing shorter calls. Welch *et al.* (1998) conducted an experiment in which eggs were removed from some ten female frogs. Half of the eggs from each female were fertilized with sperm from a short-calling male and half from a long caller. Offspring were compared for survival and growth rate under conditions of

Fig. 7.6. (a) Female mallard ducks (center) are drab and somewhat smaller than males. (b) Bull elk grow antlers each year while females (foreground) are antlerless (courtesy of Agustin Orihuela).

both scarce and plentiful food resources. Descendants of the long callers were superior to descendants of the short callers in every case where a difference was found, suggesting that the long-calling males which females prefer are naturally more vigorous. Males of this species have no contact with their offspring, so the only contribution they can make to their progeny is genetic. Both calling and enhanced survival of the young may be controlled by the same genes or linked genes, perhaps by increasing stamina or by more efficient metabolism.

Sexual selection is sometimes influenced by non-genetic mechanisms, such as when females copy the mate choices of other females. Swaddle *et al.* (2005) found that female zebra finches (*Taeniopygia guttata*) preferred males they had observed housed with another female over males seen housed with another male. Galef and White (1998) tested the preference of female Japanese quail (*C. japonica*) for two males and then allowed each female to observe the non-preferred male mating with another female. In subsequent tests, the females switched their preferences to the previously non-preferred males, who had now proven their success as potential mating partners. Interestingly, in a similar experiment (White and Galef, 1999) in which *male* Japanese quail were allowed to observe preferred females mate or court with other males, their subsequent preference for the 'taken' females declined.

Incest Avoidance

Incest, the breeding of close relatives, generally results in reduced reproductive success and vigor of offspring due to the increased likelihood of rare deleterious gene alleles being paired and expressed (see p. 26). Animals in nature avoid incest and inbreeding by a natural tendency for one or both sexes of offspring to disperse from their birth site and, in cases where dispersal does not occur, the ability to recognize kin. A study of the breeding biology of an island population of Savannah sparrows (*Passerculus sandwichensis*) showed that only nine of 1110 pairings (0.8%) were incestuous (Wheelwright *et al.*, 2006). All but one case of close inbreeding involved 1-year-old males breeding for the first time and there were no cases of father–daughter matings. It is believed the recognition of close relatives is achieved either through early experience and familiarity with kin, or by using cues that reflect genetic similarity independent of experience and familiarity.

The ability to identify unfamiliar kin is attributed to 'phenotype matching', in which an individual learns a common phenotype for familiar kin during early development and then applies this template when discriminating between unfamiliar conspecifics (Tang-Martinez, 2001). The common phenotype may be based on visual, auditory or olfactory characteristics shared by kin but different in non-kin. Gerlach and Lysiak (2006) found that female *juvenile* zebrafish (*Brachydanio rerio*) showed preferences for familiar kin over unfamiliar kin and for unfamiliar kin over non-kin, based solely on odor. Sexually *mature* females prefer unfamiliar, unrelated males to unfamiliar brothers, suggesting that they can discriminate between unfamiliar kin and unfamiliar non-kin, thus exhibiting a mechanism to avoid inbreeding. Although male zebrafish preferred unfamiliar kin to unfamiliar non-kin as juveniles, they showed no preference for the odor of related or unrelated females after reaching sexual maturity. These results suggest that inbreeding in this species is avoided through female choice of mating partners. Breeders of zebrafish and other fish species for aquaria displays can maximize the reproductive success and vigor of their breeding populations by pairing fish from different populations only.

Animal breeders are generally aware of the deleterious effects of close inbreeding and make efforts to avoid incestuous pairings when choosing mating partners for their animals. Since most of our domestic farm animal populations consist primarily of females, males (or semen) are typically obtained from other populations where the chance of genetic relatedness is greatly reduced.

Mating Systems Used in Captive Animal Breeding

There are four basic mating systems used in the breeding of captive animals: (i) multi-sire mating; (ii) single-sire mating; (iii) controlled (hand) breeding, in which sexually receptive males and females are introduced to one another in pairs just long enough for mating to occur; and (iv) artificial insemination. Male–male competition and female choice are eliminated in all but the multiple-sire breeding system. Table 7.2 points out that most of our common domestic animals are bred using systems other than the multi-sire mating scheme.

The elimination of male–male competition and female choice in captive animal breeding systems has led to considerable variability in sexual motivation (i.e. libido), behavioral compatibility of pairs and, thus, reproductive success (Price, 2002). Its effect on mating competence (i.e. the ability to perform species-specific sexual behaviors) is less evident, except in male domestic turkeys (*M. gallopavo*) where artificial selection for large breast size has made natural mating difficult to achieve. However, its effect on the libido of breeding-age males is particularly evident (see p. 107). The loss of female choice when the animal breeder, rather than the animal itself, chooses its mate ('forced' pairings) can increase the incidence of behavioral incompatibility and breeding failure greatly. Long-term forced cohabitation of animals prior to breeding may lead to a kind of 'assumed kinship' and incest-related mating failure. Blohowiak (1987) found that when captive black ducks (*Anas*

Table 7.2. Occurrence of male–male competition in mating systems typically employed in breeding domestic animals.

Species	Common	Somewhat common	Rare
Dairy cattle			×
Beef cattle		×	
Sheep		×	
Dairy goats			×
Swine			×
Horses			×
Laboratory rodents			×
Rabbits			×
Dogs			×
Cats			×
Chickens	×		
Turkeys			×

rubripes) were randomly paired, less than a third of the pairs mated. When maintained in large groups and allowed to select their own mates, the majority of them bred. Likewise, captive female canvasback ducks (*Aythya valisineria*) must choose their own mates in order to obtain fertile eggs (Bluhm, 1988). Gonadal development in unpaired female canvasbacks was seldom stimulated by courtship from randomly assigned males. In many species, males must dominate females for successful matings to occur. Captive rhinoceros (*Diceros bicornis*) breed better if the female is dominant over the male (Carlstead *et al.*, 1999). Such findings highlight the importance of knowing the behavioral biology of animals before establishing a breeding plan.

Courtship (Pre-mating) Behavior

Courtship, or pre-mating, behaviors in our common domestic animals tend to be short-lived and sometimes difficult to distinguish from mating behavior, per se. This is not surprising considering that most of our domestic animal species are promiscuous and do not form stable male–female pair bonds. As stated earlier, males primarily serve the role of inseminator and are not involved in parental care. Courtship is typically more prolonged and elaborate in monogamous species.

There are four primary functions of pre-mating behaviors.

1. Reproductive isolation

Matings between individuals of different species is typically prevented by pre-mating behaviors, which allow for a period of social 'testing' of potential mates. Incompatibilities due to species differences in behavior preclude mating, thus minimizing gamete wastage and preserving species integrity. Cross-species matings are uncommon among domestic animals, largely because of human control over the breeding process and relatively large species differences.

2. Advertisement of sex and sexual receptivity

Pre-mating behaviors advertise sex (gender) and degree of sexual receptivity. These are important functions served by pre-mating behaviors in promiscuous species where contact between male and female can be fleeting and periods of sexual receptivity for the female are often very short. Male pre-mating behaviors include an active search for females and, once females are found, they include behaviors which advertise the male's sexual readiness and behaviors which test female sexual receptivity. Male turkeys (*M. gallopavo*) and peacocks (*Pavo cristatus*) strut in front of females with their tail feathers spread out in the shape of a fan (Fig. 7.7), bulls (*Bos*) approach cows and rest their chins on the female's back or rump (Fig. 7.8), male sheep (*O. aries*) and goats (*C. hircus*) raise their forelegs to make physical contact with females, often while uttering grumbling sounds and nudging the female with their head (Fig. 7.9), and male pigs (*S. scrofa*) nudge sows with their nose while emitting characteristic courtship grunts (Fig. 7.10). Non-receptive females will avoid and move away from males exhibiting these behaviors.

Fig. 7.7. Male wild turkeys display their tail feathers to females during the spring breeding season.

Fig. 7.8. Bulls often rest their chins on the back or rump of females when testing for sexual receptivity.

Females often solicit copulation. Female rats (*R. norvegicus*) solicit attention and copulation by approaching males and then darting away. Receptive female rats exhibit the 'lordosis' posture when mounted by the male (Fig. 7.11). Sexually receptive female ungulates (cattle, sheep, goats, pigs, horses, etc.) stand immobile when approached and mounted (Fig. 7.12). Sexually experienced males 'understand' these postures and displays and respond accordingly.

The most obvious components of the courtship behavior of male chickens (*G. domesticus*) are: (i) 'waltzing', in which the male drops one wing and approaches the hen with short shuffling side-steps and (ii) the rear approach, in which the male holds the comb or neck of the hen or flaps his wings over her. Sexually recep-

Fig. 7.9. Ram strikes an estrous ewe with its foreleg to test her willingness to stand immobile for mounting.

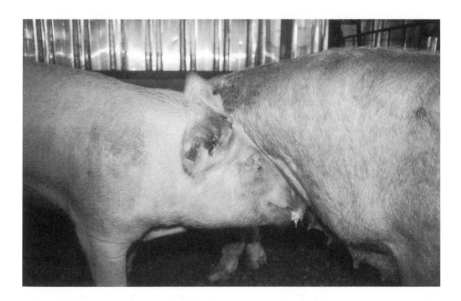

Fig. 7.10. Boar nudges sow with its snout as part of its pre-mating behavior.

tive female turkeys (*M. gallopavo*) assume a crouching posture in response to the 'strutting' behavior of the male, in which he fans out his tail feathers, drops his wings so that the tips scrape the ground and paces back and forth in a deliberate motion while 'trumpeting' a series of deep notes that have been described as 'pulmonic puffs'.

Fig. 7.11. Lordosis posture in the estrous female rat in response to being mounted.

Fig. 7.12. Estrous cow stands immobile for mounting.

3. Enhancement of libido and fertility

Pre-mating behavior can improve sexual libido and fertility in some species. Studies have shown that the libido of male goats (*C. hircus*) and cattle (*B. taurus*) is enhanced by a short (5–10 min) exposure to estrous females before mating is permitted. In pasture breeding, male goats and cattle use odor to identify females coming into estrus and will often stay in close proximity to these females, courting and waiting for them to become receptive to mounts and copulation. This pre-estrous period of courting, or 'tending', females enhances their libido and fertility when mating finally occurs.

In male cattle (*B. taurus*), the quantity and quality of sperm in an ejaculate is correlated with the amount of pre-mating stimulation (see also p. 117). Semen volume, sperm concentration, sperm motility, percentage of live sperm and conception rates (of mated females) are all improved when the bull is exposed to one or more females for a few minutes before it is allowed to copulate.

Hemsworth *et al.* (1978) studied the importance of male courtship in pigs (*S. scrofa*) on female fertility following natural mating and artificial insemination. Boar courtship typically involves anogenital sniffing, nosing (massaging the sides of the female with the nose) and 'chanting', a specific courtship vocalization. In the first study, 24 boars were each naturally mated to 20–80 estrous females and the frequency of courtship behaviors were recorded. Females receiving the most nosing from the boar had the highest conception rates. Anogenital sniffing and chanting were not significantly correlated with fertility. In the second study, 88 estrous sows were artificially inseminated in 44 pairs. One sow in each pair received 2 min of courtship by a boar prior to insemination and the other member of the pair was inseminated without boar contact. Conception rates and average litter sizes were greater for the sows that had been courted (Table 7.3). Boar courtship, particularly nosing, is believed to improve sperm transport in the uterine horns. Courtship may also advance the onset of ovulation, thus improving the timing of ovulation relative to mating.

4. Synchronization of reproductive behavior and physiology

Courtship behaviors are important in synchronizing the reproductive physiology of monogamous bird species in which both male and female participate in nest building, incubation and care of the young. Lehrman and his colleagues (Lehrman, 1961) noted that if a female ring dove (*Streptopelia risoria*) was placed alone in a cage with nesting material, it would not engage in nest-building behavior. If a single female is placed in a cage with a preformed nest, it will not lay eggs. If a male and a female are placed together in a cage with nesting material, they will build a nest, but only after a day or so of courtship in which the male struts, bows and 'coos' to the female. Studies have shown that courtship directed at the female releases follicle-stimulating hormone in the anterior pituitary of the brain, which prompts the ovary to release estrogen. Estrogen release is necessary for the female to begin nest building. Females injected with estrogen behaved as though they had recently been courted. Thus, courtship has a critical role in stimulating the release of hormones needed to initiate reproduction in this species.

Table 7.3. Effect of courting behavior of the male on the reproductive success of sows inseminated artificially (from Hemsworth *et al.*, 1978).

	Courted by male (*N* = 44)	Not courted (*N* = 44)
Conception rate	86.7%	62.1%
Litter size (mean no. of piglets)	10.3	8.6

Pre-mating behaviors are particularly important in synchronizing the release of eggs and sperm in species with external fertilization, such as aquatic vertebrates (e.g. amphibians, fish, etc.) where fertilization of eggs takes place in water. In these species, pre-mating activities bring the male and female in close proximity so that eggs and sperm are released at approximately the same time and in the same place.

Mating Behavior

Mating behaviors lead to gamete fertilization. Fertilization occurs either externally or internally. In external fertilization, courtship behaviors bring male and female in close proximity and the male deposits spermatic fluids directly over the extruded eggs. Internal fertilization is achieved in three different ways.

1. Spermatophore transfer

Male and female do not come in contact with one another in spermatophore transfer. After a period of courtship, the male deposits a gelatinous capsule containing sperm (spermatophore) somewhere near the female. The female then picks up the spermatophore with the lips of her cloaca and fertilization takes place internally. Spermatophore transfer is common among salamanders and a wide variety of invertebrates.

2. Cloacal contact

Most bird species have no intromittent organ. The male mounts the female and copulation is achieved when semen is transferred from male to female by the rapid apposition (placing together) of the male's engorged phallic folds and the female's everted (turned inside out) cloaca. Cloacal contact in chickens (*G. domesticus*) and turkeys (*M. gallopavo*) is maintained for a very brief time (1 s or less) as semen is transferred. The female's everted cloaca is immediately retracted as the male dismounts. Male chickens may copulate up to 30 times/day. Typically, only half of these copulations are accompanied by semen transfer.

3. Employment of an intromittent organ

The male deposits semen directly into the appropriate female reproductive orifice using an intromittent organ such as the penis. Male waterfowl have a penis because they commonly mate in water, where the cloacal contact method of copulation would be inefficient. As in cloacal contact, not all waterfowl copulations are accompanied by semen transfer.

In most domestic animals, successful copulations are completed in just a few seconds. Exceptions are the pig (*S. scrofa*) and dog (*C. familiaris*). Of all the farm animals, the boar has the longest ejaculation time (3–20 min; mean = 9 min) and the greatest ejaculate volume. In male canids such as the domestic dog, the *bulbus glandis* at the base of the penis enlarges rapidly within the female's vagina following intromission and the penis is 'locked' in place by the constriction of vaginal muscles, preventing withdrawal for 10–30 min. Ejaculation begins at about the time the lock is formed and continues until nearly the end of the tie (*bulbus glandis* deflates). The male typically dismounts *during* the tie and it is not uncommon for the male and female to assume positions facing in opposite directions.

Hormonal Influences on Reproductive Behavior

During embryonic development, the male testis produces testosterone but no sperm and the female ovary produces eggs but no estrogen. In mammals, the relative absence of testosterone results in the formation of the female reproductive tract and feminization of the brain. Since testosterone is converted to estrogen in the male brain by the process of aromatization, moderate levels of estrogen in the developing fetus induce development of the female phenotype, while higher levels of estrogen and possible estrogen synergism with androgens cause masculinization of the genitalia and nervous system.

The sensitive period for sexual differentiation of the genitalia and nervous system occurs prenatally in most precociously born species of domestic animals. However, domestic rat (*R. norvegicus*) pups are born in an altricial state of development, such that the sensitive period for sexual differentiation of the hypothalamus of the brain occurs in the first few days following birth, thus affording an excellent opportunity to study the role of hormones in the development of gender-specific sexual behaviors.

Hormones have both an organizing effect on anatomical structures (e.g. genitalia, brain) and an activating effect, in which they activate (or inhibit) cells of target organs (e.g. gonads) they may have helped to organize. Table 7.4 describes a series of experiments investigating the sexual differentiation of the hypothalamus of the rat brain (Whalen and Edwards, 1967; Döhler *et al.*, 1984). Both organizational effects of hormones on neonatal rat pups as well as activational effects on adult rats are summarized. The presence of testosterone (relatively high levels of estrogen; see explanation above) in the neonatal rat results in masculinization of the preoptic area of the hypothalamus (organizational effect) of the brain so that male sexual behaviors are exhibited as an adult. The absence of testosterone in the neonatal rat (moderate levels of estrogen) results in feminization of the hypothalamus (organizational effect) so that female sexual behaviors are exhibited as an adult. Injections of estrogen and progesterone do not elicit female behaviors in adult rats masculinized at birth, nor do injections of testosterone result in male behaviors in adults feminized at birth. In contrast, males castrated at birth exhibit relatively high levels of female sexual behaviors after injections of estrogen and progesterone as an adult (activational effect). Likewise, females administered testosterone at birth exhibit relatively high levels of male sexual behaviors after injections of testosterone as an adult (activational effect). Interestingly, in some litter-bearing mammalian species, female fetuses located between two male fetuses in the uterus become somewhat masculinized and are subsequently more aggressive than females which developed between two female fetuses in the womb (Ryan and Vandenbergh, 2002).

Gonadal hormones are not always responsible for gender-based differences in sexual behavior. The Caribbean bluehead wrasse (*Thalassoma bifasciatum*) is a

Table 7.4. Organizational and activational effects of testosterone and estrogen in the sexual differentiation of the rat brain (from Whalen and Edwards, 1967 and Dohler *et al.*, 1984).

Genetic sex	Organizational effects – neonate		Relatively high levels of estrogen
	Neonatal operation	Neonatal injections	
Male	Sham	None	Yes
Male	Castration	None	No
Male	Castration	Testosterone[a]	Yes
Female	Sham	None	No
Female	Ovariectomy	None	No
Female	Ovariectomy	Testosterone[a]	Yes

Adult operations	Activational effects – adults		Behaves like........
	Female behavior – lordosis[b]	Male behavior – intromissions[c]	
Castration	Low	High	Male
None	High	Low	Female
None	Low	High	Male
Ovariectomy	High	Low	Female
None	High	Low	Female
None	Low	High	Male

[a]Aromatized to estrogen in rat brain.
[b]After injections of estrogen and progesterone to induce female sexual behavior.
[c]After injections of testosterone to induce male sexual behavior.

coral-reef fish distinguished by the ability of females to change into males when males are unavailable. Semsar *et al.* (2001) found that masculinity in this fish species is determined not by the gonads but by a brain hormone, arginine vasotocin (AVT). Wrasse males are found in two forms; aggressive, frequently spawning 'supermales' and more timid yellow males. When a supermale leaves a spawning site, females and yellow males immediately try to take over the site. Both change colors temporarily and one (often the female) eventually emerges as the replacement supermale. These events occur even when the gonads of the fish have been removed. Females normally produce little AVT, but levels increase fourfold when they assume a supermale role, suggesting a direct relationship between AVT levels and behavior.

The role of hormones in the reproductive cycle of ring doves (*S. risoria*) provides an excellent illustration of the activational effects of hormones on behavior. Figure 7.13 describes the major events in this cycle. Courtship takes place on day 1, followed by nest building (over about 5 days), egg-laying and incubation (over about 2 weeks) and parental care (another 2 weeks). Both the male and female participate in nest building, with a tendency for males to bring nest material to the female, who forms the nest. The female does most of the incubating, with the male's help. Both male and female feed the young 'crop milk', a liquid secreted by a sloughing of the epithelial lining of the bird's crop, a pouch in its esophagus in which food is typically held for later digestion or regurgitation to nestlings.

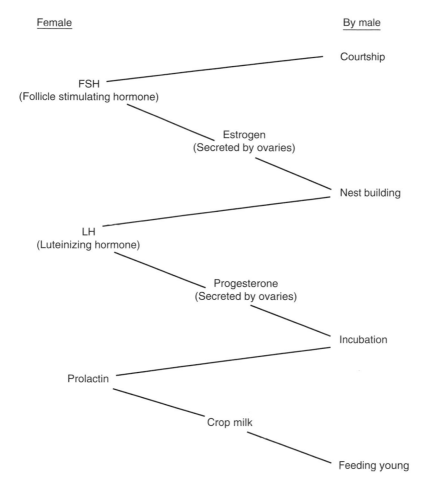

Fig. 7.13. Behavioral and hormonal events in the reproductive cycle of the female ring dove. Each behavior listed stimulates the physiological response needed to attain the phase of the cycle (Lehrman, 1961).

We have already discussed (p. 99) how the male ring dove's courtship behaviors (primed by testosterone) causes the female's anterior pituitary to release follicle-stimulating hormone, which stimulates the ovary to secrete sufficient levels of estrogen needed to prime the animal for nest-building behavior. Nest-building behavior, in turn, stimulates the anterior pituitary to release luteinizing hormone, which initiates ovulation and the secretion of progesterone from the ovaries. These hormonal changes prime the bird for egg laying and incubation. Incubation behaviors stimulate the pituitary to release the hormone, prolactin, which is needed for crop milk production. Males produce crop milk only if they have participated in nest building and spent at least 72 h with the female during incubation. Each step in the sequence primes the bird physiologically to perform the next step, once the appropriate stimulation is received. These events are not triggered merely by the perception of a mate, nesting material, eggs or young. In addition, steps cannot be omitted. If a previously isolated male and female are placed together and given a nest bowl with a preformed nest and some eggs, the pair will engage in courting, nest-building activity (a new nest is built on top of the one provided) and the female will lay a new clutch of eggs.

One can demonstrate experimentally the priming effect of hormones in each step of the sequence by taking previously isolated intact or ovariectomized females, injecting them with the appropriate hormone for that step and then exposing them to the stimulus which triggers that behavior. Doves injected with estrogen behave as though they had been courted and will immediately engage in nest-building behaviors. Birds injected with progesterone behave as though they have been courted and have built a nest and will immediately begin to incubate a clutch of eggs presented to them in a preformed nest.

These studies illustrate the complex role that hormones play in reproductive behaviors and the feedback mechanisms involved. Behavior influences hormone secretion and hormone secretion influences behavior.

Sexual Motivation

Sexual motivation at any given time is determined by the capacity to become sexually aroused and the quality and quantity of sexual stimuli perceived, as well as the negative effects of other competing and distracting stimuli. The capacity to become sexually aroused is affected not only by maturational and physiological factors, such as hormonal effects on neural centers, as we have just discussed, but also by experience (i.e. learning). As the young animal matures and gains sexual experience, it becomes sexually aroused by an increasing number of sexual stimuli associated with positive sexual experiences. A sexually receptive female is more attractive to an aroused male than a non-receptive female. Males develop sexual preferences for certain females and vice versa. Stimuli associated with negative sexual experiences lose their ability to arouse, or may inhibit sexual arousal completely. A male may encounter a sexually receptive female but have his arousal limited by the presence of a dominant or aggressive competitor. Thus, sexual motivation is dynamic. Animals can be classified based on their general propensity to become sexually aroused, but the degree of sexual motivation actually exhibited at any given time is dependent on a host of relevant internal and external factors. The next two chapters provide a number of examples to illustrate these principles.

8 Male Sexual Behavior

Definitions of Sexual Performance

Sexual behavior has both motivational and performance components. The motivational component of sexual behavior if often referred to as 'libido'. The performance component is based on both libido and the motor patterns that characterize each sexual behavior exhibited. It is possible to quantify the frequency, duration or intensity of most male sexual behaviors but, in doing so, it is performance that is being measured, not just libido. For example, the number of ejaculations attained in a given period of time, often referred to as 'serving capacity' (Blockey, 1976), is based not only on the sexual motivation of the animal but also on its ability to execute the behavior patterns associated with copulation. An individual must have relatively high libido to attain a relatively high serving capacity score, but an animal that attains a relatively low serving capacity score will not necessarily do so because it lacks libido. A low score may be due to locomotor problems resulting from structural deficiencies, pathological conditions, genital abnormalities, or lack of mating experience. For example, small males may have difficulty copulating with larger females. Horse breeders will sometimes place small stallions (*E. caballus*) on a mound of dirt behind taller mares so the male will be high enough to gain intromission (penetration of the vagina). A bull (*B. taurus*) we raised for sexual behavior research could scarcely support its weight (on its hind legs) when mounting a female and would fall back on its rump following an ejaculation. It soon showed a reluctance to mount, even though it had ample libido. Another bull had a condition called 'corkscrew penis', in which the erected penis assumed a corkscrew shape just prior to intromission (Fig. 8.1), thus preventing intromission. In a study on the sexual behavior of male goats (*C. hircus*), we noted that sexually inexperienced males would often 'charge' (run at) estrous females, kick them vigorously with their forelegs and immediately attempt to mount, usually causing the females to exhibit avoidance behaviors. In contrast, our experienced males would typically approach females at a slow pace, followed by 'gentle' contact with forelegs, nudging and anogenital sniffing and only mount when females demonstrated willingness to stand (for mounting). Sexually inexperienced male sheep (*O. aries*) were less proficient when approaching estrous ewes for the first time and, after a few sexual encounters, would attain the same serving capacity scores as more experienced rams.

Inability to measure sexual libido accurately is no great loss to animal breeders since their primary interest is in sexual performance. Mating efficiency is also a topic of interest to animal breeders. Some definitions of mating efficiency are: (i) the total number of females impregnated by a male during a restricted breeding period; (ii) the proportion of estrous females in the group that were inseminated at least one time (or impregnated) by a male; and (iii) the number of mounts per ejaculation. For the livestock producer, the most efficient males are those which impregnate the largest

Fig. 8.1. Corkscrew penis condition in the bull, preventing intromission.

number of females in the shortest period of time with the fewest total number of mounts and ejaculations. Needed energy is saved if a male successfully copulates on the first two or three mounts and then moves on to mate with other receptive females. Maximum conception rates in cows are obtained with two successful copulations ('services') by a bull. The most efficient males providing semen for artificial insemination are those which provide a quality semen sample in the shortest period of time after exposure to the mount stimulus.

Measuring Libido

Quantification of male libido is possible by devising tests which remove performance (i.e. copulatory behavior) as a confounding variable. One approach is to measure their motivation to seek out and (or) remain in close proximity to potential mates. Libido scores based on 'time near females' where physical contact cannot be made are of limited usefulness because males soon learn that contact and mating are not possible and will wander away from the females. Another approach is to measure the frequency with which males engage in pre-mating and mounting when intromission is prevented. In one such study with male sheep (*O. aries*) (Price *et al.*, 1992), we tested rams exposed to estrous females under each of the following three treatments: (i) only when pre-mating activities were allowed (e.g. foreleg kicks, anogenital sniffs, nudging); (ii) pre-mating activities plus mounting allowed (copulation prevented); and (iii) pre-mating activities, mounting and copulation allowed. Libido scores in the first two treatments were found to be correlated with ejaculation rate when copulations were permitted (third treatment), thus validating the use of pre-mating behaviors and mounting (when copulation is precluded) as a measure of the animal's underlying sexual motivation or libido.

Evolutionary Considerations

Not all male farm animals exhibit adequate sexual performance when exposed to females. Among the causes of substandard sexual performance are methods of genetic selection and rearing management practices. In captivity, males are typically selected for breeding programs with little concern for sexual performance. Natural selection for the ability of males to compete for females is eliminated when single males are exposed to females, or when collecting semen for artificial insemination. In essence, natural selection for sexual libido is relaxed under such breeding programs. As long as matings eventually occur and females are impregnated, the animal breeder is not likely to cull a slow-working male if it proves desirable for other reasons. McDonnell (1986) reported that 8 (17%) of 47 stallions (*E. caballus*) examined for breeding competency (for the first time) had inadequate sexual behavior. It is estimated that 20% of male pigs (*S. scrofa*) produced for breeding in the USA exhibit inadequate sexual performance. Many of these males lack sexual libido. In some cases, it is necessary to assist boars manually to achieve intromission with receptive sows. Furthermore, sexual performance in meat animals may be suppressed inadvertently as a correlated effect of artificial selection for other phenotypic characteristics, such as body size and conformation. Increased body size may reduce the physical dexterity and agility of animals, thus affecting their ability to copulate, as illustrated by the clumsiness and copulatory inefficiency of domestic male turkeys (*M. gallopavo*).

The effects of relaxed selection on libido and artificial selection for traits that may have a detrimental effect on male sexual performance are reflected in relatively high levels of variability in male sexual performance. Statistical measures of variation, such as the 'coefficient of variation' in livestock species, illustrate that individual differences in sexual performance are considerably greater than changes in such variables as body weight (same-age animals), milk production in dairy animals and rate-of-gain in meat-producing animals. For example, in one of our studies at UC Davis, the mean number of ejaculations attained by 16 rams (*O. aries*) when exposed to estrous females for two 60-min periods ranged from 0.0 to 6.5, with a coefficient of variation (CV) of 0.54. In another study, the mean number of ejaculations attained by 14 bulls (*B. taurus*) in six 40-min sexual performance tests ranged from 1.8 to 5.5 (CV = 0.33). Considerable variability has also been observed in the sexual performance of male pigs (*S. scrofa*). A study by Boyd and Corah (1986) offers proof that the variability in sexual performance currently observed in male livestock is influenced by genetics. Table 8.1 illustrates

Table 8.1. Mean number of ejaculations attained by the male offspring of four beef bulls. Each male was exposed to four females for 20 min/test (from Boyd and Corah, 1986).

Bull	Number of sons tested	Mean number of ejaculations	
		Test 1	Test 2
A	42	4.2[a]	4.7[a]
B	9	3.5[ab]	3.2[a,b]
C	11	2.4[bc]	2.5[b]
D	8	0.6[c]	1.2[b]

[a,b,c]Ejaculation frequencies with different superscripts within columns differ ($P < 0.01$).

the number of successful copulations attained by the male *offspring* of four of their beef bulls (*B. taurus*). Livestock breeders can no longer assume that individual males selected for breeding programs will exhibit levels of sexual performance needed to impregnate very many females in a timely manner.

Genetic Influences on Male Sexual Behavior

The tendency of animal breeders to ignore sexual libido and sexual performance in artificial selection programs has left huge pools of genetic variation ready to be exploited by artificial (or natural) selection. Blockey *et al.* (1978) calculated a very high heritability estimate (see p. 28) for an explanation of heritability) of 0.59 for sexual performance in beef bulls (*B. taurus*), while Snowder *et al.* (2002) reported a more modest heritability estimate of 0.22 for sexual performance in rams (*O. aries*). Both of these estimates suggest considerable genetic variation for this trait. Bench *et al.* (2001) found a significant difference in the sexual performance of the sons of high- and low-performing rams (*O. aries*) in just one generation of artificial selection (Fig. 8.2). Bernon and Siegel (1983) reported that after 20 generations of selection for sexual performance in male chickens (*G. domesticus*), the high-performing line averaged 24.4 completed matings over eight 10-min test periods, while the low-performing line averaged 0.3 successful copulations (see p. 27) for more details). Bane (1954) compared the sexual performance of six pairs of identical twin bulls (*B. taurus*) (Fig. 8.3). Note the lack of variability in sexual performance *within* sets of genetically identical twins compared to the differences *between* twin sets. In summary, there is ever-increasing evidence for a significant genetic component to sexual libido and performance and an untapped source of genetic variability, which could be used to improve the efficiency of farm animal breeding.

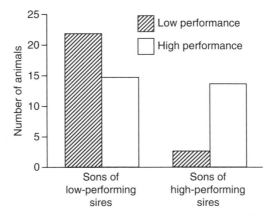

Fig. 8.2. Sexual performance of the sons of high- and low-performance rams. High-performance rams sired a significantly higher proportion of high-performance sons than did their low-performance counterparts (Bench *et al.*, 2001).

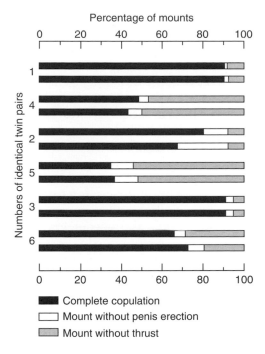

Fig. 8.3. Sexual performance of six identical twin bulls. Note the similarity in the sexual behaviors of twins relative to the variation between twin pairs (Bane, 1954).

Effect of the Rearing Environment

It is a common practice on farms to isolate females from males after weaning to prevent unwanted pregnancies. On some farms, males may be reared singly in individual stalls or enclosures to control disease, prevent injury from aggressive interactions with other males and to ensure that nutritional requirements are being met. The study with pigs (*S. scrofa*; described on p. 45–46) demonstrates how male pigs reared in social isolation exhibited significantly poorer sexual performance with estrous females than boars reared in either all-male or heterosexual groupings (Hemsworth *et al.*, 1977). An interesting aspect of this study was that the isolate-reared pigs did not show an improvement in sexual performance with age or sexual experience over the 6-month test period.

Our work with young beef bulls (*B. taurus*) demonstrated that individual rearing (i.e. physical isolation) from birth did not affect their sexual performance as adults (Price *et al.*, 1990). The only difference between our isolate-reared bulls and group-reared counterparts was that the isolate-reared animals initially exhibited a significantly higher number of disoriented mounts directed at the side or front of the stimulus animal (Silver and Price, 1986). It did not take very many exposures to estrous females for them to learn to mount from the rear. Group-reared males learn to mount from the rear early in life, either by mounting other youngsters and (or) trying to mount their mothers. Mount attempts oriented toward the side or front of another animal usually result in the mounter being butted or avoided.

Approaches from the rear are not as easily detected by the recipient animal and, by being in the rear, mounters can more easily avoid retaliatory butt attempts by the mount recipient.

Hemsworth's study (cited above) also showed that rearing boars in *all-male* groups did not impair their subsequent sexual performance. We obtained this same result with male cattle (*B. taurus*) (Price and Wallach, 1990b). However, our work on the development of sexual behavior in sheep (*O. aries*) revealed that rearing rams in all-male groups could contribute to the development of substandard levels of sexual performance (Katz *et al.*, 1988). One group of rams was reared in traditional all-male groups following weaning, while our experimental rams were reared in heterosexual groups. As adults, the rams reared with females exhibited significantly more mounts and successful copulations when exposed to estrous females than control rams reared in all-male groups (Table 8.2).

Rearing rams in all-male groups also encourages the development of male (same-sex) sexual orientation and lack of sexual interest in females. In one of our studies (Price *et al.*, 1988), we found that 4 of 44 (9.1%) rams reared in all-male groups after weaning developed a sexual preference for males. Their male sexual orientation persisted even after many exposures to females in the absence of other males. In another (unpublished) study, we obtained offspring from male-oriented rams by collecting their semen and artificially inseminating ewes. Sons of the male-oriented rams were no more likely to be male-oriented than the sons of female-oriented rams produced in the same manner, suggesting that their atypical sexual orientation was a result of being reared in an all-male social environment rather than inheriting the trait from their sires. (Of course, they could have inherited the trait from their mothers.) Follow-up studies supported the rearing environment hypothesis. Male sexual orientation almost never developed in rams reared with females after weaning (Katz *et al.*, 1988). Young rams preferred to direct their sexual behaviors toward females when given a choice between males and females but they would frequently mount other males in the absence of females (Price *et al.*, 1988). Sexual experience in all-male rearing groups may condition certain predisposed individuals to seek out male stimuli toward which to direct their sexual behaviors. Interestingly, we found that male–male sexual preferences can sometimes be changed by long-term exposure to females (in the absence of males) and that, once female-oriented, rams would not revert back to being male-oriented, even when housed in all-male groups.

Table 8.2. Mean (± SE) ejaculation frequencies of rams reared in all-male or heterosexual groups during four 30-min sexual performance tests (from Katz *et al.*, 1988).[a]

Rearing group	No. of rams	Test no.			
		1	2	3	4
Heterosexual	24	4.7 ± 0.6[b]	4.4 ± 0.5[b]	4.7 ± 0.6[b]	3.6 ± 0.4[b]
All-male	23	2.4 ± 0.5[c]	2.6 ± 0.5[c]	2.3 ± 0.4[c]	2.3 ± 0.4[c]

[a]Means include values for rams which failed to copulate.
[b,c]Ejaculation frequencies with different superscripts within columns differ ($P < 0.05$).

Table 8.3. Effects of the early social environment on the sexual performance of male pigs, sheep and cattle.

Species	Individual rearing (physical isolation)	Reared in all-male group
Boar	Detrimental	Normal[a]
Ram	Normal	Detrimental
Bull	Normal	Normal

[a]Normal refers to sexual performance as good as that obtained when males are reared with females from weaning.

The studies cited above point out that there are important species differences in how the social environment of the animal can influence the development of male sexual behavior and performance. The results for male pigs, sheep and cattle are summarized in Table 8.3.

Advantages of Social Dominance

The advantages of social dominance for male sexual performance and reproductive success have been well documented. Hulet *et al.* (1962) confirmed that dominant rams were more successful than subordinate males in mating with ewes (Table 8.4).

The effects of dominance on male sexual behaviors are more obvious in mixed-age groups of males than when young males of similar age are used in breeding programs. Blockey (1979) compared the reproductive success of groups of mixed-age bulls (one 5-year-old and two 2-year-old males) and same-age bulls (three 2-year-old bulls) when placed with 114 heifers for a 6-week period. Approximately five heifers were in estrus each day in each group during the first 3 weeks (first estrous period). With this many sexually receptive females, the dominant bull in each group could not prevent some subordinates from gaining access to females and mating. During the second 3 weeks of the study (second estrous period), approximately three females were in estrus on any given day. Subordinate bulls in the mixed-age groups were interrupted in their mating activities 87% of the time compared to 20% of the mating attempts of subordinates in the same-age groups. It was clear that the larger 5-year-old bulls in the mixed-age groups suppressed the sexual activities of the subordinate males more so than the dominant males in the 2-year-old groups.

Table 8.4. Mean number of mounts and ejaculations per ram over a 24 hr period when rams were mated singly or in groups (from Hulet *et al.*, 1962).

	Mounts	Ejaculations
Single male	23.8	5.4
Multiple male groupings		
Dominant ram	26.5	4.6
Intermediate	10.7	0.8
Subordinate ram	6.8	0.0

Interestingly, the greater dominance displayed by the older males in the mixed-age groups was not a reproductive advantage for these groups as a whole. Conception rates during the second estrous period were 64% for mixed-age groups and 81% for the same-age bull groups, and pregnancy rates after 6 weeks were 85% for mixed-age groups and 94% for same-age groups. Overall, reproductive success was better with the younger same-age groups. There are two additional reasons why younger males should be used in breeding programs. First, younger males should be genetically superior to older males in any artificial selection program. Secondly, younger males should have fewer reproductive problems, such as structural, locomotor and genital abnormalities.

Dominant males can inhibit the sexual activities of subordinate animals merely by their presence. A study by Lindsay *et al.* (1976) showed that the sexual behaviors of subordinate rams were inhibited when dominant rams were restrained in full view of the subordinates. They referred to this phenomenon as the 'audience effect'. Subordinate males become conditioned to fear and avoid dominant conspecifics.

Sexual performance is not necessarily correlated with aggressiveness or social dominance. For example, we compared the aggressive behaviors of male sheep differing in levels of sexual performance when competing for females and for food (Erhard *et al.*, 1998). Rams were tested in pairs of one high- and one low-performing individual. High and low rams did not differ significantly in the number of aggressive interactions initiated in either context. Guhl *et al.* (1945) found no relationship between the social (dominance) rank of male chickens (*G. domesticus*) and their sexual performance when placed *alone* with hens.

Sexual Performance Effects on Population Fecundity

Just about every captive-animal breeding plan has guidelines for maximizing reproductive success. Unfortunately, many males used in breeding programs fail to reproduce, or do so at unacceptably low levels. How does one predict substandard sexual performance before a great deal of time, effort and money are invested? The Australian veterinarian, Mike Blockey, noted that many beef bulls (*B. taurus*) did not show adequate levels of sexual performance in pasture matings. He subsequently devised a simple barnyard serving capacity test for bulls and then compared their performance in the barnyard with performance in the field (Blockey, 1978). The barnyard test simply counted the number of successful copulations (ejaculations) each animal achieved in a 40-min period. Each bull was then placed in a pasture with 35–40 heifers for 10 weeks and the pregnancy rate of the heifers at 3 weeks (end of first estrus) and after 10 weeks was compared with their barnyard serving capacity scores. Table 8.5 shows that, as a group, bulls with serving capacity scores of zero or one (number of ejaculations in 40 min) were clearly inferior in the field tests as well. Medium-rated bulls impregnated their heifers but it took them longer than the bulls with relatively high serving capacity ratings. Beef cattle breeders prefer to use high-serving capacity bulls because they benefit financially when young are born early in the calving season and during a relatively short period of time. Early births permit early sales (of weanlings) at higher prices. Also, a shorter calving season saves money on extra labor and other resources needed for the birthing period.

Table 8.5. Relationship of serving capacity of bulls to sexual performance in pasture mating (from Blockey, 1978).

Serving capacity[a]	No. of bulls	Percent heifers pregnant at end of first cycle[b]		Percent heifers pregnant after 10 weeks[b]	
		Average	Range	Average	Range
Low (0, 1)	6	21%	4–40%	33%	4–67%
Medium (2, 3)	10	60%	55–68%	92%	89–96%
High (4+)	7	72%	70–78%	97%	90–100%

[a]Based on number of ejaculations in 40 min pen tests prior to the breeding period.
[b]When placed with 35–40 heifers for 10 weeks.

A similar barnyard serving capacity test for male sheep (*O. aries*) was used by Perkins *et al.* (1992a) to select rams for pen breeding. Four high-performing rams and four low-performing rams were each placed in a pen for 9 days with approximately 30 estrous synchronized ewes. Observations confirmed the greater sexual activity of the high-performing rams. Nearly three times as many ewes were observed being inseminated by high-performing rams versus low-performing rams. Thirty-six percent of the ewes exposed to low-performing rams eventually gave birth to lambs, while 89.5% of the ewes exposed to high-performing rams lambed. Interestingly, sire group did not significantly affect the number of lambs born to those ewes that lambed. A similar study was conducted by Stellflug *et al.* (2006) using one high- and one low-performance ram in each of five breeding pens containing 200 ewes. In a 3-week period, high-performing rams impregnated almost twice as many ewes and sired more than twice as many lambs as low-performing rams.

The above examples demonstrate that males with relatively high libido and sexual performance are capable of impregnating more females than low-performing individuals in a given period of time. This can be achieved not only by a faster rate of ejaculation but also by fewer repeated matings with the same sexually receptive females. We tested the repeated mating hypothesis (Price *et al.*, 1996) by exposing 24 sexually experienced rams (*O. aries*) to ten estrous ewes on three occasions to determine the relationship between ejaculation rate (ejaculations per unit time) and the incidence of repeated matings. Rams that attained six ejaculations in a relatively short period of time (high-performing rams) copulated fewer times with the same ewes than low-performing rams. We concluded that high-performing rams not only mated faster than low-performing rams but they distributed their matings among a greater number of different ewes. Interestingly, we did not get this same result with beef bulls (*B. taurus*) (de Araujo *et al.*, 2003). High-performing bulls did not mate with a greater number of females than low-performing bulls when they were observed for the same number of ejaculations. One possible explanation for this species difference relates to differences in the sexual behavior of females in the two species. Estrous ewes typically exhibit stronger male seeking and sexual soliciting behaviors than estrous cows. Estrous ewes often compete for the attention of rams by wedging themselves between males and the other females the males are courting. This behavior was not observed in female cattle. Such intense solicitation behavior by ewes could encourage rams to switch females more frequently than bulls.

Maintaining Libido

Maintaining libido is essential to an efficient breeding program, whether we are dealing with natural mating on pasture, hand breeding or collecting semen. It is normal for breeding males to exhibit a temporary waning of libido after each ejaculation. This period of relative sexual inactivity (recovery period) typically increases in length with each successive ejaculation with a given female.

Male farm animals experiencing sexual satiation with one female often show an immediate surge in libido when exposed to a new, unfamiliar receptive female. Pepelko and Clegg (1965) demonstrated this resurgence in sexual interest in sheep (*O. aries*) and found that the surge was greatest when rams were exposed to a new unmated female. In a study by Jakway and Sumption (cited in Craig, 1981), it was found that male pigs (*S. scrofa*) would mate only once with a given estrous sow in a 40-min period. However, when placed in a pen with a number of receptive sows, they averaged eight ejaculations in 135 min. The surge in libido when exposed to novel females encourages polygamous males to shift their attention from one female to another when breeding (Dewsbury, 1981). Polygamous males that investigate and inseminate many females will potentially leave more offspring for the next generation than males that remain with a single female. Interestingly, male montane voles (*M. montanus*) exhibit a surge in copulatory activity when exposed to novel females, but male prairie voles (*M. ochrogaster*) do not (Dewsbury, 1973; Gray and Dewsbury, 1973). This is in keeping with the fact that montane voles are polygamous and prairie voles are monogamous (see p. 83).

Sexual satiation is reflected in the male's physiological responses to females (Coquelin and Bronson, 1979). Male mice (*M. musculus*) normally release luteinizing hormone (LH) when first exposed to a sexually *non-receptive* female. In Coquelin and Bronson's study, continuous exposure of male mice to non-receptive females was accompanied only by spontaneous, random elevations in LH. Likewise, successive presentations of the same female at 2 h intervals resulted in gradually diminishing LH responses. In contrast, exposing unresponsive males to a novel female dramatically stimulated the release of this hormone, demonstrating how socially induced endocrine responses are subject to habituation and recovery.

'Hand breeding', in which the male and female are brought into contact solely for the purpose of mating, is commonly practiced with horses (*E. caballus*), swine (*S. scrofa*), dairy goats (*C. hircus*) and companion animals (e.g. dogs and cats). Hand breeding can be a time-consuming task if many pairings are involved and if male libido is relatively low. One of my former graduate students and I (Mader and Price, 1984) conducted a study with beef bulls (*B. taurus*) to determine how we could maximize libido and sexual performance in the context of hand breeding. Twelve 22-month-old, sexually experienced Hereford bulls were tested three times under each of the following four conditions: (i) *spectator*: bulls were allowed to observe another bull mate with a cow for 15–30 min immediately prior to being tested themselves; (ii) *watched*: no pre-test stimulation but bulls were observed by another male (spectator) while being tested; (iii) *restrained*: confined next to a stimulus female (without a male) for 15 min before being tested; and (iv) *control*: no pre-test stimulation and tested in visual isolation from other bulls. Bulls attained 87% more completed copulations in the spectator treatment than when they served as controls. Sexual performance was also improved, albeit to a lesser degree, by being watched by another male and by restraint in the presence of the stimulus female. We repeated this study with male dairy goats (*C. hircus*) and obtained similar results (Price *et al.*, 1984/1985); 75% more successful copulations were obtained in the spectator treatment than as controls. Interestingly, when we conducted a companion study on male sheep (*O. aries*) (Price *et al.*, 1998a), no significant differences were found between the spectator and control conditions. Furthermore, direct competition between rams for estrous females did not enhance their sexual performance either (Price *et al.*, 2001).

In seeking an explanation for the 'spectator effect' in bulls and male goats, we noted that: (i) cows (*B. taurus*) and female goats (*C. hircus*) commonly participated in female–female mounting when in estrus but ewes (*O. aries*) did not; and (ii) bulls and male goats are attracted to females engaged in mounting. We surmised that natural selection in the ancestry of cattle and goats has likely favored those males most sexually stimulated by being spectators of conspecific mounting. These males may be the first to contact and mate with sexually active females. There is a distinct advantage to being first to mate with sexually receptive cows and female goats. In both of these species, sexual receptivity can be terminated by as few as two or three successful copulations, whereas ewes typically remain sexually receptive for many services. In addition, cows and female goats are sexually receptive for a significantly shorter period of time than ewes. In terms of offspring left for the next generation, there is no 'penalty' for rams that are slow to contact sexually receptive ewes. The relatively large testes of rams (rivals that of male cattle many times their size) produce relatively large quantities of sperm. Large sperm reserves accommodate repeated mating. Multiple inseminations (of females) with relatively large amounts of sperm increase the likelihood that a male will impregnate many females. High sperm-producing males have a selective advantage over males with lesser amounts of available sperm.

Nowhere is sexual fatigue more obvious than in semen collection. The efficiency of a semen-collecting program is dependent on maintaining adequate sexual libido. Variables of particular interest are 'reaction time' (time between exposure to the teaser stimulus and ejaculation) and number of ejaculations (semen samples) obtained per collecting session (day). Individual males differ in reaction times and ejaculation rates. Males also differ in recovery time after a sexually satiating experience with a given stimulus animal.

Most dairy cows (*B. taurus*) are now inseminated artificially. Hale and Almquist (1960) studied the relationship of bull sexual behavior with the efficiency of semen collection. Figure 8.4 plots the reaction time to ejaculation for three bulls exposed to the same stimulus animal over successive weekly collection periods. The data indicate that: (i) initial reaction times to the stimulus animal are different for the three bulls; (ii) recovery times vary; (iii) the initial reaction time is not necessarily correlated with recovery 1 week later (compare A versus B). Both initial reaction times and recovery rate must be considered in evaluating the sexual behavior of bulls used in artificial insemination programs. The frequency of collection is determined, in part, by recovery time. Figure 8.5 plots the rate of ejaculation over time for the three bulls. These data show: (i) individual differences in ejaculation frequency during initial 1h exposures to a single stimulus animal; (ii) individual differences in recovery 1 week later when exposed to the same stimulus animal; and (iii) an increase in ejaculation rate when the bulls are exposed to a new stimulus animal.

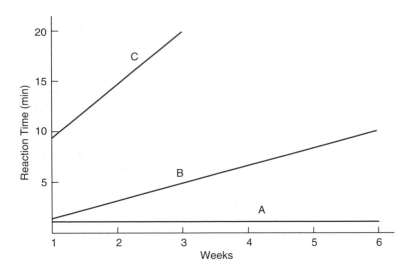

Fig. 8.4. Reaction time (latency) to ejaculation of three bulls exposed to the same stimulus animal for six successive weekly semen collections. Bull A recovers completely between ejaculations, bull B exhibits incomplete but good recovery and bull C shows poor recovery (Hale and Almquist, 1960).

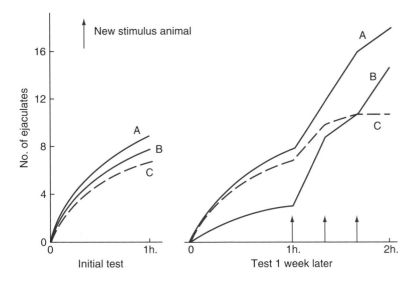

Fig. 8.5. Number of ejaculates obtained from three bulls during initial 1 h semen collection periods and during tests 1 week later. Note the enhanced response when the bulls are exposed to a new stimulus animal (vertical arrows) (Hale and Almquist, 1960).

Changes in the stimulus animal's environment can also improve the efficiency of semen collection. Hale and Almquist (1960) found that changing the location of stimulus animals or moving them back and forth in front of the bull could reduce reaction times by 73–84%. Changes in location by just a few meters appeared to be sexually stimulating. The faster recovery from sexual fatigue which accompanies exposure to a new stimulus animal or collection site suggests that sexual satiation is specific to a particular stimulus animal or situation (stimulus complex), rather than a manifestation of a general sexual fatigue, per se.

Both male and female conspecifics can be used as teaser (mount) stimuli when collecting semen. Hale and Almquist (1960) noted that male stimulus animals were just as effective as females when collecting semen from bulls (*B. taurus*). Work at Colorado State University on semen collection with horses (*E. caballus*) showed that estrous females, diestrous mares and geldings were equally effective in stimulating stallions for semen collection (Anderson *et al.*, 1996). In some cases, male teasers are preferred to lessen the possibility of venereal disease transmission.

In addition to maintaining libido, semen collectors are also concerned with obtaining the highest quality semen. One technique which improves a bull's semen quality is to allow two or three 'false' mounts on the stimulus animal (the penis is manually deflected away from the mount stimulus) prior to semen collection. When collecting one to two ejaculates per week from a bull, a single false mount at collection time can increase sperm output by about 50% (Hale and Almquist, 1960). Two additional false mounts can increase sperm output by another 50%. Another technique is to restrain the bull (i.e. prevent it from mounting) after exposure to the stimulus animal. Restraint for 10 min can improve semen output by as much as several false mounts. These 'sexual preparation' techniques maximize sperm output with minimal demands on the animal's libido. Bulls with the poorest semen quality show the greatest improvement in quality in response to false mounts and restraint.

Bulls, boars, goats, rabbits and dogs all show an increase in sperm output with sexual preparation, particularly when females are involved in the preparation. In contrast, the sperm output of stallions (*E. caballus*) is not improved (Ionata *et al.*, 1991). Perhaps the degree of sexual stimulation needed for erection in the stallion is sufficient for harvesting the maximum number of sperm.

Sensory Influences on Male Sexual Behavior

In topographically diverse environments where vision is precluded, female olfactory stimuli serve as important sex attractants in locating estrous females. Olfactory cues associated with urine from an estrous female or from different parts of the female's body (e.g. external genitalia, muzzle, groin area) can also aid males in locating and identifying sexually receptive females and stimulating pre-mating and copulatory behaviors. Vaginal flora increases the attractiveness of females. Ewes given daily intravaginal treatment with antibiotics were less stimulating to rams than ewes given the antibiotic vehicle alone (Ungerfeld and Silva, 2005). Sniffing or touching the vulva or urine of an estrous female with the tongue often stimulates a 'flehmen' response by the male, an upward curling of the upper lip (Fig. 8.6a–e). Flehmen is exhibited by males of all of our common farm mammals, including swine, in which the lip curl is very subtle due to the construction (stiffness) of the snout. It is also observed in male domestic cats (*F. domestica*), but not dogs (*C. familiaris*) or rodents. Flehmen appears to facilitate the movement of chemical substances through the nasopalatine ducts and into the vomeronasal organ, where olfactory receptor neurons relay information to the hypothalamus of the brain (see also p. 178–179).

Fig. 8.6. Flehmen in the bull (a), ram (b), male goat (c), stallion (d) and tiger (e) (tiger photo courtesy of Patrick Martin-Vegue, Marine-World Africa USA).

(d)

(e)

Fig. 8.6. continued

In a study with domestic goats (*C. hircus*), Ladewig *et al.* (1980) found that diestrous female goat urine elicited flehmen by males more than twice as frequently as estrous goat urine. It is possible that the sexual status of females in estrus is readily determined through activation of receptors in the nasal epithelium and that the diestrus condition requires more thorough examination using the vomeronasal system.

At close range, vision is the more important sensory modality. In males of most domestic animal species, visual cues associated with immobility of the female (when approached by a male) provide the greatest single stimulus for sexual arousal, mounting and copulation. This was clearly demonstrated in one of our studies (Wallach and Price, 1988), where we allowed beef bulls (*B. taurus*) the choice of *restrained* estrous and diestrous (non-estrous) cows. The bulls showed no preference. Immobility, in this case, was the critical stimulus for mounting and copulation. Another study by Anderson *et al.* (1996) showed that estrous mares (*E. caballus*) provide so many visual cues signaling their sexual condition that olfactory stimuli are not important in the elicitation of sexual arousal. Estrous mare urine appeared to be a sexual attractant, not a sexual stimulant.

Domestic ungulates can be trained to mount inanimate objects for the purpose of semen collection (Fig. 8.7). Hale (1966) found that bulls (*B. taurus*) readily mounted bales of hay covered with a tarp once they learned to associate mounting with ejaculation. The salient connection is that mounting *anything* leads to ejaculation, not mounting object X. Dummy ('phantom') mount stimuli (Fig. 8.8) are routinely used to collect semen from stallions (*E. caballus*) who might otherwise be injured by the 'teaser' stimulus animal. Training to mount inanimate objects involves simple conditioning processes (Conboy, 1992). At first, the animal may need to be coaxed to mount the dummy by placing a sexually receptive female in a position in front of or next to the dummy. Once the male has ejaculated into an artificial vagina while mounting the dummy, it will come to associate the collection site and equipment (e.g. dummy, artificial vagina, etc.) with a rewarding experience and routinely mount for semen collection. Stallions are typically given a 10–15 min exposure to the dummy in the presence of an estrous mare on the first training session. If the stallion does not mount the dummy, he should be allowed to mount the mare at the end of the session. Approximately 40% of stallions with normal libido successfully mount the phantom stimulus on the first day. About 90% will mount on the second day of training.

Fig. 8.7. Boar mounting 'dummy' for the purpose of semen collection.

Chapter 8

Fig. 8.8. Stallion mounting 'phantom mare' for the purpose of semen collection (at Colorado State University).

Physiological Influences on Male Sexual Behavior

Sexual activity is expressed in many male animals before reaching puberty. For example, we found that 10 of 39 (25.6%) male beef calves (*B. taurus*) mounted estrous heifers at 3 months of age and 29 of 74 (39.2%) at 6 months of age (Price and Wallach, 1991). Such observations beg us to ask what factors initiate and maintain prepubertal mounting when testosterone levels are so low and before sensory feedback from ejaculations can be experienced? One hypothesis is that prepubertal mounting is maintained by tactile and thermal stimulation to the penis as it rubs against the relatively immobile penile sheath. Second, the animal may experience deep pressure sensations as the abdomen of the male rubs against animate (or inanimate) objects when mounting. Katz and Price (1986) sought to examine this hypothesis by studying the role of afferent input from the penis and copulatory experience in the development of sexual activity in the bull (*B. taurus*). This was done by comparing the libido of intact bulls with bulls whose penises had been altered at 6–7 months of age, either by lesioning the dorsal penile nerves or by surgically deviating (deflected) their penises approximately 45 degrees from the midline of the body so they were incapable of gaining intromission when mounting females. At 18–24 months of age, the neurectomized bulls exhibited lower levels of sexual activity than both the intact bulls and those with deviated penises. The deviated bulls, which had never had a copulatory experience but had neurally intact penises, mounted estrous females just as frequently as sexually experienced intact bulls when exposed to estrous females whose perineums had been covered with cloth patches to prevent intromission (by the intact bulls). The results suggested that sensory stimulation from the genitalia rather than copulatory experience was important in maintaining sexual libido in the bull. Metzler *et al.* (1988) conducted a similar study with male dairy goats (*C. hircus*) and obtained the same results. Stimulation of the genitalia during mounting appears to be critical for the maintainance of sexual activity in these species.

An alternative hypothesis to explain prepubertal mounting is that the testes of the developing male produces sufficient quantities (albeit very small) of androgen, such as testosterone (TP), to support mounting at this stage of development. This hypothesis is supported by the fact that sexual differentiation of the mammalian brain takes place very early in development by the action of androgen (see p. 101). Young males become 'hard wired' to exhibit male sexual behavior based on the action of what might seem like insignificant levels of androgen.

It is also of interest that adult males produce more androgen, such as TP, than is actually needed to support their inherent levels of libido and sexual performance. This explains why individual differences in base-line levels of TP and luteinizing hormone (LH) in intact males (when not engaged in sexual activity) do not correlate with individual differences in their libido or sexual performance. Levels of TP and LH collected from sexually experienced beef bulls (*B. taurus*) over a 24 h period (13 samples per animal) did not correlate with their sexual performance at 18 and 24 months of age (Price *et al.*, 1986). Research has also shown that the sexual performance of intact adult males is usually not improved by injections of TP or LH. This makes sense only if a portion (threshold amount) of the androgen produced is needed to maximize an individual's sexual performance. D'Occhio and Brooks (1982) castrated male sheep before puberty and, once mature, injected them with graded doses of TP while observing them for sexual behaviors. Circulating plasma levels of TP required for complete mating activity (mount, intromission and ejaculation) were lower than plasma levels of TP typically found in intact rams. Roselli *et al.* (2002) found that basal androgen concentrations in adult male sheep (*O. aries*) were not responsible for low libido or male–male sexual preferences and proposed that functional differences in sexual performance and orientation could exist in the brain. They later identified a cell group within the medial preoptic area/anterior hypothalamus of adult sheep that was significantly larger in rams than ewes. Interestingly, the volume of the ovine sexually dimorphic nucleus (OSDN) was twice as large in female-oriented rams as in male-oriented rams (Roselli *et al.*, 2004). Because the medial preoptic area/anterior hypothalamus is known to control the expression of male sexual behavior, these results suggested that differences in sexual preference might be related to differences in brain anatomy and its capacity for estrogen synthesis. This hunch was confirmed in a recent study (Roselli *et al.*, 2007) which demonstrated that the OSDN was organized prenatally and that exogenous TP administration to lamb fetuses between days 30 and 90 of gestation (term is 150 days) resulted in the development of male-like OSDNs in ewes. It appears that TP acts on the organization of aromatase-expressing neurons in the OSDN during a critical prenatal period of brain development.

As stated previously, increased levels of TP and LH have been reported in response to sexual arousal and mating activity in males of a wide variety of domestic animal species (e.g. cattle, sheep, pigs, rabbits, mice and rats). Adult male mice (*M. musculus*) housed with an adult female over an extended period have the same plasma TP levels as males housed with other males. However, if the resident female is replaced by a new female, TP levels become markedly elevated in 30–60 min (Macrides *et al.*, 1975). If the resident female is replaced by another male, TP levels do not change. Post *et al.* (1987) demonstrated that beef bulls (*B. taurus*) showing relatively large increases in TP to a single injection of gonadotropin-releasing hormone (GnRH) exhibited higher levels of sexual performance in pasture matings than bulls with a low testosterone response to GnRH. GnRH, a decapeptide secreted in the hypothalamus, stimulates the production of LH in the anterior pituitary. LH travels, via the systemic circulation, to leydig cells in the testes, which produce testosterone. Perkins *et al.* (1992b) found that basal TP and LH concentrations in sexually active rams (*O. aries*) were elevated by exposure to estrous ewes but not in their sexually inactive counterparts. Such results demonstrate the important feedback effect of sexual behavior on hormone secretion in the male.

The effect of castration on the sexual behavior of males varies with the species, the individual and the animal's age at the time of castration. The ejaculatory response is usually the first component of sexual behavior that deteriorates following castration (Beach, 1948). In mammals, castration *during adulthood* typically results in a decline in sexual libido and performance but, in some species (e.g. dogs, goats), certain sexually experienced individuals continue to exhibit mounting and behavioral ejaculation for years after the surgery (Fig. 8.9). Castration *prior* to sexual maturity usually results in reduced libido and impaired sexual performance as an adult but, again, it depends on the species and individual. Prepuberally castrated rabbits (*Oryctolagus cuniculus*) and cats (*F. domestica*) exhibit greatly impaired sexual function, whereas rats (*R. norvegicus*), hamsters (*M. auratus*), guinea pigs (*C. porcellus*), bulls (*B. taurus*) and chimpanzees (*P. troglodytes*) display more elements of the complete sexual pattern. Le Boeuf (1970) demonstrated that early castration of beagle dogs (*C. familiaris*) had little effect on their *libido* or *attractiveness* to estrous females. The most noticeable deficiency was a loss of the ejaculatory reflex, in most cases, and the inability to attain a copulatory lock with females, presumably due to underdevelopment of the penis resulting from the lack of androgens during maturation. This is in stark contrast to the male domestic cat (*F. domestica*), which typically loses all interest in females following prepuberal castration. Rosenblatt and Aronson (1958) reported that only 1 of 13 male cats castrated at 4 months exhibited mounting when exposed to an estrous female and that individuals failed to show 'stepping' and

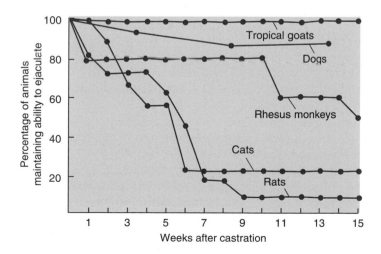

Fig. 8.9. Species comparison of differences in the persistence of copulatory behavior following castration during adulthood (Hart, 1985).

thrusting characteristic of intact males. An intermediate effect of castration is typically seen in horses (*E. caballus*). As many as half of castrated horses (geldings) show stallion-like behavior to mares, with many exhibiting herding behavior and mounting. Interestingly, stage of sexual maturation does not affect the response of cats and horses to castration. These observations support the fact that while androgens are critical for the expression of normal adult male sexual behavior, we cannot predict the effects of castration on individual libido and sexual performance without knowledge of the species and individual response to the loss of androgen production.

Restoration of mating behavior in males castrated after achieving mating competence is attained relatively soon following injections of androgen such as testosterone propionate (TP), a process referred to as 'replacement therapy'. For example, Clegg *et al.* (1969) castrated sexually experienced rams (*O. aries*) and subsequently administered daily injections of TP after copulatory behavior had ceased or 1 year had passed, whichever came first. All of the rams exhibited pre-castration levels of sexual performance after 9 days of replacement therapy. In a second experiment, rams were castrated shortly after birth and then administered daily injections of TP in adulthood. The full complement of copulatory behaviors was observed in these animals after 2–6 weeks of therapy. Copulatory behaviors in the wethers subsequently disappeared within 4–7 weeks after hormone withdrawal.

Mate Preferences

Animal breeders readily agree that most males show preferences for certain females. In multi-female groups, some females receive ample inseminations to attain pregnancy, while other females in close proximity are ignored. Rams prefer older ewes to younger ones. Color or odor may influence the preferences of stallions for certain mares. Pickett *et al.* (cited in Anderson *et al.*, 1996) reported that stallions spent more time investigating passive mares whether or not they were in estrus. Also, I remember one study in which our bulls (*B. taurus*) continued to ignore a specific estrous cow during test after test, even after other cows in the group had lost their sexual receptivity due to repeated matings.

Male–Male Mounting and the Buller Steer Syndrome

Male dogs, horses, sheep, cattle, pigs and many other species will mount other males when females are not present. There is a long-standing debate about whether male–male mounting is an expression of general stimulation, sexual libido, or that certain males use mounting to reinforce their social status. In reality, all of these motivating forces are likely involved at one time or another, with one being more important than the others, depending on the species, maturational state of the animals involved, season of the year, etc. I have noted that moving bulls (*B. taurus*) to a new location will often stimulate male–male mounting, suggesting that general stimulation is sufficient to stimulate this behavior. While the causes of male–male mounting are a matter of intellectual curiosity for most people, male–male mounting is a matter of great economic concern for commercial feedlot operators. Castrated male cattle (steers) (*B. taurus*) are typically placed in feedlots as yearlings, where they are fed a special 'finishing' diet in preparation for slaughter. Relatively large numbers of steers (200+ individuals) are sometimes placed in each pen. About 2.5% (typical range:1–4%) of the steers in these feedlot enclosures will either stand like estrous females to be mounted by other males or be followed and mounted when the opportunity arises (Blackshaw *et al.*, 1997). Steers being mounted are referred to as 'bullers' and the animals performing the mounts are called 'riders'. Buller steers are often injured from persistent mounting by pen mates, or debilitated to the point that they must be removed from the group and housed separately. The added labor and pens required to manage buller steers cuts into profit margins. Losses are even greater when buller steers become debilitated. Feedlot operators claim that economic losses due to 'bulling' are second only to respiratory diseases.

It is not clear why certain steers submit to being mounted. The normal response of steers to receiving mounts is to exhibit avoidance behaviors (run away) or to turn around and threaten or butt the mounter. Rider steers rarely exhibit an erection or attain rectal intromissions when mounting and do not show aggression toward other riders when pursuing bullers. Riders often take turns mounting bullers without interacting with each other. Riding typically stops when the buller is removed from the pen. Bullers do not mount one another when housed together in pens.

Different hypotheses have been proposed for the development of buller behavior. One report suggests that the frequency of bulling increases when steers are given growth-promoting hormones such as diethylstilbestrol. Hormone implants containing estrogens could partially feminize the odor and (or) behavior of steers. This idea was supported by another study, which found that blood estrogen levels were elevated for buller steers but not riders. In addition, riders may have been castrated late or incompletely, or their growth-promoting implants (if any) may be metabolized differently than in other steers. A second idea is that buller steers are simply individuals which lack an aversion to being mounted and that riders learn to take advantage of their submissive behavior. Mounting is highly subject to social facilitation in which the activities of one animal stimulate conspecifics to engage in the same behavior. Observing the mounting activity of conspecifics is sexually stimulating to male cattle (see p. 115), thus encouraging other males to participate.

9　Female Sexual Behavior

Physiological Control of Female Sexual Behavior

Hormones have a critical role in the expression of female sexual behavior. The reproductive cycle begins with hormones secreted by the hypothalamus of the brain. The hypothalamus responds to external cues such as seasonal changes in light:dark cycles in many species to secrete gonadotropin-releasing hormone (GnRH). GnRH activates the anterior pituitary gland to secrete follicle-stimulating hormone (FSH), which initiates growth of one or more ovarian follicles. The growing ovarian follicle secretes ever-increasing amounts of estrogen, which prepares the reproductive tract for sperm transport, promotes changes in the epithelial tissue of the vagina to facilitate copulation, activates the preovulatory surge of luteinizing hormone (LH) from the anterior pituitary and initiates and supports female sexual behaviors (reviewed by Hurnik, 1987). The LH surge triggers ovulation. Female sexual libido wanes following ovulation and terminates as estrogen levels plummet and progesterone secretions from the corpus luteum increase. Progesterone prepares the uterus for implantation of the fertilized egg if mating is successful and facilitates the maintenance of pregnancy.

Cyclicity of Female Reproductive Behavior

Sexual receptivity in females is cyclic in nature. Adult female cattle (*Bos*) and swine (*S. scrofa*) exhibit continuously repeated estrous cycles. They are referred to as 'polyestrous' because they cycle until they conceive. Female domestic sheep (*O. aries*), goats (*C. hircus*), horses (*E. caballus*) and cats (*F. domestica*) are 'seasonally polyestrous' because they exhibit repeated estrous cycles during specific seasons of the year only. Female sheep and goats cycle during short days (fall season in the northern hemisphere), female cats cycle when day length is increasing (late winter and spring) and mares cycle during long days (spring and early summer). Estrous cycling in these species continues during their respective breeding seasons, unless terminated by conception. Female dogs (*C. familiaris*) are 'monestrous' because they have only one estrous period at a time. They may have one to four estrous periods per year but two is most common, about every 6 months. These parameters are summarized in Table 9.1.

Estrous cycles in female cattle, swine, goats, horses and cats are about 21 days in duration. Cycles in female sheep average 16–17 days. The duration of 'true estrus' during these cycles, when females will permit copulation by the male, varies from a few hours to 10 days, depending on the species. 'Proestrus' immediately precedes true estrus. During this period, females exhibit signs of sexual interest in conspecifics (e.g. approaching and attempting to mount) and produce olfactory cues that stimulate males to remain in close proximity. In female cattle (*B. taurus*), proestrus may begin 40 h before the onset of true estrus. 'Metestrus' is the period immediately following true estrus when sexual interest is waning and females no longer permit copulation.

Table 9.1. Some female reproductive parameters and behavioral indicators of estrus in selected domestic animal species (adapted from Craig, 1981 and Hurnik, 1987).[a,b]

Item	Cow	Sow	Ewe	Doe (goat)	Mare	Queen (cat)	Bitch (dog)
Mean age first estrus	9 mo	7 mo	9 mo	5 mo	24 mo	6 mo	10 mo
Mating season and estrus cyclicity	All year polyestrous		Seasonally polyestrous (short days)	Seasonally polyestrous – fall	Seasonally polyestrous – winter to early summer	Seasonally polyestrous – late winter to early summer	Monestrous[c]
Length of estrous cycle (mean)	21 days	21 days	16.5 days	21 days	21 days	21 days[d]	One cycle, twice a yr
Estrus duration	6–30 hr	1–3 days	1–2 days	18 hr–2 days	2–10 days	6–10 days	7–10 days
Duration of receptive phase of estrus	1–12 hr	12–32 hr	18–36 hr	16–24 hr	1–3 days	1–4 days with copulation	
Primary criterion in absence of males	Mounting by other females	Mating stance to pressure on back	None	Mounting by other females, tail wagging	Clitoris exposed, tail raised, urination stance	Lordosis, treads to pressure on back	
Duration of copulation	3–12 sec	3–12 min	2–4 sec	2–4 sec	20–90 sec	6–10 sec	10–30 min ('lock')

[a]In absence of males.
[b]Under moderate to high levels of nutrition in temperate climates, a few breeds may differ from these descriptions.
[c]Large breed variation.
[d]Highly variable.

Metestrus in cattle may last for 20 h. Estrous odors produced by cows are present up to 3 days prior to and 1 day after true estrus. Thus, the duration of estrus includes that period of time from the beginning of proestrus to the end of metestrus. 'Diestrus' is the period between the end of metestrus and the beginning of the next proestrus in polyestrous species. 'Anestrus' refers to the period of time when seasonally polyestrous females are not cycling. A flock of sheep is considered in anestrus when 80% of the ewes are not ovulating. 'Estrus' is spelled *without* an 'o' when used as a noun (e.g. 'during estrus') and *with* an 'o' when used as an adjective ('estrous cycle').

Female rodents (e.g. mice, rats, gerbils) frequently experience a post-partum estrus and become pregnant within days of giving birth. In captivity, some female rodents are engaged in a repeated cycle of pregnancies, giving birth and nursing. Interestingly, a recent study (Clark *et al.*, 2006) showed that female Mongolian gerbils (*Meriones unguiculatus*) who were repeatedly nursing and pregnant at the same time produced female offspring which were less likely as adults to socially bond with their future mates, and which produced fewer offspring than females who had dams who were not nursing when they were gestating.

The Effect of Males on Estrous Cyclicity

The introduction of a male, or male secretions (e.g. urine, saliva, scent gland secretions), can facilitate the onset of estrus in many species (e.g. cattle, sheep, goats, swine, horses, dogs, cats, etc.). In anestrous female sheep (*O. aries*), the onset of estrous cycling (as the breeding season approaches) is advanced 2 weeks to 2 months by the introduction of an adult ram (Knight, 1983; Ungerfeld *et al.*, 2004). The 'ram effect', as it is called, also serves to synchronize estrous cycles among the ewes in the flock. About 50% of the ewes in the flock will experience their first ovulation by the third day after ram introduction. This ovulation is typically not accompanied by estrous behaviors (i.e. it is 'silent'). The second ovulation, which is accompanied by estrous behaviors, usually occurs about 17 days later (Fig. 9.1). The stimulative effect of the ram is believed to be

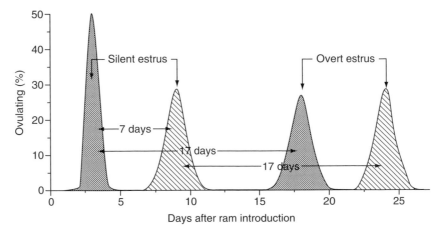

Fig. 9.1. Percentage of anestrous ewes ovulating over a 4-week period after introduction of a ram. The first ovulations are accompanied by a 'silent' estrus (absence of estrous behavior) (Signoret, 1990).

Chapter 9

Table 9.2. Percentage of anestrous ewes that ovulated in response to various ram stimuli (from Knight and Lynch, 1980).

Treatment	No. of ewes	Percentage (no.) that ovulated
No male stimuli	30	7 (2)
Water spray	14	0 (0)
Male urine spray	18	22 (4)
Muzzle smeared with wax and wool	29	48 (14)
Intact male	26	50 (13)

mediated largely by olfactory (chemical) substances on the ram's wool (Table 9.2) since the ram effect can be obtained by exposing ewes to rams' wool in a face mask. The periodic introduction of a new ram(s) results in a stronger ram effect than continuous exposure to a single ram. Apparently, stimulus novelty is also an important component of the ram effect.

Synchronization of first behavioral estrus via the ram effect requires large sperm reserves since breeding males will potentially service many females in a relatively short period of time. Fortunately, rams have evolved larger testicles and sperm production for their body size than almost any other mammal. Estrous synchronization also results in a relatively short lambing period, 5 months later, which helps to reduce losses due to predation when ewes are maintained on pasture and exposed to predators. Also, lambs born early in the breeding season reach market weight earlier and command a better price than late-born lambs.

Many years ago, Whitten noted that the introduction of male mice (*M. musculus*) into a cage of females induced and synchronized their estrous cycles (Whitten, 1956). This phenomenon is known as the 'Whitten Effect'. When female mice are placed in large groups, their estrous cycling is often suppressed. If these females are subsequently exposed to male mouse urine for 24–48 h in the diestrous condition, estrous cycling is initiated and incidentally synchronized so that most of the females come into estrus 3–4 days later. The active chemical substance in the urine of males which stimulates this effect on females is a secretion of the preputial gland and is an androgen metabolite, or the product of androgen-maintained tissues, since the effect of the urine is lost following castration and restored with testosterone therapy (injections). A recent study (Marchlewska-Koj *et al.*, 1990) found that the salivary glands of male mice (*M. musculus*) also secreted the estrus-stimulating substance.

In the beef industry, cows (*Bos*) are usually bred before their calves are weaned to maintain a 12-month reproductive cycle. Nursing and caring for calves slows their return to ovulation and behavioral estrus following parturition. The introduction of a bull in a herd of nursing post-partum beef cows stimulates them to ovulate and exhibit behavioral estrus sooner than they would otherwise in the absence of the male. Dairy cows (*B. taurus*) are often slow to return to behavioral estrus after parturition, in spite of the fact that their calves are normally hand reared by humans. This is likely due to the physiological demands of milk production for human consumption. Interestingly, studies have shown that exposure of high milk-producing, post-partum dairy cows to bulls does not advance the onset of their return to estrous cyclicity (Shipka and Ellis,

1998). Perhaps the high energy demands of milk production overrides the stimulating effect of the bull, or possibly the response has been lost through relaxed selection resulting from decades of artificial insemination.

Behavioral Manifestations of Ovulation and Sexual Receptivity

Table 9.1 describes some of the behaviors typically exhibited by females when in true estrus. As stated in the last chapter, the single, most important female behavior signaling sexual receptivity is willingness to stand immobile (or crouch) to be mounted and served by the male. Sexual receptivity normally occurs just before and (or) during the period of ovulation but, sometimes, females will ovulate and not exhibit sexual behaviors. These periods of 'silent ovulation', as they are called, are most common during the first one or two ovulations after reaching sexual maturity, the first ovulations following parturition and at the beginning of the breeding season in seasonally breeding species. Table 9.3 provides data on the incidence of silent ovulation in dairy cattle (*B. taurus*) at puberty and after giving birth.

In nature, males are not always present when females become sexually receptive. Consequently, behaviors have evolved which facilitate finding suitable mates. In general, females locate males either by seeking them out, by attracting males via visual displays advertising sexual readiness, or by auditory or olfactory cues. Females of most domestic farm animal species exhibit male seeking to some extent during periods of sexual receptivity but, of all these species, male seeking is expressed most strongly by estrous ewes (*O. aries*). Both ram seeking and other estrus-related behaviors are estrogen dependent (Fig. 9.2) (Lindsay and Fletcher, 1972). The term 'proceptive behavior' has been given to female behaviors associated with seeking out and soliciting males (Beach, 1976).

Female–female mounting is one of the more obvious visual displays advertising sexual readiness in cattle (*Bos*), goats (*C. hircus*) and swine (*S. scrofa*). Estrous females may form sexually active groups (SAGs) and take turns mounting one another (Fig. 9.3). Males are attracted to these SAGs from a distance; seeing other animals mounting is sexually stimulating (see p. 115). Estrous females sometimes try to mount males in close proximity. Sexually active males often sniff the anogenital region (perineum) of estrous females (Fig. 9.4) and touch this area with their tongue. Non-receptive females typically run away from the male in response to anogenital sniffing and then often stop to urinate. Pursing males may sniff their urine and then

Table 9.3. Incidence of 'silent ovulation' (absence of estrous behaviors) during the first three ovulations following puberty and parturition in female dairy cattle (from Morrow, 1969 and King *et al.*, 1976).

Ovulation number	Following puberty – percentage of heifers ($N = 53$)	Following parturition – percentage of cows ($N = 36$)
First	74%	50%
Second	43%	6%
Third	21%	0%

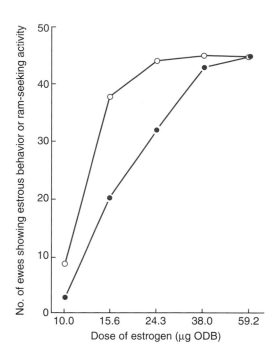

Fig. 9.2. Number of ovariectomized ewes (*N* = 48) exhibiting estrous behavior and ram-seeking activity after receiving replacement therapy with different doses of estrogen (estrous behavior: ○—○; ram-seeking: ●—●) (Lindsay and Fletcher, 1972).

Fig. 9.3. A sexually active group of Holstein heifers.

Fig. 9.4. Ram sniffs anogenital region of an estrous ewe. Diestrous ewes normally run away from rams in response to this behavior.

exhibit flehmen. The odor of urine and vaginal secretions from females convey information to sexually experienced males about the female's state of receptivity. If the female does not run away from the male, he tests her readiness to stand immobile (for mounting) by either placing his chin on the female's back or rump (cattle, goats) (Fig. 9.5), contacting the female with foreleg kicks (sheep and goats), nudging the sides of the female (swine, sheep, goats) (Fig. 9.6), or showing an intention to mount

Fig. 9.5. Male goat rests its chin on the shoulders of an estrous doe. She was bred shortly after I took this photo.

Fig. 9.6. Ram nudges side of estrous ewe. Diestrous ewes normally attempt to escape from rams in response to this behavior.

by shifting his weight to his hind legs and raising his head while making contact with the female's back or rump (nearly all species) (Fig. 9.7). In response to these male behaviors, the sexually receptive female stands immobile (all species) (Fig. 9.8), the most important visual signal to the male that she is ready to be mounted. In some species (sheep and goats), the courted female will wag her tail periodically. The mare typically raises her tail, exposing the clitoris (Fig. 9.9), and often assumes a urination stance. The receptive sow becomes immobile when pressure is applied to her back.

Fig. 9.7. Mount intention by a bull on a non-receptive cow. The animal shifts its weight to its hind legs and raises its head but both front feet never leave the ground.

Fig. 9.8. Female–female mounting in domestic goats.

Fig. 9.9. An estrous mare raises her tail, exposing the clitoris (at Colorado State University).

Humans can demonstrate the immobility response of the estrous sow (and detect true estrus) by placing their hands on either side of the sow's rump and applying pressure. This test is 97% effective if the sow can see, hear and smell a boar. Only about 50% of estrous sows will stand immobile to the backpressure test in the absence of boar stimuli. The estrous cat (*F. domestica*) and rat (*Rattus*) assume a 'lordosis' posture (back arches downward with raised pelvis) when mounted. The estrous cat may also tread the ground with its rear legs. Sexually receptive female chickens (*G. domesticus*) and turkeys (*M. gallopavo*) solicit copulations by assuming a sexual crouch in response to approaches and courtship by the male.

Female Mate Preferences

In Chapter 7, I stated that females largely controlled mate choice in sexually dimorphic species. Since most of our common domestic animals are sexually dimorphic to some degree, how does this play out in captive animal breeding systems? Can female choice be exercised only in beef cattle (*Bos*), sheep (*O. aries*) and chickens (*G. domesticus*) where multi-sire mating is still practiced? In single-sire mating and hand breeding, only one male is available to females at any given time, and in artificial insemination programs, the female may never see a male. Thus, females subjected to single-sire mating and hand breeding are faced with either accepting or rejecting their male suitor. Females may avoid mating with young, inexperienced males. They may also reject males (often young males) who exhibit overly vigorous pre-mating behaviors. And then, some females are just 'plain picky' for reasons not apparent to humans. Domestic female cats (*F. domestica*) and dogs (*C. familiaris*) tend to fall into this category. In some cases, a lack of social experience (i.e. proper socialization) with conspecifics can result in responses toward certain males (e.g. fear, aggression) which compete with sexual activity. Certain female dogs (bitches) will not mate with subordinate males, while other females are indifferent to social status. Females of some breeds of sheep (*O. aries*) prefer to mate with males of their own breed.

Several years ago, we conducted a series of experiments on the mate preferences of female sheep (*O. aries*) (Estep *et al.*, 1989). In one experiment, we gave estrous ewes the choice of young versus older males. They consistently chose the older males. In another experiment, we presented rams of similar size that exhibited different rates of sexual solicitation (grunting, licking the ewe, nasonasal contact, nudging the ewe with the head, kicking and pawing the ewe with a foreleg). The ewes preferred the rams with higher rates of solicitation. A third experiment paired rams which differed in levels of sexual performance (serving capacity). Both ram-lambs and older rams were used, keeping age and body size constant in each test. Ewes consistently chose the rams with higher sexual performance ratings, regardless of age and size. The last experiment showed that males were more attractive to some estrous ewes if the rams were engaged in interactions with other estrous ewes. These studies, taken together, illustrate the proactive role of female choice in this species and how female preferences may account, in part, for the greater reproductive success of older, larger and more sexually active rams.

Female choice has also been demonstrated in other farm animal species. Estrous sows (*S. scrofa*) prefer adult intact males to younger boars and castrated males. Female goats (*C. hircus*) also show preferences for older, more sexually active males, particularly those with a strong buck (caprine) odor. Avoidance of young, overly aggressive and (or) socially subordinate males is characterized by moving away from the male when approached, tightly covering the vulva with the tail and, in some cases, aggressive threats directed at the pursuing male.

Horse breeders are well aware that estrous mares (*E. caballus*) have preferences for certain stallions, particularly if the mare is older and more sexually experienced. In free-ranging groups of horses, preferences may be based on the stallion's position in the social hierarchy. Pickerel *et al.* (1993) found a preference among estrous mares for stallions exhibiting the highest rate of vocalizations.

Methods of Estrous Detection

Reproduction in domestic dairy cattle (*B. taurus*) and turkeys (*M. gallopavo*) nowadays is almost totally dependent on artificial insemination. The efficiency of hand breeding and artificial insemination is based largely on the accuracy of estrous detection techniques. Each time an estrus is missed, or breeding (insemination) is attempted at the wrong time of the cycle, the animal breeder loses time and money. In addition, hand-bred animals can be injured when attempting to mate with non-receptive females. It has been estimated that half of the estrous periods go undetected on the average dairy farm in the USA. A study on eight California dairies revealed that 13% of the artificial inseminations were on cows *not* in estrus (cited by Stevenson, 1996). The importance of efficient estrous detection techniques is reflected in the number of techniques that have been developed to increase accuracy while reducing costs.

Monitoring of mount interactions

Table 9.4 lists some of the estrous detection techniques that have been used and researched in the dairy industry. Over the years, the dairy industry has relied most heavily on direct observation of female–female mounting for setting the time of artificial insemination. Females that stand immobile when mounted by other females are temporarily taken out of the herd and inseminated by trained technicians, veterinarians, or the dairy farmer. Unfortunately, direct observation is not always reliable. Hurnik *et al.* (1975) found that the periods of true estrus in some dairy cows (*B. taurus*) can be so short that they go unnoticed using direct observation programs. Direct observation of cows for sexually active females is typically scheduled twice a day (12 h apart), either just before or after each milking. The researchers found that some females would go in and out of true estrus in less than a 12 h period and that this was most likely to happen during the hours of darkness (Fig. 9.10). Furthermore, estrous females sometimes will not be mounted frequently enough to be observed. They may be experiencing a silent ovulation, as discussed above, or there may be too few sexually active females in the herd at the time. Hurnik *et al.* (1975) found that the number of mounts received by estrous females averaged 11.2 with one cow in estrus to 52.6 mounts per cow when three cows were in estrus on the same day. This corresponds with the finding that the duration of true estrus for an animal is longer when there are a number of cows in estrus, thus offering more time for mounts to be exhibited and observed. These results

Table 9.4. Some traditional and 'experimental' methods used to detect estrus in female cattle (*B. taurus*) and other large domestic farm animals.

Commonly used techniques:
- Direct visual observation of female–female interactions (e.g. mounting) and response to a male.
- Visual inspection of the anogenital region (perineum) for vaginal discharge and tone.
- Tail-head or rump painting/marking (paint is smeared off when animal is mounted).
- Pressure-sensitive mount detector devices used with or without radiotelemetry (e.g. Heat-Watch®, KaMar® heat detection devices, respectively).
- Use of vasectomized males or androgenized females fitted with 'chin-ball' halters (cattle), chest marking harness (sheep and goats), or chest paint (sheep and goats) which leaves paint marks on mounted females.

'Experimental' techniques:
- Rectal palpation for mature follicles.
- Measures of locomotor activity (e.g. pedometers).
- Progesterone levels in milk or blood (decline in progesterone levels to < 1.0 ng/ml).
- Electrical conductivity of reproductive tissue (probe or implant measures changes in electrical conductivity of vaginal mucus).
- Increase in vaginal temperature (> 1.0 °C).
- Ultrasound scanning of ovaries (disappearance of fluid-filled pre-ovulatory follicle).
- Increase in blood levels of luteinizing hormone.
- Detection of estrous odors by dogs.
- Changes in milk yield (increases in later stages of estrus).
- Changes in aggressive behavior (increases in early stages of estrus).
- Elevation of milk temperature (0.3–0.4 °C).
- Artificial electronic nose to detect volatile perineal odors associated with estrus (measures changes in the electrical resistance of sensors exposed to soiled cotton swabs).

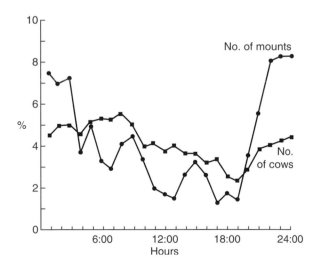

Fig. 9.10. Distribution (%) of mounts and number of cows in true estrus over a 24 h period. Estrous activity is greatest during the hours of darkness (Hurnick *et al.*, 1975).

suggest that dairy cows should be kept in large enough groups so that at least two or three cows will be in estrus on any given day.

On my family's dairy farm, we always assumed that only the cows *being* mounted were in estrus. Hurnik *et al.* (1975) found that nearly 80% of the cows *performing* mounts in their study were also in estrus. This is more likely to be true if there are multiple females in estrus at any given time. Furthermore, there is no relationship between social dominance and the tendency to be mounted or to mount, even though the frequency of aggressive interactions increases just before and during estrus.

The substrate on which animals are housed can influence the efficiency of estrous detection. Research at North Carolina State University (Britt *et al.*, 1986) has shown that estrous cows (*B. taurus*) housed on dirt exhibit 65–130% more mount interactions than animals housed on concrete. Concrete does not provide good footing for hoofed animals, especially when wet or when manure accumulates. Unsure footing can inhibit mounting activity.

Season of the year and weather patterns at any given locality can influence the frequency of mounting. Rain, high winds and hot weather, particularly accompanied by high humidity, tend to suppress mounting and other estrus-related motor activities.

Mount detection techniques other than direct observation have become widely used in the livestock industries. Tailhead painting is commonly used to determine if cows (*B. taurus*) have been mounted (Xu *et al.*, 1998). A paint or crayon mark is placed on the back of the cow at the base of the tail, which is smeared or obliterated when the cow is mounted. KaMar® pressure-sensitive mount detectors (KaMar Marketing, Inc., Portland, ME) consist of a tube of gel attached to a small patch glued to the animal's tailhead (Fig. 9.11). When the cow is mounted, the weight of the animal performing the mount turns the color of the gel from white to red, signaling that a mount has occurred. The HeatWatch® system (DDx Inc., Denver,

Fig. 9.11. KaMar® mount detection device which has been activated (tube of clear gel turns red) from the cow being mounted.

CO) consists of a pressure-sensitive device with radio transmitter attached to material glued to the cow's tailhead. When the cow is mounted, the radio transmitter sends a signal to a computer, which records the mount and the time it occurred (see examples of use in Xu *et al.*, 1998 and White *et al.*, 2002). The advantages of these techniques are that they operate 24 h/day and save time and money otherwise spent in direct observation. Disadvantages include the cost of the equipment and materials and labor costs associated with marking or attachment of the devices and monitoring their use. In addition, estrous cows will sometimes mount diestrous females, creating false positive readings.

Marker animals

Male conspecifics are sometimes used to monitor estrus in females and to determine when they have been mounted and likely inseminated. Some beef cattle (*B. taurus*) breeders fit 'chin-ball' marking halters on their bulls (Fig. 9.12) to mark cows that have been mounted. The chin-ball marker works like a ballpoint pen under the bull's chin, leaving a telltale ink line on the cow's back when the bull lowers its head while mounting. A study at Cornell University (Foote, 1975) found that 87% of the cows in estrus were marked by a bull wearing a chin-ball device. As noted earlier, bulls sometimes ignore certain females, even when they are in estrus.

It is estimated that about 20% of the beef cows (*B. taurus*) in the USA fail to breed in any given year. Monitoring estrus and mounting activity helps the beef cattle breeder determine which cows have likely conceived during the ranch's designated breeding season and which animals should be culled. Dairymen occasionally use in their estrus

Fig. 9.12. Chin-ball marking harness. The ink-filled metal receptacle under the bull's lower jaw acts like a ballpoint pen when the bull mounts the cow, leaving ink marks on her back.

detection programs males whose penises have been surgically deflected (see p. 121) to prevent intromission (conception) and the spread of venereal disease. Beef bulls are typically used in this role because they are less dangerous to handlers than dairy bulls. Many sheep breeders use chest-mounted marking harnesses on their rams (*O. aries*) to monitor which ewes have been mounted. Each harness has a colored crayon attached between the animal's front legs (Fig. 9.13), which leaves a colored mark on the ewe's rump each time she is mounted. Different colored ink or crayons can be used in multi-sire mating programs to determine which males mounted each estrous female.

Steers or cows (*Bos*) treated repeatedly with testosterone are also used in estrous detection programs. They can be fitted with chin-ball markers to aid in identifying estrous females and they are easier to handle than intact bulls. Androgenized marker animals vary in their effectiveness and should be monitored closely, especially when first used.

Fig. 9.13. Marking harness on a ram, which leaves a crayon mark on the ewe's rump when she is mounted. In multi-sire mating, different colored crayons indicate which ram(s) has marked each ewe.

One advantage of marker animals is that they often stimulate mounting activity among the estrous females in the herd or flock, increasing the likelihood they will be visually observed. In fact, the most efficient estrous detection is obtained using a combination of marker animals with visual observation. In one study (Fulkerson *et al.*, 1983), only 50% of estrous cows (*B. taurus*) were identified using twice-daily visual observation (at milking time). Adding tailhead painting to visual observation increased the rate of estrous detection to 80%, nearly the same percentage detection rate (83%) as 12 h of continuous daily visual observation.

Monitoring increases in general activity

Sexually receptive females of most species are more active during estrus due to male seeking and greater interest in conspecifics. In fact, the general activity of dairy cows (*B. taurus*) increases two to fourfold during estrus (Kiddy, 1977). Some dairymen have placed pedometers on the legs of their dairy cows (Fig. 9.14) and used distance travelled per day as part of their estrous detection program. Activity scores are read electronically into a computer as the animal enters the milking parlor and the computer is programmed to identify those animals whose activity scores have increased by a set amount (e.g. three times normal). These animals are separated from the herd after milking and are artificially inseminated.

Fig. 9.14. Pedometer on a cow's leg. An electronic sensor 'reads' the pedometer (number of steps taken) as the cow enters the milking parlor and sends this information to a computer. Cows several times more active than normal are flagged and temporarily removed from the herd for artificial insemination.

Experimental techniques

A number of other estrous detection techniques have been researched and proven reliable but have not been adopted by the livestock industries because of cost, time factors and practicality. Many of these techniques are listed in Table 9.4. One of the more interesting experimental techniques for estrous detection in cattle (*B. taurus*) is the use of trained dogs (*C. familiaris*). Kiddy *et al.* (1978) tested the responses of four German shepherds and two Labrador retrievers to vaginal fluids obtained from cows in estrus and diestrus. (The dogs had been trained previously for olfactory detection of explosives.) Estrous samples (swabs) were taken a few hours after the cows became sexually receptive and diestrous samples were obtained from the same cows 6–12 days later. The dogs were also tested for their ability to discriminate between urine samples from estrous and diestrous cows. Testing consisted of presenting each dog with a single estrous sample and one or two diestrous samples, all 90 cm apart on a 'discrimination board'. Dogs were trained to sit down next to the sample of their choice and were rewarded with praise and bits of food each time they chose the estrous sample. The results from hundreds of trials revealed that the dogs were about 80% accurate in selecting the estrous samples, including trials in which the stimuli were estrous and diestrous urine samples from different cows. The dogs were 87% accurate when asked to choose between live cows in estrus and diestrus. (A platform was provided behind the cows to put the dogs at a convenient working height.) Kiddy *et al.* (1984) also reported that the dogs could discriminate between estrous and diestrous cows based on milk and blood plasma samples. Unfortunately, dairymen have been reluctant to use dogs in estrous detection programs, perhaps because of the intensive training required.

Other mammals have shown the same ability as dogs to discriminate estrous odors. Domestic rats (*R. norvegicus*), for example, can discriminate between estrous and diestrous cow (*B. taurus*) urine, even when diluted with water to a ratio of 1:25 (Dehnhard and Claus, 1988). Furthermore, rats retain their ability to discriminate between estrous and diestrous urine for up to 12 months.

Scientists are trying to develop an 'artificial nose' which can accurately detect odors associated with estrus in cows (*B. taurus*). One such device (Neotronics Scientific Ltd., Stansted Mountfitchet, UK) measures changes in the electrical resistance of 12 conducting polymer sensors when exposed to swabs of bovine perineal secretions (Lane and Wathes, 1998). The 'nose' sensors respond to estrous cycle changes in perineal odors with changes in resistance. An efficient electronic nose would be practical for animal breeders if it was marketed at a reasonable cost and was 'user-friendly'.

10 Maternal and Neonatal Behavior

Behaviors Prior to Birth or Hatching

Preparturient behavior refers to those behaviors preceding and associated with the birth of offspring (parturition) by female mammals. Females of most species show an increase in general activity as parturition approaches. Female swine (*S. scrofa*), cats (*F. domestica*), dogs (*C. familiaris*) and rodents (e.g. *R. norvegicus*) build nests, starting 1 or 2 days before giving birth. Nest sites are important for these species because of the difficulty in moving large numbers of offspring (litters), particularly when the young are born in a relatively helpless (altricial) state and lack mobility (rodents, cats and dogs). On pasture, sows construct a bowl-like depression in the ground and cover it with branches, twigs and grass some 7–14 h prior to parturition. Female cattle (*Bos*), sheep (*O. aries*), goats (*C. hircus*) and horses (*E. caballus*) housed on pasture often isolate themselves from conspecifics in the hours preceding parturition, which offers them privacy during the birthing process and less interference from inquisitive conspecifics as the mother bonds to her offspring and encourages the young to stand and suckle.

Nest building and incubation in chickens (*G. domesticus*) and turkeys (*M. gallopavo*) is triggered by release of the peptide hormone, prolactin, from the anterior pituitary gland. Free-living or free-range chickens and turkeys form bowl-shaped nests with grass and small twigs. Eggs are laid in the nest and are incubated by the female until the young have hatched, usually after 21 (chicken) or 27 (turkey) days.

Behaviors Associated with Birth or Hatching

In female mammals, parturition involves the rupture of the amnion and expulsion of the fetus(es) and placenta, in that order. Parturition in many mammals is characterized by a rapid increase in the ratio of estradiol to progesterone and a dramatic increase in the synthesis, storage and release of the hormone, oxytocin, from the posterior pituitary gland. Distension of the uterus, cervix and vagina during the expulsion of the fetus stimulates the release of oxytocin into the peripheral circulatory system and, at the same time, oxytocin is released in the brain to stimulate maternal behavior. Endogenous opioid peptides in the brain also increase at parturition to facilitate maternal behavior and raise the animal's pain threshold. Labor and expulsion of the fetus is accompanied by restlessness, lying down for periods of time, followed by brief periods of standing and circling. Offspring are usually born while the mother is lying down. Birth of the offspring is typically followed by standing and licking (grooming) the offspring, often with soft vocalizations. Grooming stimulates the newborn of precocious species (e.g. most large domestic ungulates) to stand and engage in teat seeking. Mother cats (*F. domestica*), dogs (*C. familiaris*) and rodents (*Rattus*) lick the anogenital region of their young to stimulate defecation and urination.

The odor and taste of amniotic fluids on the newborn stimulates grooming by the mother. Licking the young removes amniotic fluids, which would otherwise cause a loss of body heat in cold environments. Amniotic fluid becomes attractive to the female several hours before parturition (10 h in the cow) and its attractiveness fades within several hours following parturition. Amniotic fluid becomes repulsive to the ewe (*O. aries*) starting at about 4 h post-partum (Lévy *et al.*, 1983).

Figure 10.1a–f shows the series of events associated with parturition in the ewe (*O. aries*). After the onset of intense labor, it takes her about 30 min to give birth to her firstborn. The second lamb (if she is carrying one) is born about 20 min later.

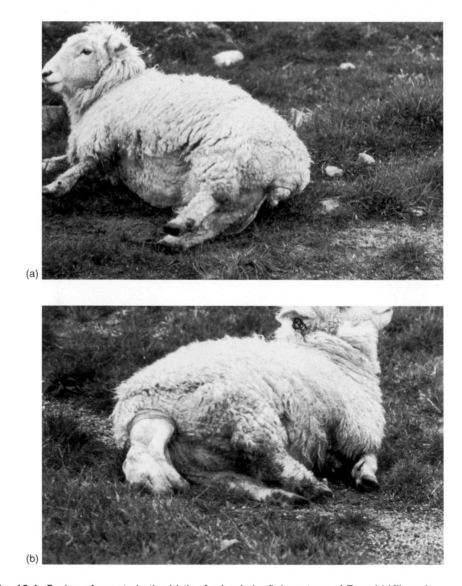

(a)

(b)

Fig. 10.1. Series of events in the birth of a lamb (a–f) (courtesy of Ronald Kilgour).

(c)

(d)

Fig. 10.1. continued

Problems can arise for the firstborn while the second lamb is being born. Preparturient females are often interested in newborn young. A preparturient ewe can inadvertently, or purposefully, lead a neonate away from its mother while she is in labor with her second young, especially under crowded conditions where many pregnant females and newborn young are housed in close proximity. These 'adopted' young are sometimes abandoned once the alien female gives birth to her own offspring.

In sheep (*O. aries*), the birth of twins rather than single lambs results in a higher incidence of mother–young separation. Alexander *et al.* (1983) found that breed affects the probability that ewes will become separated from one or both twins.

(e)

(f)

Fig. 10.1. continued

Merino ewes were by far the poorest mothers, with 80% becoming temporarily or permanently separated from one or both twins compared to 55.5% of Dorset Horn ewes and 9.6% of New Zealand Romney mothers. When ewes give birth on steep hillsides (slope > 30 degrees), firstborn young can tumble down the slope while attempting to stand, never to be recovered by their mother after the birth of the secondborn (Knight *et al.*, 1989).

Females giving birth to altricial young cuddle or crouch over their newborn young, which keeps them warm. Altricial young are not able to control their body temperature fully for some time after birth. Young kittens (*F. domestica*) begin to regulate their body temperature by 3 weeks of age but complete temperature control is not attained until they are 7 weeks old (Olmstead *et al.*, 1979). Even precocious young seek their mother's warmth. Newly hatched chicks (*G. domesticus*) are not able to generate normal body temperature fully until 3 weeks after hatching, and thus periodically nestle under their mother's wings or body.

Although the female bird lays her eggs over a period of many days (1 egg/day), the young birds hatch at about the same time due to 'synchronized hatching'. Synchronized hatching is adaptive for birds with precocious young since late-hatching young would be left behind if the mother bird led the earlier hatched birds away from the nest. Vince (1973) found that if an egg was in physical contact with an egg that had been incubated 24 h longer, the younger fetus in the 'stimulated' egg would hatch about 10 h sooner than it would otherwise. Vince also noted that artificial 'clicks' administered at a rate of 3/s to isolated eggs in the later stages of incubation can advance hatching by more than 24 h. Interestingly, the fetuses of many birds emit peeping and clicking noises (associated with breathing) as they approach hatching. These clicking noises promote hatch synchrony among eggs in physical contact.

The fetal chick (*G. domesticus*) and turkey (*M. gallopavo*) cuts its way out of its eggshell using a hard, sharp tubercle on the tip of its bill (egg-tooth) and its large neck muscle. Once the precocious young have hatched, they follow their mother about (see discussion of filial imprinting, p. 35) and quickly learn that nestling under the mother's wings and body provides warmth and protection from the elements.

At birth, precocious mammalian offspring will attempt to approach and suckle any adult female in their immediate proximity. Social attachment of young to their mother and avoidance of strange or alien females are learned through simple associative (i.e. operant) conditioning. Mother becomes associated with positive reinforcers (e.g. milk, warmth), while alien mothers thwart the youngster's suckling attempts and often show aggression if the unfamiliar neonate invades her space.

Sensitive Period for Maternal Responsiveness

Females become primed to behave maternally by hormonal changes occurring at the time of parturition or hatching. In mammalian females, increases in estrogen in the days preceding parturition act as an essential primer for the induction of maternal behavior once contact with a young neonate is made. The release of oxytocin at the time of parturition facilitates the expression of maternal behavior by its action on the medial preoptic area and the paraventricular nucleus of the hypothalamus (Lévy *et al.*, 1996). The timing of these hormonal events and critical feedback from the young neonate profoundly influence the quality of maternal care the mother offers.

Domestic ungulates

Female ungulates are strongly motivated to exhibit maternal behavior at the time of parturition and to direct that interest to *any* young neonate. However, maternal responsiveness to *any* young neonate does not last for an extended period of time.

Rather, it is confined to the first 24 h period following parturition and may only be a few hours in some species. Klopfer *et al.* (1964) allowed 14 domestic female goats (*C. hircus*) no contact with their newborn young for 1–3 h following parturition. When their young were returned, only 2 of 14 mothers accepted their own offspring. Hudson and Mullord (1977) conducted a similar study with cows (*B. taurus*), returning the young 1–5 h after parturition. Only 10 of 21 cows accepted their own young under these conditions. Smith *et al.* (1966) prevented 21 ewes (*O. aries*) from contacting their newborn young for up to 8 h at parturition. All but one ewe accepted their offspring on reinstatement, suggesting that sheep are more resistant to the effects of separation (at parturition) than goats and cattle. Other research has shown that alien lambs are just as acceptable as own lambs during the first few hours of parturition and that newborn aliens are actually preferred over older (12 or 24 h old) own young.

Maternal experience facilitates the expression of maternal responsiveness when young are presented to mothers for the first time following a brief period of post-partum separation. Lickliter (1982) separated 12 female goats (*C. hircus*) from their offspring for 2 h immediately after parturition. Six of the goats were first-time (primiparous) mothers and six were experienced (multiparous). Only one of the primiparous females accepted her own young, while all of the multiparous mothers showed acceptance behaviors. An additional six primiparous and six multiparous females all accepted their own young when separation was not administered (control group). Multiparous females were used in the cattle and sheep studies cited above; the authors of the goat study did not state whether their does were experienced mothers.

The sensitive period for maternal responsiveness to *any* young is terminated once the mother has interacted with a young neonate. The interaction may be very brief (5–20 min) and it initiates the establishment of a discriminative bond with that neonate. In the Klopfer *et al.* (1964) study cited above, 15 additional mother goats were given 5 min of contact with their young at parturition and then were separated for 1–3 h. On reinstatement, 13 of the 15 mothers accepted their own offspring. Five of these 13 mothers were exposed to alien young as well and all aliens were rejected. In the Hudson and Mullord (1977) study cited above, an additional nine multiparous cows were given 5 min of contact with their own calves at parturition and then separated for 1–12 h. All nine cows accepted their own young at reinstatement and all rejected alien calves. Smith *et al.* (1966) noted that if a ewe had licked any lamb for about 20 min or more, she would subsequently reject all *new* lambs presented to her, even if the new lamb was her own.

Continued interaction with the young neonate is necessary to maintain maternal reponsiveness. Ramírez *et al.* (1996) gave 30 maternally experienced goats (*C. hircus*) 5 min of contact with their own young at parturition and then rejoined mother and young 1, 8 or 24 h later. All of the females reinstated with their offspring 1 or 8 h later accepted their offspring, while all of 10 females reinstated 24 h later rejected their young. Similarly, Hudson and Mullord (1977) found that cows (*B. taurus*) given 5 min of contact with their young at parturition and then separated for 24 h would consistently reject their offspring on reinstatement. Lévy *et al.* (1991) gave ewes (*O. aries*) 4 h of contact with their own lambs following parturition and then separated them for 6, 12, 24 or 36 h. All of the ewes were maternally responsive when rejoined with their young following 24 h of separation but more than a third had lost interest after 36 h.

Role of the Senses in Maternal Responsiveness and Offspring Discrimination

Olfactory cues

Olfaction is not a requirement for the manifestation of maternal responsiveness in domestic ungulates but can be very important in offspring discrimination. Baldwin and Shillito (1974) surgically removed the olfactory lobes of the brain of eight female sheep (*O. aries*) when they were 3 months pregnant (gestation is 5 months). Seven of the eight ewes successfully reared their young and six of the seven nursed alien lambs as well. A similar result was found by Meese and Baldwin (1975) with female pigs (*S. scrofa*) whose olfactory lobes had been removed prior to breeding. Their bulbectomized sows readily accepted both their own and alien piglets.

Klopfer and Gamble (1966) gave 16 female goats (*C. hircus*) 5 min of contact with their young at parturition. Nine of the mothers had been temporarily rendered anosmic (could not smell) at parturition by spraying a 10% cocaine hydrogen chloride solution into their nostrils 20–90 min prior to parturition. These females could smell when reinstated with their offspring 3 h later. The remaining seven mothers could smell at parturition but were rendered anosmic for the reinstatement 3 h afterwards. Eight of the nine females who were anosmic at parturition but could smell at reintroduction accepted their own offspring, as well as alien young during reinstatement. They were maternally responsive based on the 5 min contact at parturition but had not yet formed a maternal bond based on olfactory cues. In contrast, only three of the seven females who could smell at parturition but were anosmic at reintroduction showed acceptance behaviors toward their young. The four who rejected their offspring at reinstatement continued to reject them, even after the anosmia had worn off. The lack of 'own young' odor perceived when their own young were first reintroduced compromised their ability to recognize and accept their own young.

Romeyer *et al*. (1994) noted that visual characteristics of the young could not compensate for a loss of olfactory recognition cues. Nine female goats (*C. hircus*) were rendered permanently anosmic 3 weeks before parturition by irrigating the olfactory mucosa with zinc sulfate. Eight goats were left untreated. Observations following parturition revealed no differences in the maternal behavior of the anosmic and intact mothers. The two groups of mothers were then individually exposed to alien young of similar and dissimilar hair color and pattern. Most of the anosmic mothers accepted all of the 'strange' young presented, while the intact mothers rejected all aliens. All of these studies on the role of olfaction in maternal responsiveness and offspring discrimination suggest that the post-partum female quickly stores a memory of the odor of the first neonate contacted following parturition and any change replacing 'own offspring' odor, whether it be the odor of an alien young or no odor at all (as in anosmia), will lead to rejection behaviors. This finding is supported by the fostering research reported in the next section of this chapter.

Prior to parturition, offspring odors have little effect on the release of neurotransmitters or the electrical activity of neurons in the olfactory bulbs of the brain, the brain area that receives sensory information from the nose. At parturition, olfactory processing in the brain is altered dramatically to accommodate olfactory recognition of offspring. Kendrick *et al*. (1992) point out that there are three basic types of neurons in

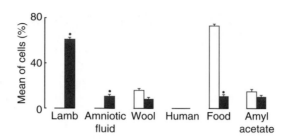

Fig. 10.2. The effect of parturition on the percentage of sheep mitral cells that preferentially responded to different classes of odors. Open bars represent data from 105 cells recorded before parturition; solid bars represent data from 83 cells recorded after parturition (Kendrick *et al.*, 1992).

the olfactory lobes. Of these three, mitral cells are most important in the processing (receiving and transmitting) of offspring odors. Physiological events associated with parturition result in a marked increase in the proportion of mitral cells that respond preferentially to offspring odors. Figure 10.2 illustrates how lamb (*O. aries*) odors become the most potent olfactory stimulus for mitral cells at parturition, largely displacing food odors in importance. Seventy percent of the mitral cells responding to lamb odor responded indiscriminately to the mother's own lamb and alien lambs. Thirty percent of the cells responded preferentially to the odor of the mother's own young. In addition, a small proportion of mitral cells (11%) responded preferentially to amniotic fluid odors at parturition, a finding which may help to explain why amniotic fluid is so attractive at parturition and so repulsive at other times.

Maternal 'labeling' of young through milk intake or by licking (e.g. mother's saliva), or both, has been proposed as an explanation for the rejection of alien young by post-partum females. Gubernik (1980, 1981), working with goats (*C. hircus*), and Alexander *et al.* (1989), working with sheep (*O. aries*), found that unmothered young (human fed) had a higher acceptance rate when fostered than young which *had* been mothered. However, Lévy *et al.* (1991) and others have not been able to replicate these findings.

Visual and auditory cues

While the maternal acceptance of young is ultimately based on olfactory stimuli, both visual and auditory cues are used by domestic ungulates to discriminate between own and alien young, typically at greater distances than olfaction. Alexander and Shillito (1977) studied the importance of visual cues from various body regions on offspring recognition by sheep (*O. aries*). Various regions of their own lambs' bodies were blackened and ewes were scored for acceptance and rejection behaviors. Lambs were 3–14 days of age at testing and each ewe was tested only once so that test experience would not be a confounding variable. Figure 10.3 illustrates the various body areas that were blackened and the proportion of ewes initially rejecting their own offspring. The results indicated that the head of the lamb was the single most important body area used by the ewe in offspring discrimination by visual cues.

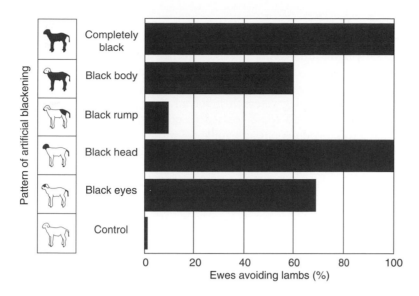

Fig. 10.3. Changing the visual appearance of lambs interferes with the mothers' ability to identify her offspring. Blackening the head, either partially or completely, resulted in a relatively high percentage of ewes avoiding their lambs (Alexander and Shillito, 1977; figure taken from Hart, 1985).

Poindron and Carrick (1976) investigated whether ewes (*O. aries*) could discriminate between the vocalizations of their own and alien lambs at a distance. Tape recordings were made of the bleats of 18 lambs. When the lambs were 6–8 weeks old, their mothers were exposed to bleats of their own young and the bleats of an alien lamb, alternately broadcast over two loudspeakers 15 m apart in an enclosure and each 13 m from the point of the ewe's release. The experimenters recorded time spent by the ewe on the side of the enclosure from which own-lamb bleats were broadcast versus time spent on the side emitting alien-lamb vocalizations. Half of the trials were run with own-lamb recordings on the left and half with own-lamb recordings on the right to avoid the possibility of a 'position effect' confounding the results (e.g. always going left, regardless of the stimulus presented). Fourteen of the 18 mothers (78%) showed a preference for the side of the enclosure broadcasting their own lamb's vocalizations, confirming that ewes can discriminate between own and alien offspring using auditory cues at 6–8 weeks of age. Searby and Jouventin (2003) confirmed that ewes and lambs recognized one another by audition and noted that individual differences in sound frequency and timbre (quality) were most relevant.

Fostering Techniques

Breeders of companion animals and livestock are often faced with caring for orphaned young. Mothers can die from complications associated with giving birth, or illnesses incurred shortly thereafter. Young may be abandoned or rejected by their mothers. In other cases, females may have insufficient milk to support multiple offspring effectively. In any of these situations, the caretaker must either care for the young or find a surrogate mother to do so. Hand rearing is costly and labor-intensive, but doable, particularly if there are not many orphans and time is available. Fostering orphans on surrogate mothers is usually more time-saving and cost-effective in the long run, but mothers must be found who can be enticed to accept the orphan and care for it. That can be difficult, particularly if the mother has already formed a maternal bond with her own offspring.

The bulk of my fostering research involved sheep (*O. aries*), since the greatest need for information on this topic seemed to come from the sheep industry. Most well-nourished ewes produce enough milk to support two lambs. However, some ewes give birth to single lambs and some bear more than two. When this happens, it makes sense for sheep breeders to entice ewes bearing single lambs to adopt one of these 'extra' lambs, or possibly an orphan that had lost its mother. Fostering (grafting) lambs from one mother to another is not especially difficult if it is done immediately at parturition before the ewe has contacted her own newborn. However, as we noted in the previous section, once she has groomed her young and bonded with it, it is extremely difficult to entice her to accept an alien lamb. In all likelihood, she will butt the alien lamb or move away from it if it tries to suckle. Such rejection behaviors are the result of hundreds of generations of natural selection on mothers to care only for their own offspring. This gives their young the best chance of survival, thus increasing the probability that the mothers' genes will be passed on to the next generation.

A number of techniques have been developed to facilitate fostering in sheep and other species of domestic farm animals. I will divide these techniques into 'traditional' and 'experimental' for the purpose of our discussion (Table 10.1). Fostering involves either substituting an alien neonate for the mother's own young ('substitution fostering') or adding a second offspring to a mother with one of her own ('add-on fostering').

Two traditional fostering techniques used with domestic sheep (*O. aries*) involve transferring odors associated with the mother's own young to the alien being fostered. Slime grafting involves rubbing the alien's body with amniotic fluids or placental membranes from the birth of her own young. This procedure is obviously used at the time of parturition, or very soon thereafter. Skin grafting is used when the ewe's natural young has been stillborn, or died after birth. Since own-lamb odor is carried on the lamb's wool, the dead lamb is skinned and the hide ('jacket') is fitted over the

Table 10.1. Traditional and experimental fostering techniques used with sheep (*O. aries*) in both substitution and add-on modes.

	Substitution[a]	Add-On[b]
Traditional:		
Slime grafting	×	×
Skin grafting	×	
Restraint fostering	×	×
Experimental:		
Odor transfer	×	×
Use of novel odorant		×
Vaginal stimulation	×	×

[a]Alien lamb substituted for own lamb.
[b]Ewe with single offspring is given an additional lamb.

individual to be fostered. Portions of the alien not covered by the hide (head, legs and tail) may be rubbed with the hide to transfer as much odor as possible. Restraint fostering can be used if slime or skin grafting is unsuccessful, or if fostering is attempted a day or so after parturition. In this technique, the foster mother is placed in a small enclosure and tethered or restrained in such a manner that she cannot butt or avoid the suckling attempts of the alien young. If the ewe has a natural lamb with her, an effort should be made to match the alien in regard to size and vigor. The ewe is untethered periodically to test her willingness to accept the alien. Once the ewe shows acceptance behaviors toward the alien (e.g. allows fostered lamb to suckle; no aggression), she should be left untethered in the enclosure for 1 or 2 days before she and her newly accepted young are returned to the flock. Our studies on restraint fostering (Price *et al.*, 1984a) revealed that ewes generally accepted an alien lamb substituted for their own lamb after 4–5 days of cohabitation.

There have been three general *experimental* approaches to fostering research. One has been to alter the identity of alien young so they are perceived as own (natural) offspring. Of the work done in this area in recent years, odor transfer has gained the greatest acceptance by sheep breeders. Like slime and skin grafting, the odor transfer technique recognizes that fostering is best facilitated if the alien lamb's odor can be altered to match that of the ewe's own lamb. Slime is not always available when a fostering decision is made. Furthermore, skin grafting requires that the mother's dead lamb is recovered and not everyone is amenable to skinning out a deceased animal. We surmised that a cloth jacket worn by the ewe's own lamb might acquire enough own-lamb odor to fool an experienced ewe when the jacket was placed on an alien lamb. We used orthopedic stockinette made of an elastic knitted fabric for our jackets (Fig. 10.4). When a jacket worn by a ewe's natural lamb was transferred to a substitute alien lamb, the alien was accepted more than 85% of the time and 40% of the acceptances were immediate (Price *et al.*, 1984b). In contrast, it was very rare for a ewe to accept an alien

Fig. 10.4. Lamb fitted with stockinette jacket (Price *et al.*, 1984b).

wearing a clean jacket (control condition). A greater challenge was to entice ewes with single lambs to accept a second alien lamb (add-on mode) since their own lambs would be present for comparison. Separate jackets, which had been worn by a ewe's own lamb and an alien lamb, were exchanged at the time of fostering (initial introduction of the alien lamb) so that her own lamb now carried alien-lamb odor and the alien carried own-lamb odor. Under these circumstances, approximately 70% of the ewes adopted the alien lamb in addition to their own lamb and, again, 40% accepted the alien immediately (Martin *et al.*, 1987). It was found that, once acceptance occurred, the stockinette jackets could be removed without the ewe reverting to rejection behaviors. Interestingly, washing alien lambs to remove their natural odors does not facilitate fostering using the odor transfer technique or any other fostering method, a fact which suggests that it is the presence of 'own-lamb odor' rather than the absence of 'alien-lamb odor' that is important in offspring acceptance. The odor transfer technique has gained acceptance by the sheep industry because it is relatively easy to administer and costs are minimal. Also, the jackets can be reused after washing.

The odor transfer fostering technique also works with cattle (*B. taurus*). We found that 9 of 10 cows adopted alien calves substituted for their own young within 48 h of fostering, while only 1 of 12 cows in the control group (i.e. given an alien calf wearing a clean stockinette jacket) met the adoption criterion during the same period (Dunn *et al.*, 1987).

We also examined the efficacy of using an artificial odorant, 'neatsfoot oil', to facilitate fostering in the add-on mode (Price *et al.*, 1998b). Neatsfoot oil is a natural animal triglyceride often used to preserve leather. Alien and own young were smeared with neatsfoot oil shortly after birth and the alien was fostered 20–24 h later. Only 50% of the ewes accepted the alien with this treatment. However, 80% of those adopting the alien did so immediately. This was twice the immediate acceptance rate obtained using odor transfer with stockinettes. Because of the high percentage of immediate adoptions, a subsequent experiment was performed using both stockinettes *and* neatsfoot oil. An additive effect was observed; 80% of the ewes accepted the alien as well as their own lamb and, of those ewes adopting aliens, 83% did so immediately. It was concluded that the main advantage of using an artificial odorant like neatsfoot oil is that it increases the percentage of immediate acceptances.

A second general experimental approach to fostering research has been to control the maternal responsiveness to alien young by artificially stimulating the experiences and physiological events at the time of parturition. Notable among these studies is the work on uterus/cervical/vaginal stimulation. In one study, Keverne *et al.* (1983) found that inflating (for 5 min) a rubber balloon in the uterus of ewes (*O. aries*) 2 h after giving birth facilitated acceptance of alien young. Each of the 12 ewes had developed a selective bond with their natural offspring during the 2 h period prior to stimulation and shown rejection behaviors toward alien young. After uterine stimulation, 10 of the 12 ewes accepted alien lambs. A similar result was obtained when the investigators mechanically stimulated the vagina of non-pregnant ewes which had been hormonally treated to simulate the physiological events at parturition. Eighty percent of the ewes receiving vaginal stimulation accepted alien lambs within 60 min of stimulation, in contrast to only 20% of the unstimulated controls. They hypothesized that mechanical stimulation of the uterus and genitalia caused a resurgence of oxytocin secretion, which renewed the ewes' interest in *any* lamb. The investigators also found that genital stimulation 2 and 4 h after parturition enhanced the olfactory attraction of ewes to amniotic fluid at a time when attractiveness was normally fading (Lévy *et al.*, 1990). However, this same effect was not seen 8 h after giving birth, when estrogen levels had declined significantly.

A third general experimental approach to fostering research involves incapacitating the sensory capabilities of the mother so that she cannot discriminate between her own and alien young. In my opinion, studies designed to eliminate the mother's olfactory sense temporarily to facilitate fostering have proven impractical, inconclusive and questionable from an animal welfare standpoint.

Fostering should be attempted as soon after parturition as possible. The stronger the mother's bond to her natural young, the more difficult it will be to substitute or add alien young. This also applies to species bearing large litters. On a sabbatical in Australia, I fostered piglets (*S. scrofa*) between sows within 2–9 h of birth or at 2, 4 or 7 days of age (Price *et al.*, 1994). Eleven of 12 piglets fostered within 9 h met the criterion for successful suckling within 6 h after fostering. Fostering success within 6 h at the three older ages was much lower, 25–50% (Fig. 10.5).

Not only did sows tend to be more aggressive toward older fostered piglets, but piglets fostered at 2, 4 or 7 days showed a greater reluctance to engage in suckling, higher rates of ambulation and more frequent vocalizations than piglets fostered from 2 to 9 h after birth. It appeared that the older piglets were reluctant to accept their new mother and maternal environment, possibly because familiar visual and olfactory cues were missing (see pp. 181). This was quite different than what was observed in our fostering studies with sheep and cattle.

Dairy cows (*B. taurus*) whose milk is no longer being sold commercially are sometimes used as foster mothers for orphaned beef calves or young of other species (Fig. 10.6). Dairy cows appear to be more accepting of alien young than beef cows or females of other livestock. This may be a consequence of removing natural selection

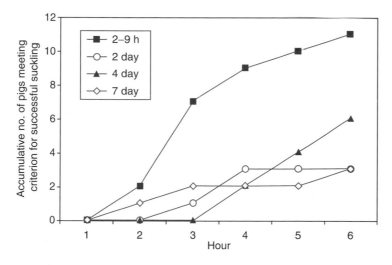

Fig. 10.5. Accumulative number of piglets that met the criterion for successful suckling (5 s or more of sucking during two successive milk ejection periods) within 6 h of being fostered at 2–9 h, 2, 4 and 7 days following birth. Fostering was facilitated by introducing piglets to their new mother within 9 h after birth (Price *et al.*, 1994).

Fig. 10.6. Dairy cow nursing lambs (courtesy of Ronnie Gardner).

pressure on fidelity to offspring resulting from the standard industry practice of separating dairy cows from their calves at birth.

Follower versus hider species

Following birth, the precocious offspring of some species (e.g. sheep, horses) follow their mothers wherever they go (Fig. 10.7), while the offspring of other species go into hiding for 1–3 days (cattle) to 1–2 weeks (goats) and only contact their mothers when the latter

Fig. 10.7. Triplet lambs following mother.

return for nursing (Ralls *et al.*, 1987). When the mothers (hider species) want to nurse, they approach the area where they last saw their young and vocalize. The young often respond by standing and approaching their mother. The young of hider species eventually become socialized to other young in the population and typically spend much of their time with them. 'Nursery groups' of calves (*B. taurus*) can often be seen together in a field, while their mothers are somewhere else grazing. Lambs (*O. aries*) are seldom seen very far from their mothers and are more attracted to her than to their non-sibling peers.

The young of nesting species stay in the nest until they become mobile enough to follow their mothers. Piglets (*S. scrofa*), being relatively precocious at birth, begin to leave the nest for short periods after several days, while kittens (*F. domestica*) and puppies (*C. familiaris*), being relatively helpless and immobile (altricial) at birth, do not leave the nest on their own for several weeks. Young domestic rat pups (*R. norvegicus*), also born in an altricial state, do not leave the nest until they are about 14 days of age. Interestingly, the mother rat begins to produce an odor at that same time ('maternal pheromone'), which is attractive to rat pups and encourages them to stay close to the nest (Leon, 1974).

Females of hider species (e.g. cattle, goats) and species which make nests (e.g. dogs, cats, pigs) tend to eat the placenta, which is expelled following parturition. More than 80% of cows (*B. taurus*) eat their placenta if given the opportunity (Machado *et al.*, 1997). Follower species (e.g. horse, camels) tend not to eat the placenta but, with their precocious newborn young, move away from the birth site. One theory to explain placentophagia in hider and nesting species is that it removes odor from the birth site, which could attract predators. For nesting species, it could also help to keep the nest area clean and free from material that could harbor bacteria. It also has nutritive value in the form of protein and water, which may permit the mother to stay with her newborn young longer before leaving to feed.

Neonatal Feeding Behavior

Teat seeking and suckling in the newborn mammal

Newborn mammals are born with a well-developed sucking behavior. The 'first milk', or colostrum, obtained from the mother provides nutrients needed for thermoregulation and early locomotor activities, and antibodies needed to counter disease. Once precocious young are on their feet, they approach the mother and locate a teat (nipple) by following a thermal, olfactory and textural gradient with their nose and tongue (Vince, 1993). Of course, the underside of the mother's body is warmest, particularly the area under the front and hind legs (axillary and inguinal regions, respectively), which is devoid of hair or wool. The sheep's udder produces an inguinal wax, which is attractive to lambs. Tactile stimulation provided by any protruding appendage in these areas triggers intense suckling behavior. The location of the udder on the mother is quickly learned.

Female cattle (*Bos*), sheep (*O. aries*), goats (*C. hircus*) and horses (*E. caballus*) nurse their offspring in a standing position. Udders with two (sheep, goats, horse) or four (cattle) teats are located in the inguinal region. Sows (*S. scrofa*) usually nurse lying down. There are typically seven pairs of teats on the sow's ventrum to accommodate their relatively large litters (mean: 10.5 young). Female cats (*F. domestica*) and dogs (*C. familiaris*) also nurse lying down and have four (cat) and four to six (dog) pairs of teats to accommodate their litters, averaging about five young.

Kittens and puppies are born with their eyes closed and first locate teats on their mother's body primarily by tactile and thermal stimulation, as do the young of many other mammals whose young are born in an altricial state (Freeman and Rosenblatt, 1978). Kittens often tread with their paws against the mother's ventrum as they suckle, which is thought to stimulate milk ejection (Turner and Bateson, 2000).

Young ungulates typically access the udder (i.e. suckle) from the side with their rump angling toward the female's head (Fig. 10.8). Butting the udder is often observed during suckling, particularly at the beginning and end of a suckling bout. Butting is believed to increase mammary pressure in the udder and the flow rate of milk through the teats at the beginning of and during milk ejection. Suckling and butting often continues after all or most of the milk has been extracted, providing the mother remains immobile. Suckling during and after a meal aids the digestive process by increasing levels of the metabolic hormones, insulin and cholecystokinin (de Passillé *et al.*, 1993; de Passillé and Rushen, 1997). When milk is consumed very quickly, as when young are fed milk or milk 'replacer' from a bottle or pail, the animals will subsequently suck on nearly anything that protrudes in their environment, such as the appendages of nearby animals (e.g. ears) and inanimate objects. Research points to the fact that once a young neonate gets the taste of milk and begins suckling, its motivation to continue suckling is reduced more by the duration of the suckling bout than by the ingestion of milk (Rushen and de Passillé, 1995).

If a young ungulate is not getting enough milk from its mother or is orphaned, it may learn to sneak between the hind legs of another female and suckle, often while the female's own young is suckling. 'Sneakers' can sometimes be identified by feces on their heads and necks (Fig. 10.9).

Mammalian mothers nurse their young at various intervals throughout the day and night, depending on the species, age of the young and environmental conditions at the time (e.g. temperature). The one exception is the female European rabbit

Fig. 10.8. Characteristic positioning of lambs while suckling.

Fig. 10.9. Lamb that had been rejected by its mother and was stealing milk from other ewes by rear suckling. Feces on the head is usually a sign that a lamb is experiencing maternal rejection.

(*O. cuniculus*), which returns to the nest to nurse her young for 3–5 min only once a day (see Gonzales-Mariscal and Rosenblatt, 1996). Lactating female rabbits emit from their ventrum a chemical substance ('nipple pheromone') that allows the neonate to find a nipple quickly and begin suckling. After nursing, the mother rabbit deposits several fecal pellets before leaving the nest. These pellets contain bacteria from the mother's intestinal flora, as well as traces of the mother's diet. The young pups nibble on the feces as they get older (coprophagia), which helps them establish their own intestinal flora as they start eating solid foods. In addition, the odor of the feces may lead the young to the same appropriate and safe foods consumed by the parent.

Establishment of a teat order in newborn piglets

Milk is continually available to piglets (*S. scrofa*) at the time of birth but, starting soon thereafter, the sow offers milk to her piglets at only 45–60 min intervals. The sow signals these periods of nursing by grunts, which the young soon associate with feeding. The piglets attach to a nipple and massage the udder by moving their snouts up and down. This stimulates oxytocin secretion in the sow and milk ejection after about 2 min. Milk ejection lasts for only 15–20 s, on average, during which time the piglets exhibit rapid suckling (2–5 sucks/s).

Newborn piglets are very precocial. Within minutes following birth, piglets stand and move about their immediate environment until they contact a vertical surface, whether this is the sow, a littermate, or a wall. Once a vertical surface is contacted, it is explored by nose contact, coupled with lateral movement of the snout. The prostrate mother provides the most prominent, relatively warm vertical surface in the piglets' environment at this time and the piglets eventually find a teat while exploring the mother's body. This phase has been referred to as *teat seeking* and is accompanied by little or no aggressive behavior (Hartsock and Graves, 1976; Hartsock *et al.*, 1977).

Rather than remain at the first teat contacted, piglets will tend to move from teat to teat for a while, a process referred to as *teat sampling*. During this phase, the piglets remain oriented to the udder. Data indicate that the anterior teats are preferred, perhaps because of their higher milk yield (15% greater, in one study) relative to the posterior teats. Contact with littermates at this stage often results in aggressive behavior. Aggression may consist of biting or pushing with the nose or shoulder and is commonly directed at other piglets which are engaged in suckling. Suckling piglets respond to this kind of interference by positioning their bodies so that their teat position is inaccessible to the challenger. If this defensive technique is unsuccessful, the defender may resort to aggressive behavior as well.

Aggression appears to be the primary activity during the teat-sampling phase. Testosterone levels are relatively high during this stage. Aggression between piglets is most frequent at 2–3h following parturition and then declines sharply from 3 to 7h of age. Aggression becomes infrequent by 14–18h following birth and will remain relatively low until just before puberty.

Clipping the 'eye teeth' (temporary canines and lateral incisors) of the piglets shortly after birth dramatically reduces the incidence of facial wounding incurred from aggressive interactions during the establishment of the teat order. Fraser (1975) demonstrated that only 16% of clipped litters showed evidence of facial wounding, in contrast to 67% of unclipped litters.

The *teat defense* phase follows, during which time the piglets tend to confine their activities to a specific area of the udder. Their interactions during this phase become limited to adjacent littermates, primarily because of their developing preferences for specific teats. Figure 10.10 plots the mean percentage of different littermates fought

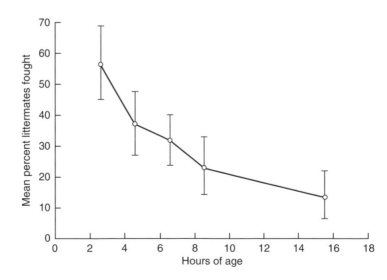

Fig. 10.10. The mean percentage of littermates fought declines over hours following birth as aggression becomes increasingly confined to piglets on adjacent teats (Hartsock *et al.*, 1977).

during the first 16 h following birth. The curve indicates a sharp decline over time corresponding to the increasing tendency for piglets to interact primarily with littermates on adjacent teats. The more productive anterior teats on the sow are contested most vigorously and are usually won by the larger piglets of the litter (Fig. 10.11). Interestingly, the first piglets to be born are usually the heaviest and the ones most likely to defend an anterior teat.

Data from three litters of piglets (in the Hartsock study) monitored for 40 consecutive hours following birth indicated a decided advantage in defending the 'home' teat. Piglets successfully defended their home teats about 80% of the time, while only about 40% of the fights initiated at other teats were won. From this standpoint, teat defense resembles a form of territoriality (i.e. defense of a feeding area). Hartsock et al. (1977) reported that 51% of the piglets in 13 litters nursed consistently from a specific teat or pair of teats by 72 h. De Passillé et al. (1988) found that 75% of 55 piglets had attained teat specificity by 72 h and Wesley (1967) reported 45% specificity by 72 h in eight litters.

Following the successful defense of a given teat, piglets go into a fourth phase of suckling and sleeping, referred to as the phase of *teat maintenance*. Little fighting occurs and piglets have relatively uncontested access to their own teat or pair of teats. Hartsock and Graves (1976) noted that 10% of the piglets maintained two teats throughout lactation, whereas Wyeth and McBride (1964) reported that 17% of the piglets in their studies suckled two teats. Whether one or two teats are used, a stable teat order results in faster growth of the piglets.

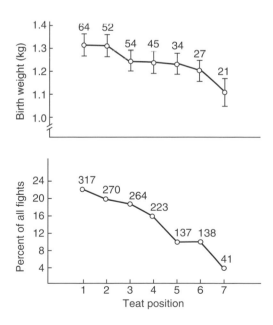

Fig. 10.11. Top: Mean (± SE) birth weight and total number of piglets occupying the seven teat positions (1 = anterior-most teat). Bottom: Percentage and number of fights involving piglets at the seven teat positions (Hartsock *et al.*, 1977).

Weaning

Mammalian mothers gradually reject their offspring's suckling attempts as the young progress to a solid food diet. Natural weaning of piglets (*S. scrofa*) in a semi-natural environment begins at about 6 weeks and is completed by 14–15 weeks. Beef cows (*Bos*) naturally wean their calves at 7–14 months, depending on the breed and availability of feed. Animal management practices typically favor early weaning of mammalian young. Piglets are typically weaned from 2 to 6 weeks of age and beef calves at about 6 months. Cessation of nursing (i.e. lactation) facilitates a return to estrous cycling in females of some mammalian species. Also, early weaning of young allows more of the mother's energy to be channeled into her subsequent pregnancy.

Early weaning can be stressful to both the mother and her offspring. They are not only separated from one another but the young are prematurely and abruptly denied the opportunity to suckle. Their distress is typically expressed by an increase in activity (e.g. pacing back and forth along enclosure walls and fences), frequent vocalizations and a temporary reduction in food intake. There is also evidence for immunological compromise in some species. The digestive physiology of piglets (*S. scrofa*) goes through a rapid period of maturation from 3 to 5 weeks of age. Thus, piglets weaned before or during this period cannot benefit fully from a solid food diet. McGlone and Anderson (2002) reported that piglets weaned at 3–4 weeks of age and placed on a solid food diet gained weight faster when their feeders or snouts were painted with a synthetic fatty acid mixture similar in composition to sow skin secretions (i.e. 'maternal odor'). The increased weight gain in treated piglets was believed due to better feed:gain ratios (i.e. feed utilization efficiency), since feed intake was similar in treated and untreated animals. Perhaps the presence of sow odor reduced the stressfulness of weaning and transition to dry feed.

Weaning in young piglets is often accompanied by increased aggression, resulting in injuries and other stress-induced behaviors (e.g. tail biting, belly nosing, etc.). Dudink *et al.* (2006) found that placing newly weaned piglets in an area with fresh straw bedding and seeds on the floor reduced the incidence of weaning-induced aggression. Interestingly, the investigators also found that announcement of being placed in this area using an auditory signal (door bell) as a conditioned stimulus reduced piglet aggression at weaning even further and increased play behaviors.

As discussed in Chapter 1, fenceline weaning, in which cows (*B. taurus*) and their calves are placed on opposite sides of a fence, can alleviate many of the stress-related behaviors associated with abrupt weaning (Price *et al.*, 2003b). Mother and young can gain close proximity to one another (social contact is maintained) but the young cannot obtain milk. Haley *et al.* (2005) studied the relative importance of social separation versus the inability to suckle on the stress-related behaviors exhibited by early-weaned beef calves (*B. taurus*). The experimental subjects were fitted with plastic antisuckling devices (nose flaps) in the days prior to weaning so they could remain with their mothers but not suckle. After 3 (short-term) or 14 (long-term) days of wearing nose flaps, they were physically separated from their mothers (at a considerable distance), along with control calves who had been able to suckle up to the time of separation. The results showed that after physical separation from their dams, calves weaned in two stages vocalized less, walked less and spent more time eating and resting/lying down than control calves. Similar results were obtained with two-stage weaning involving sheep (*O. aries*) (Schichowski *et al.*, 2008) and foster cows (*B. taurus*) each nursing four calves (Loberg *et al.*, 2007). These results are similar to those obtained in fenceline weaning in which social contact is permitted (albeit across a fence) but suckling denied, starting on the first day of weaning. Since both techniques use the same control treatment (abrupt denial of both social contact and suckling at the time of weaning), it appears that simultaneous denial of both social contact and suckling are responsible for the stress-related behaviors observed at weaning in this species and that permitting social contact while denying suckling provides a more welfare-sensitive approach to the weaning experience.

Early weaning may encourage the development of atypical behavior patterns (see also p. 280). Würbel and Stauffacher (1998) found that premature weaning of caged laboratory mice (*M. musculus*) resulted in the development of stereotyped wire-gnawing behaviors. Callard *et al.* (2000) reported that captive-born roof rats (*R. rattus*) began backflipping in their cages immediately following early weaning. Nicol (1999) noted that two-thirds of 14 horses (*E. caballus*) that developed stereotyped behaviors did so within 1 month of weaning, even though they were weaned at different ages. These findings are especially interesting since the significance of early weaning on the development of atypical behaviors has not been fully appreciated.

Young subjected to early weaning and then socially isolated may be denied important socialization with littermates or conspecifics. We discussed in Chapter 4 how the period of socialization in domestic puppies (*C. familiaris*) ranged from 3 to 12 weeks of age. Puppies taken from their littermates early in this period and isolated

Fig. 10.12. Calf drinking milk from a pail with an artificial teat attached at the bottom.

from other dogs frequently become overly aggressive toward conspecifics and exhibit other atypical social behaviors.

An extreme example of early weaning (from the mother) is in dairy calves (*B. taurus*) that are typically removed from their mothers within hours of birth. Care is given to provide these calves with colostrum from either their natural mother or another female during the first 24 h. Colostrum provides the newborn calf with immunoglobulin, a protein that binds to infectious agents and either neutralizes them, kills them or makes it easier for the body to kill them. After receiving colostrum, dairy calves are typically fed an artificial milk ('replacer') diet directly from a bucket or from a pail or bottle with a nipple (teat) (Fig. 10.12).

11 Communication

General Principles

Wilson (1975) defined communication as: 'an action on the part of one organism that alters the probability of a behavior occurring in another organism in a manner that is adaptive to either one or both participants.' It is tempting to think of communication as simply information exchange. But, as Wilson's definition states, communication is more than just information exchange. The exchange can alter the behavior of the parties involved. It can even be used to manipulate or deceive.

Communication is dependent on the structure and functioning of the sense organs, which differ from species to species. Development of the sensory systems has been studied in a wide array of mammalian and avian species. Interestingly, the order in which the major sensory systems begin functioning is the same in each of the species analyzed. In mammals and birds, development proceeds in the order, tactile, vestibular, chemical, auditory and visual, regardless of the stage of development at birth (Gottlieb, 1971).

Behaviors almost always accompany the dissemination of communicatory signals. Behaviors associated with the reception of communicatory signals are typically more subtle. For example, the courtship or pre-mating behaviors of male birds often include rather dramatic visual cues (e.g. strutting, feather displays, acrobatic flights, etc.). Females generally receive these visual signals in a passive manner. Active behaviors are not required to receive visually based messages. Mammals mark their territories with chemical signals using highly ritualized behavior patterns (e.g. raised-leg urination, rubbing scent glands on objects, etc.). Receiving these chemical signals (i.e. messages) requires little action on the part of the receiver.

Some distinct patterns are observed when animals communicate. These patterns generally have adaptive significance, considering the nature of the messages conveyed and the environment in which communication takes place. The following section discusses discrete versus graded signals, redundancy used in communication and how the environment dictates the relative usefulness of the major sensory modalities used in communication.

Discrete versus graded signals

Behaviors used in communication can be categorized roughly into two groups, discrete and graded. Discrete signals are those that present messages in a simple on-off manner, signifying 'yes' or 'no', 'present' or 'absent', or similar dichotomies. For example, the sexual receptivity of a female mammal or bird is often communicated to the male by whether or not she stands immobile when approached or mounted (Fig. 11.1a,b). The 'standing posture' is a relatively discrete signal in that the female either does or does not do it.

Fig. 11.1. Female ungulates either stand immobile to be mounted (a) or they do not (b). The message is discrete. Note that the same two heifers are in both photos. The lighter-colored heifer is in estrus and the black one is not, even though she mounts her penmate.

Discrete signals show little variability in intensity and duration. Graded signals are more variable. Graded signals vary in intensity and (or) duration as the relative arousal or motivation of the animal changes. Graded signals communicate a continuum of arousal or motivational states.

Graded communication is often achieved by adding a succession of related behavioral responses to the display. Horses (*E. caballus*) signal aggressive intent by flattening their ears on the back of their head. More intense aggression is communicated by adding an open mouth with teeth bared. Domestic dogs (*C. familiaris*) use similar graded facial signals to communicate aggressive intent and submission (Fig. 11.2).

Certain communicatory signals may dominate or supersede others. Domestic dogs signal intent to play by bowing to a conspecific (Fig. 11.3) and sometimes emitting a growl. Growls are normally associated with aggression but, when coupled with a bow, the two signals solicit playful behavior. The visual component (bow) takes precedence over the auditory cue.

Redundancy

Animals frequently communicate the same message using multiple sensory modalities. Redundancy in communication is particularly common in highly social, group-living animals and can facilitate communication under a wide variety of circumstances. It increases the likelihood that the message will be transmitted regardless of the circumstances at the time. In domestic goats (*C. hircus*), alarm may be communicated by directing visual attention to the fear-provoking stimulus, running away from the stimulus, stomping on the ground with the forefoot, snorting and the release of odoriferous chemical substances from certain scent glands. Visual, auditory and olfactory modalities are all involved. Multimodal signaling usually, but not necessarily, results in an enhanced response.

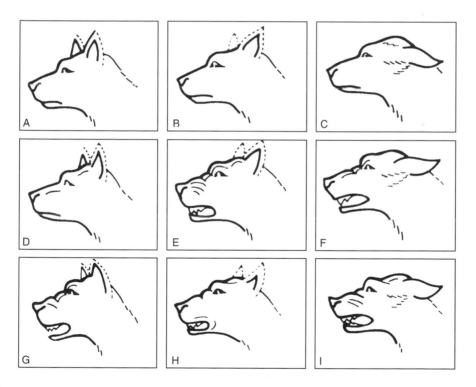

Fig. 11.2. Graded communicatory signals reflect a continuum of motivational states, as illustrated by successive intensities of aggressive and fearful facial expressions in the dog. Aggressiveness increases from top to bottom and fearfulness from left to right (after Lorenz, 1981).

Fig. 11.3. Invitational bow of the dog (right) inciting play (Hart, 1985).

Comparison of the sensory modalities used in communication

Table 11.1 compares the four major sensory modalities (visual, auditory, chemical and tactile) with respect to signal range, rate of transmission, ability to flow around a topographical barrier, fade-out time (i.e. time from transmission until the signal can no longer be perceived), locatability of sender and risk of exploitation by an illegitimate receiver such as a predator. The advantages and disadvantages of a signal vary depending on the environment in which it is used, the physical properties of the signal, the sensory acuity of the receiver and the presence (or absence) of potentially harmful receivers.

The main advantages of visual communication are the instantaneous rate of transmission, short signal fade-out times, particularly when messages are quickly changing, and the instantaneous locatability of senders. Showy visual signals can be transmitted instantly over long distances; less conspicuous visual signals work only at close range. The tail position of a dog (*C. familiaris*) communicates something about the animal's social status at a distance, whereas facial expressions and piloerection (erection of hair on the neck and shoulders) communicate best at close proximity to the receiver. Visual signals are not as effective in the hours of darkness and in topographically diverse

Table 11.1. Relative advantages and disadvantages of the four principal sensory modalities used in communication (from Alcock, 1989).

Modality	Signal range	Rate of transmission	Flow around barrier?	Fadeout time	Locatability of sender	Risk of exploitation
Visual	Medium	Fast	No	Fast	Easy	High
Auditory	Long	Fast	Yes	Fast	Fairly easy	Medium
Olfactory	Long	Slow	Yes	Slow	Difficult	Low
Tactile	Very short	Fast	No	Fast	Easy	Low

Chapter 11

environments where physical barriers prevent reception. In addition, the risk of exploitation by illegitimate receivers is high because of the instantaneous locatability of the sender. For example, instantaneous visual locatability can be a disadvantage for prey species when traversing areas harboring predators.

Auditory signals communicate information over both short and long distances and around physical barriers. They are effective in darkness and the risk of exploitation is not as great as with visual cues. Locatability of the sender is not as good as in visual communication, particularly when high frequency sounds are emitted. Alarm calls often consist of high frequency sounds for this very reason.

As in auditory communication, chemical (olfactory) signals provide information over both short and long distances; they are not impeded by physical barriers and they are useful in the hours of darkness. Fade-out time is slow and locatability of the sender can be relatively difficult compared to visual and auditory signals. Slow fade-out time can be advantageous in certain circumstances, such as marking territories and signaling sexual status. Chemical cues have communication value even in the absence of the sender. However, lingering olfactory cues can be exploited by predators when searching for prey.

Tactile signals have the advantage of being transmitted quickly day or night, fade-out time is fast, locatability of the sender is easy and the risk of exploitation is very low. The main disadvantage is that they are only useful when conspecifics are in physical contact.

Darwin's principle of antithesis

Darwin (1872) in his book, *The Expression of the Emotions in Man and Animals*, was perhaps the first to document that when an animal reverses its motivation, the associated communicatory signals are also reversed (Fig. 11.4). When one dog (*C. familiaris*) threatens another animal, it maximizes the size of its profile. It stands erect, with neck extended. Ears are up, its tail is raised and piloerection may occur. It may walk broadside (sometimes stiff-legged) to the other animal, while staring directly at its opponent. In contrast, submissive displays are just the reverse. The submissive dog generally adopts a low profile. Ears are flattened, its tail is drawn between its legs and it lowers its body in a crouch. It may even lie down on its side or back. It also avoids looking directly (i.e. staring) at its opponent.

Visual Communication

Impediments to visual communication become less important when animals are maintained in captivity. The amount of space typically provided for domestic and captive wild animals is usually less than they would use in nature, thus increasing the likelihood that communication takes place at shorter distances. In addition, this space (e.g. cages, corrals, pastures) typically has fewer topographical features to impede visual communication. Also, the risk of visual cues being exploitated by illegitimate receivers such as predators is greatly reduced in captivity. At close range, visual signals may exploit changing patterns of form, movement and color. Body posture and movements serve as major communicatory stimuli in captive wild and domestic animals.

Fig. 11.4. Threatening and submissive postures in the dog. Note the features that are opposites (e.g. ear and tail positions, contour of the spine, general posture) (Darwin, 1872).

Visual acuity

Visual acuity involves the degree of detail that can be seen. Humans have very good visual acuity relative to most domestic animals. This is because our retinas have proportionately more cone cells to rod cells. Cone cells in the retina of the eye enable us to perceive both well-defined and color images. Rod cells are more sensitive to light, which favors night vision, but they provide images that are coarse and poorly defined. The images are not out of focus, they just lack detail.

Nocturnal animals like domestic cats (*F. domestica*) have retinas with proportionately more rods than cones. In addition, the pupils of the cat's eyes can enlarge 170% in the dark to capture more light for the retina. As a consequence, they have excellent night vision at the expense of visual acuity and color vision. Cats can see well at light intensities approximately ten times lower than the ideal light intensities for humans.

Dogs (*C. familiaris*) are visual generalists; their visual acuity is reasonably good at both high and low light intensities. Some breeds of dogs are myopic (near-sighted). One study showed that 53% of German shepherd dogs could not focus well at distances over 0.5 m. Rottweilers also tend to be myopic, as are about 10% of Labrador retrievers. The visual acuity of dogs and horses (*E. caballus*) is about the same (approximately 20/60 to 20/70 on an optometrist's scale for humans).

Color vision

Colored objects do not *emit* their characteristic color, per se. Rather, they absorb visible light of all frequencies impinging on them except for light in the frequency range perceived as the color observed. Grass is 'green' to us only because: (i) light in other portions of the color spectrum is absorbed by grass; and (ii) we have 'green' cones in the retina of our eyes which send signals to our brain that are interpreted as 'green'. Animals which do not have 'green' cones see grass as a different color than we do.

Color vision in humans and many other primates is highly developed due to a proportionately large ratio of cones to rods in the retina. Primate vision, including our own, is trichromatic. Humans, Old World primates and apes and some New World monkeys possess three types of cones in the retina capable of processing wavelengths in the blue, green and red range of the color spectrum, respectively. Trichromatic color vision occurs when the signals from these three cone types are compared in the retina and brain. For example, 'blue' light stimulates the 'blue' cones a lot; the 'green' cones a little; and the 'red' cones not at all. This sensory mixture (in the brain) yields what we call 'blue'. 'Green' light stimulates the 'green' cones a lot and the 'blue' and 'red' cones a little. This mixture yields 'green'. 'Red' stimulates the 'red' cones a lot, 'green' cones a little and 'blue' cones not at all.

With the exception of the primates, most mammals have dichromatic vision; that is, they have only two types of color-related cone cells (Jacobs, 1993). Domestic dogs (*C. familiaris*) lack 'green' cones and thus have a color vision deficit in the green/yellow/orange portion of the visible spectrum (Neitz *et al.*, 1989; Jacobs *et al.*, 1993). Light from the green/yellow/orange portion of the spectrum yields a color that we could call 'whitish' since the remaining 'blue' and 'red' cones are stimulated equally by light in that wavelength range. Light of wavelengths in our red end of the spectrum appears 'yellowish' due to the dominant role of the 'red' cones and the absence of 'green' cone stimulation. Consequently, dogs have difficulty discriminating between objects which, to us, are green, yellow and orange. Even red and orange are difficult to tell apart. Nevertheless, red is easily distinguished from blue, even though red does not look the same as we see it. Not surprisingly, dogs are quite good at distinguishing between various shades of blue.

Pigs (*S. scrofa*), cattle (*B. aurus*), sheep (*O. aries*), goats (*C. hircus*) and horses (*E. caballus*) have dichromatic color vision similar to that of dogs (Neitz and Jacobs, 1989; Jacobs *et al.*, 1998; Carroll *et al.*, 2001). Horses have 'blue' cones and cones which peak in the orange range of the color spectrum. Red- and green-colored objects look washed out. For example, grass looks yellowish. Rats (*R. norvegicus*) and mice (*M. musculus*) also have dichromatic vision, but their color vision peaks in the ultraviolet and middle range of the color spectrum (Jacobs *et al.*, 2001; 2004). Guinea pigs (*C. porcellus*) are unusual for rodents in having color vision similar to the dog and ungulates (Jacobs and Deegan, 1994).

White-tailed deer (*O. virginianus*) and fallow deer (*Dama dama*) differ from dogs, other ungulates and rodents by having cone receptors which peak in the short-wave (blue) range and the middle range (green/yellow/orange) of the color spectrum (i.e. they lack red cones) (Jacobs *et al.*, 1994). Whales and seals are the only known mammals with only one color-sensitive type of cones in the retina. For all practical purposes, this makes them color-blind.

The color vision of diurnal bird species surpasses that of mammals (see review by Varela *et al.*, 1993). Birds that are active by day possess up to five times the concentration of cone cells in their retinas than humans. Many bird species also have a fourth cone type, which allows them to see in the ultraviolet range of the color spectrum. In addition, there are colored oil droplets in the cone cells of birds through which light must pass before reaching the visual pigment. These factors allow a greater range of color discernment than is possible by humans.

Stereoscopic vision

Most of our large domestic animals have a wide field of vision due to the placement of eyes on the head. For example, sheep (*O. aries*) have a 270-degree field of vision; they cannot see in the 90-degree arc behind themselves. Dogs (*C. familiaris*) average a 150-degree visual field.

Stereoscopic or binocular vision is possible when both eyes see the same thing from slightly different angles, providing a degree of depth perception. The binocular field of view for sheep is about 50 degrees, between 55 and 65 degrees for horses (*E. caballus*), 60 degrees for dogs and about 25 degrees for chickens (*G. domesticus*).

Auditory Communication

Sound production

Sound production by terrestrial vertebrates is usually associated with the animal's respiratory system. Mammalian sounds originate in the larynx by the passage of air across a pair of vocal cords, causing them to vibrate. The sounds produced are then amplified by resonating chambers, such as the throat and mouth. Variation in sound is produced by: (i) varying the tension on the vocal cords; (ii) varying the rate of air flow across the vocal cords; and (iii) varying the size of the resonating chamber(s).

Birds also possess a larynx, but they lack vocal cords. Sounds are produced in the syrinx, which is similar to the larynx in structure but located where the trachea divides into the two major bronchi.

Sound reception

The ability to perceive sound varies with species, particularly at the higher frequencies. Humans can hear sounds up to about 17 kHz (kilohertz). The upper limit is about 30–35 kHz for cattle (*Bos*) and horses (*E. caballus*), 45 kHz for dogs (*C. familiaris*),

90 kHz for cats (*F. domestica*), 100 kHz for deermice (*Peromyscus*) and 150 kHz for bats (e.g. *Myotis lucifugus*). It is difficult to appreciate the role of auditory communication in animals fully when our hearing is so underdeveloped in the high-frequency range.

Development of auditory communication

There are several key points relevant to the acquisition of auditory communication. First, the acoustic morphology (structure) of animal sounds is relatively fixed (innate) for each species. Most of these sounds develop during maturation and, with few exceptions, experience is important only in refining the objects or events that elicit these sounds. There is little evidence for the production of new sounds for new situations, except in a few species such as wild and captive chimpanzees (*P. troglodytes*). In addition, most auditory cues are context specific, signaling, for example, the presence of a predator or the discovery of food. Some species (e.g. chickens, vervet monkeys) even have different calls for different predators (Evans *et al*., 1993).

Many bird species produce both songs and calls (e.g. songbirds). Other birds (e.g. chickens) produce only calls. Interest in the auditory communication of male zebra finches (*T. guttata*) and canaries (*S. canaria*), two species commonly seen in captivity, has led to some interesting findings on the neurophysiology of song development and production (Barinaga, 2003). Males of both species sing rather elaborate songs. Zebra finches learn a single song during adolescence and sing the same song throughout the year for their entire lives. Canaries sing elaborate songs during the annual breeding season and learn new song elements every year. Their singing is less frequent and less precise during the non-breeding season. Interestingly, the brains of zebra finches and canaries reflect this difference.

The high vocal center (HVC) is the most prominent part of the network of brain nuclei essential for song learning and production in male songbirds. Males have a well-developed HVC, whereas females have a smaller version which does not support song production. The female HVC is necessary for females to recognize and respond to the songs of males of their species.

Zebra finches add large numbers of neurons to their HVC only when they are young, whereas the HVC of male canaries enlarges each breeding season due to the addition of new neurons and shrinks in size during the non-breeding season (Nottebohm, 1989). This explains why zebra finches never change their song, while canaries do. The finding that the brain of some adult songbirds is capable of producing new neurons (neurogenesis) was an important discovery for neurophysiologists working in this area and for biologists, in general.

Some species are better than others at imitating sounds and this ability does not seem to be related to their cognitive abilities. Although the social life of many non-human primates is highly complex, their vocalizations only roughly reflect these complexities. The most intensively trained chimpanzees (*P. troglodytes*) can learn several hundred hand signs but can articulate only a few human words, and then only poorly. In contrast, parrots can learn and recite a relatively large repertoire of human words. Many songbird species possess geographically distinct dialects, which are learned. These birds readily copy songs they have heard, even before being able to sing, themselves (see p. 39).

Functions of acoustic signals

Two of the primary functions of acoustic signals in domestic animals are advertisement and arousal. Chickens (*G. domesticus*) have different calls to signal the presence of ground and aerial predators. Animals advertise their presence and individual identity with songs, calls and other sounds. Acoustic signals such as barking in dogs (*C. familiaris*) and other canids can reinforce an animal's dominance over other animals, including the intimidation of potential territory intruders. Dogs also learn to bark to get attention from humans and conspecifics (Yin, 2002). Auditory signals advertise age, sex and sexual receptivity. They are used to attract mates and frequently have a role in courtship and sexual arousal. The cackling of female jungle fowl (*G. gallus*), ancestors of our domestic chickens, advertises ovulation and attracts males, inciting competition between males for the female cackler. Such competition increases the probability that the female's eggs will be fertilized by males of superior genetic quality. Sounds help youngsters stay in contact with their parents, particularly in topographically diverse environments where visual communication is impeded. Wolves (*C. lupus*) and domestic dogs howl when alone, either to seek contact with 'pack' members (Mech, 1970) or to attract conspecifics during the breeding season (Klinghammer and Laidlaw, 1979).

McConnell (1991) has pointed out that animal trainers often use short, rapidly repeated broadband notes to stimulate motor activity and longer, continuous narrowband notes to inhibit or slow down activity. As an example, horse riders use repeated clicking noises or 'giddy-up' to stimulate the horse to move, while 'whoa' is used to slow the horse down. Clicks and 'giddy-up' are normally emitted sharply and abruptly, while 'whoa' is spoken with softer, more prolonged tones (i.e. 'whooooa'). When you tell your dog (*C. familiaris*) to 'sit' or 'come' (both actions), the words are spoken in a rising, abrupt and commanding voice, while 'stay' (inaction) is spoken in a prolonged manner (i.e. 'staaaaay'). Herding-dog trainers typically use two to four short whistles in rapid succession and rising in frequency to initiate approach and herding of livestock, whereas a single prolonged descending whistle is most often used to slow or terminate an action. Horse (*E. caballus*) and dog handlers in different cultures and using different languages have developed this same pattern of communicating with their animals. McConnell's point is that some sounds are particularly effective in influencing the internal state and behavior of animals. Perhaps the animals have taught their handlers what works best. Not surprisingly, there are many examples of non-human animals using these same patterns of repetition and duration of acoustic signals to influence the behavior of conspecifics. The saying, 'It's not just *what* you say but *how* you say it', applies to many animals, as well as humans.

Auditory signals can convey the emotional state of the sender. In some cases, the message is transmitted at frequencies above our normal hearing range. For example, adult laboratory rats (*R. norvegicus*) produce two types of vocalizations in the ultrasonic range, one at about 22 kHz and another at about 50 kHz. The 50 kHz vocalization generally reflects a positive emotional state in the sender, while the 22 kHz vocalization denotes a negative emotional state (Burman *et al.*, 2007). When recordings of the 22 kHz vocalization were played to rats, it induced a state of increased vigilance and emotional reactivity.

Information on the emotional state of domestic dogs (*C. familiaris*) can be conveyed to human listeners by the tonality, pitch and intervals of barks (Pongrácz *et al.*, 2006). Low-pitched barks from dogs of the Mudi breed (Hungarian herding dogs) were correctly described by a panel of people as 'aggressive', while tonal and high-pitched barks were correctly scored as expressing 'fear'. The pitch of the bark was more important to the human listener than the bark's tone (harmonic to noise ratio imparting clarity or 'roughness' to the sound). Bark sequences with relatively short interbark intervals were considered aggressive and the intensity of aggressiveness was believed to increase as the bark interval shortened. Regardless of tonality, high-pitched bark sequences with relatively long interbark intervals conveyed a 'happy' or 'playful' demeanor to the human listener. In general, dog barks function according to the same structural-motivational rules that apply to acoustic signals in other species.

Olfactory Communication

Olfactory communication is highly developed in most of our domestic mammals. Olfactory signals that have a specific behavioral or physiological effect on conspecifics are often called 'pheromones', a word derived from the Greek word 'pherein', meaning 'to carry', and 'hormaein', meaning 'to excite'. It was first thought (based on early work with insects) that pheromones were single chemical compounds and that these single compounds were responsible for the information coveyed. We now know that chemical signals important in communication usually consist of more than one chemical compound (Wyatt, 2003). In addition, information transmitted by a chemical signal may depend more on the relative mixture of certain chemical components than on the presence or absence of certain compounds. For example, Natynczuk *et al.* (1989) reported that seven different chemical substances were found in the anal sac secretions of beagle dogs (*C. familiaris*) and the relative quantities of these seven substances differed in the male and female (Table 11.2). The sebaceous glands of the male goat (*C. hircus*) secrete 4-ethyl-octanoic, -decanoic, -dodecanoic and -tetradecanoic acids, which serve as female attractants (Sugiyama *et al.*, 1981). Of these 4-ethyl fatty acids, 4-ethyl-octanoic acid is the substance which produces a strong 'goaty odor' at relatively low concentrations. Cervico-vaginal mucus from estrous cows (*B. taurus*) contains over 20 chemical compounds, including alcohols, diols, alkenes, ethers, diethers, ketones, primary amines and aromatic alkanes (Klemm *et al.*, 1987). Nine of these 20 compounds were found to be sexually stimulating to bulls. Such findings have

Table 11.2. Quantities of all volatile components identified from the anal sac secretions of male and female beagle dogs (*C. familiaris*) expressed as a percentage of the total (from Natynczuk *et al.*, 1989).

	Male	Female
Hydrocarbons	4.8[a]	24.6
Ketones	15.5	6.8
Nitrogen-containing	22.5	0.0
Sulfur-containing	3.2	1.6
Alcohols	11.4	50.1
Aldehydes	42.7	16.5
Acids	0.0	0.3

[a]Percentage of the total quantity of all volatile components.

discouraged use of the word 'pheromone' in favor of 'chemical signals', although 'pheromone' is still commonly used in the popular literature.

Chemical signals are classed as volatile (disseminated in air as a vapor) or non-volatile. Volatile signals are carried by air movement. The animal must come in physical contact with non-volatile chemical signals for them to be useful in communication. Many secretions from sebaceous and sweat glands have both volatile and non-volatile properties.

Sources of chemical signals

Odors useful in social communication normally originate in specialized skin glands located in different parts of the body (Mykytowcyz, 1970). Some chemical signals are dispersed in urine or feces. Two types of skin glands are found in mammals, sebaceous glands and sweat glands. Regions of these skin glands have been modified to produce odoriferous substances. Some of the body regions possessing scent glands include musk glands (e.g. goat) or chin glands on the head (e.g. rabbit), ventral sebaceous glands on the abdomen (e.g. gerbil), inguinal glands on the groin (e.g. rabbit), tarsal and interdigital glands on the limbs (e.g. cloven-hoofed ungulates) and preputial glands (e.g. rodents and male pigs) and anal glands (many species) in the anogenital region.

Several species of deer (e.g. *Odocoileus*) have a raised tuft of hair on the medial aspect of each tarsus (Fig. 11.5a,b). Underlying the tuft of hair is exocrine glandular tissue (sebaceous and apocrine glands). This area is collectively called the 'tarsal gland'. Whitetail deer (*O. virginianus*) urinate on the tarsal gland in a behavior referred to as 'rub urination'. Urine is trapped on the specialized scent-disseminating hairs and mixes with glandular secretions. Osborn *et al.* (2000) found that the bacterial decomposition of urine and glandular secretions adds to the mix of volatile odors produced which convey information regarding age, sex, dominance, reproductive condition and individual identity. Some 18 different species of bacteria have

(a)

Fig. 11.5a. White-tailed deer (buck) pawing the ground in the act of making a 'scrape'. Pawing is followed by urinating over the tarsal gland (dark stained area on hind leg) on to the scrape (courtesy of Karl Miller).

(b)

Fig. 11.5b. Close-up view of the tarsal gland of a mature white-tailed buck during the breeding season (rut) (courtesy of Karl Miller).

been cultured and identified from the tarsal tufts of whitetails. Interestingly, more species of microbes were found on the tarsal glands of dominant breeding males than subordinate males or females.

Secretions of the anal gland are dispersed with feces. Anal glands have been found in over 100 species of mammals. The anal glands of the European rabbit (*O. cuniculus*) consist of two clusters of brownish tissue forming a saddle-like mass around the end of the rectum (Mykytowycz, 1968). Secretions flow from the gland into the rectum, where fecal pellets are coated as they pass out of the anus. Preputial glands are modified sebaceous glands located on the tip or inner side of the sheath or skin covering the penis (preputial canal) (Odend'Hal *et al.*, 1996). Secretions from these glands are frequently disseminated with urine. Sex hormone (e.g. estrogen, testosterone) and protein odors may also be disseminated in animal urine. Cheek rubbing by members of the cat family may serve to deposit chemical substances from saliva or from urine picked up from rubbing the sides of their face on a urine-soaked substrate (Fig. 11.6). Novotny *et al.* (1985) reported that aggression in male mice (*M. musculus*) was triggered by two chemical compounds in their urine, dehydro-*exo*-brevicomin and 2-(*sec*-butyl)-dihydrothiazole. Both of these compounds need to be present together in mouse urine to have their effect. The effect of the two compounds on aggressive behavior is not seen when one compound is present without the other and when both compounds are dissolved in water.

Fig. 11.6. Captive bobcat rubbing the side of its face on urine-soaked concrete.

Perception of chemical signals

Terrestrial vertebrates possess two anatomically distinct systems to detect and analyze chemical stimuli. The main olfactory epithelium is located in the posterior recess of the nasal cavity and transmits olfactory information to the main olfactory bulb of the brain. Further sensory processing of chemical information in centers of the primary olfactory cortex and in multiple cortical and neocortical areas of the brain generates the sense of smell.

The perception of volatile (airborne) chemical signals is initiated in the olfactory epithelium of the nasal cavity. Yoshihara *et al.* (2001) have described how hundreds of different odorant molecules are recognized by specific receptors on the surface of sensory neurons in the nasal epithelium. Each neuron carries only one receptor, which is activated by a range of odorants of similar chemical structures. The axons of these neurons send olfactory information encoded as electrical signals to the olfactory bulb of the brain. These signals are picked up by the dendrites of mitral cells in the olfactory bulb. Mitral cells are clustered in structures called glomeruli, which pass on information to the olfactory cortex of the brain.

It is difficult for humans to appreciate the olfactory acuity of most of our common domestic animals. The nose of the dog (*C. familiaris*), for example, contains hundreds of different chemical receptors and can distinguish tens of thousands of different smells (Davenport, 2001) at concentrations many thousand times smaller than the lowest concentrations perceived by humans. Of course, they are aided by a much larger nose. The volume of the olfactory epithelium in the beagle dog (*C. familiaris*) is 25 times greater than in an adult human ($75\,cm^2$ versus $3\,cm^2$, respectively) (Albone, 1984). Dodd and Squirrel (1980) reported a range of $18–150\,cm^2$ in the size of the olfactory epithelium of various dog breeds.

The vomeronasal organ (VNO) is part of the 'accessory olfactory system' and is also used in the perception of chemical compounds, particularly non-volatile chemical substances. The VNO consists of olfactory-sensitive neurons encapsulated in a bilateral and tubular-shaped, fluid-filled structure of the ventral nasal septum, or in the hard palate. VNO neurons send fibers to the accessory olfactory bulb of the

Chapter 11

brain, which relays information to the ventromedial hypothalamus of the brain via the amygdala.

In rodents, access to the VNO is gained by a ventral groove in the nasal cavity where water-soluble chemical molecules are carried by nasal mucus. As the rodent moves about and sniffs, the VNO expands and contracts, pumping fluid over the olfactory-sensitive neural receptors, which then relay messages to the accessory olfactory bulb (AOB) of the brain. Different receptors in the VNO are sensitive to different odors (e.g. odors from males versus females) and are activated accordingly (Luo *et al.*, 2003). The AOB relays the messages to other parts of the brain, which assess the social relevance of the chemical cues and activate an appropriate response, such as attack or mate (Wysocki *et al.*, 1986). The accessory olfactory system in ungulates is similar. In our domestic ungulates, tiny openings in the roof of the mouth (anterior palate) provide an entrance to the nasopalatine ducts, which terminate in the VNO. The animal often uses its tongue to place non-volatile chemical substances (e.g. urine, vaginal secretions) at the openings of the nasopalatine ducts. These substances are drawn ('pumped') through the ducts into the VNO. 'Flehmen' (see p. 117–119) appears to facilitate this process.

VNO function is also important to females. For example, volatile chemicals from male rats (*R. norvegicus*) stimulate the VNO of females and influence their reproductive physiology and behavior.

There is some evidence that VNO neurons do not adapt readily (i.e. lose their stimulative function) under prolonged stimulus exposure (Holy *et al.*, 2000). This is a major difference from most all other sensory neurons. Sensory adaptation to odors in the main olfactory epithelium occurs quickly.

In general, primates and birds have very poor olfactory capabilities compared to non-primate mammals. Not only are the chemical processing regions of our brains poorly developed but our genes may also be contributing to our poor sense of smell. Rouquier *et al.* (2000) reported that more than 70% of the human genes encoding olfactory receptors (cell-surface proteins used to detect odors) have disabling mutations, making it more difficult for us to smell. About 50% of the 'olfactory genes' in gorillas (*G. gorilla*) and chimpanzees (*P. troglodytes*) are non-functional or greatly impaired. In contrast, not a single disabled olfactory receptor gene was found in mice (*M. musculus*).

Functions of chemical signals

Chemical signals function in a wide variety of contexts, including feeding, predator avoidance, habitat selection and a broad array of social interactions. One of the major functions of chemical signals is advertisement. The odor of an animal can reveal its species, age (especially juvenile versus adult), sex, group membership, individual identity, sexual status, social status, general vigor and the animal's home range or where it lives. For example, the territorial European rabbit (*O. cuniculus*) marks its territory with urine and secretions from its anal and chin glands (Mykytowycz, 1968). Anal gland secretions are deposited on fecal pellets as they are expelled and the chin gland is used to mark vertical objects that cannot be reached with urine or by defecating (Fig. 11.7). Dominant males mark younger animals and females in their territories with urine and chin gland secretions, which gives them group identity. If urine from a foreign male is placed on a group member, it is immediately attacked.

Fig. 11.7. Chin marking by a rabbit in the garden of one of my students. Rabbits will use their chin gland to mark objects that cannot be readily marked with urine or feces.

Females will reject their own offspring if smeared with foreign urine. Inguinal gland secretions from the groin serve as sex attractants for females when in estrus. Prior to eye and ear opening, nestling rabbits identify their mothers by anal gland secretions and, to a lesser extent, secretions from her inguinal gland. Anal gland secretions from strange females elicited noticeable avoidance reactions in the young nestlings (Mykytowycz, 1970).

Nest-mate recognition in termites is based on bacteria inhabiting the insect's gut. The species of bacteria in the gut of termites varies from colony to colony. Ewes (*O. aries*) identify their own lambs by odor on the lamb's wool. Mice (*M. musculus*) of the same gender and inbred strain can identify one another on the basis of odor alone (Bowers and Alexander, 1967). Pigs use urine in individual recognition. The identity of cloven-hoofed animals is also revealed by scent from interdigital glands (located between the two halves of each hoof), which is deposited as the animal moves about. The sex and sexual status of an animal is revealed by odoriferous secretions in its urine, which are derived from sex hormones (e.g. estrogen, testosterone). Studies with domestic rats (*R. norvegicus*), catfish (*Ameiurus* species) (Bardach and Todd, 1970) and other animals have shown that dominant animals can be identified by their odor. In many cases, androgen-based odors of dominant animals are simply more intense than in subordinate conspecifics. It is believed that the odor of testosterone and certain proteins in the urine of adult male goats (*C. hircus*) can be used by females to assess the general vigor of potential suitors (Coblentz, 1976).

Animals advertise their home range or 'residence' by deliberate marking of physical objects with urine, feces, saliva or skin-gland secretions. Object marking is like leaving a 'calling card' saying 'I have been here' or 'this is my home'. Such markings may enhance the confidence of animals subsequently engaged in aggressive interactions to defend parts or all of their home range. It is analogous to the well-known 'home field' advantage in human sporting events. Scent marking by males is androgen dependent. Dominant males mark more frequently than subordinates, males more than females and adults more than juveniles. Objects marked may become 'signposts' for information exchange. For example, the raised-leg urination of canids (e.g. wolves, domestic dogs) and urine spraying by cats (Fig. 11.8), the repeated defecation of horses (*E. caballus*) in the same spot ('stud piles') (Fig. 11.9) and chin-gland marking by rabbits (*O. cuniculus*) all function in this way. Animals are motivated to mark previously deposited marks as if it is important to renew their own mark or to place their mark on top of the mark of another animal.

Lactating female mammals secrete chemical signals ('maternal pheromones') from the nipple and areola region of their breasts, which are attractive to neonates. These and other odors (e.g. fecal odors) are used by young in identifying their mothers. Piglets (*S. scrofa*) can discriminate between the odor of their mother and alien mothers by 12 h after birth (Morrow-Tesch and McGlone, 1990a). They are most attracted to odors associated with their mother's skin secretions and feces. Young piglets also identify their preferred teat (see p. 159–161) by odor. Once a newborn has suckled, it leaves its own olfactory imprint on the nipple (e.g. saliva-based odors) and prefers that nipple in future suckling bouts. If the mother's nipples are washed to remove all odors, the young piglets become disoriented at the next suckling event (Morrow-Tesch and McGlone, 1990b). Mothers of altricial mammalian young produce a 'maternal pheromone' in their feces which helps to synchronize the mother–young relationship, attract and orient wandering young back to the nest and provide information about what the mother is eating. Leon and Moltz (1972) reported that lactating rats (*R. norvegicus*) start producing the maternal odor at about 14 days post-partum, at the same time that her young become responsive to the odor and about the time that the young pups start wandering from the nest. At about 27 days post-partum, close to the natural weaning age, the mother ceases to release the chemical substance and the young fail to show responsiveness to it. The source of the maternal odor is bacteria emanating from the cecum, a pouch at the juncture of the small and large intestine, which becomes enlarged during lactation (Leon, 1974). The odor is disseminated in

Fig. 11.8. Raised-leg urine marking in dogs and urine spraying in cats. Urine spraying is one of the most common behavioral problems for cat owners (Hart, 1985).

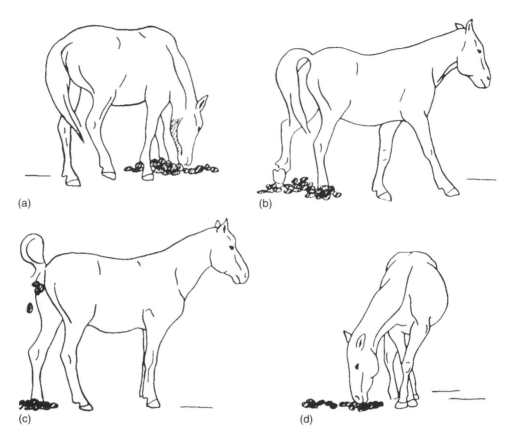

Fig. 11.9. Behavior of a stallion at a fecal pile. The stallion smells the pile (a), moves over it (b), defecates (c) and turns to smell the pile again (d) (Feist and McCullough, 1976).

the mother's feces, which advertises her location, where she has been and the location of the nest site.

A second major function of chemical signals is that of arousal. The androgenic steroid, androstenone, is secreted from the preputial gland and submaxillary salivary glands of the male pig (*S. scrofa*). Androstenone is present in the salivary foam that oozes from the mouths of boars when they are courting estrous females. Sows are very sensitive to and are sexually aroused by the odor of androstenone. In the presence of this 'boar odor', about 80% of estrous sows will stand immobile when humans apply hand pressure on their rumps. Only 50% will stand immobile in the absence of boar odor (Fig. 11.10). Interestingly, androstenone is reportedly the only single-compound pheromone discovered in ungulates. It is present in two forms, 5α-androst-16-en-3-one and 3α-hydroxy-5α-androst-16-ene (Melrose *et al.*, 1971). The former is an unpleasant urine-like odor that has been isolated from the fat of mature boars. The latter is found in high concentrations in the submaxillary salivary gland of the boar and is secreted in the saliva of males during mating. There is some evidence that boars have difficulty detecting this pheromone (Dorries *et al.*, 1991). Perhaps that is why female pigs are used to find truffles, a mushroom table delicacy possessing an aroma mimicking boar odor.

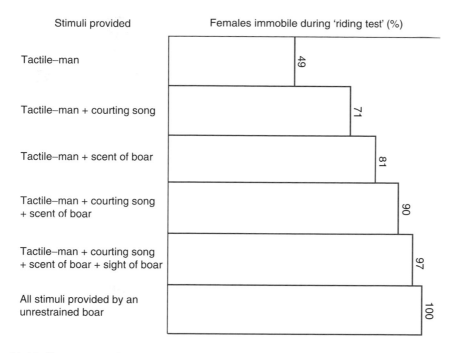

Stimuli provided	Females immobile during 'riding test' (%)

Tactile–man — 49

Tactile–man + courting song — 71

Tactile–man + scent of boar — 81

Tactile–man + courting song + scent of boar — 90

Tactile–man + courting song + scent of boar + sight of boar — 97

All stimuli provided by an unrestrained boar — 100

Fig. 11.10. Percentage of estrous sows exhibiting immobility to the backpressure ('riding') test plus various other boar stimuli (Signoret, 1971; figure from Craig, 1981).

There are many other contexts in which the arousal function of chemical signals can be displayed. During the breeding season, male goats (*C. hircus*) spray urine on their chest, front legs and chin (beard) (Fig. 11.11). It is believed that estrous females are attracted to and sexually aroused by more odoriferous males, typically the larger, more dominant and sexually active individuals (Coblentz, 1976). 'Catnip' elicits a unique combination of body rolling and face rubbing in some domestic cats (*F. domestica*) (Palen and Goddard, 1966). They report that the active ingredient in catnip is the trans, cis-isomer of the lactone, nepetalactone, which comprises 70–99% of the essential oil in the catnip plant (*Nepeta cataria*). Behavioral responses to catnip are exhibited by both sexes and even neutered animals, suggesting that catnip does not mimic estrous odors, as was once thought. Interestingly, the predisposition to respond to catnip appears to be inherited as a dominant autosomal gene. Queen bees produce a chemical substance, comprised of five chemical compounds ('queen mandibular pheromone'), which serves to arouse and attract worker bees. When a synthetic version of this 'pheromone' was sprayed on agricultural fields, researchers observed 60% more visits by worker bees. This finding has great potential value to the agricultural industry. American honeybees pollinate more than 90 different commercial crops.

Many species of vertebrates and invertebrates secrete an alarm or fear odor when confronted with fear-producing stimuli. Such chemical stimuli serve to arouse conspecifics and trigger dispersal. When some fish species are attacked by a predator, specialized epidermal cells in the injured tissure emit an 'alarm 'pheromone' ('schreckstuff'), which arouses and alerts nearby fish (Wisenden *et al.*, 2004).

Fig. 11.11. Male goat urine marking. The penis is extended and urine is sprayed on its beard, face, chest and front leg.

Studies with rats (*R. norvegicus*) and mice (*M. musculus*) have shown that stressed animals (e.g. defeated in aggressive conflict) emit a chemical substance that triggers avoidance behaviors in conspecifics (Valenta and Rigby, 1968; Carr *et al.*, 1970). The response of conspecifics is olfactory mediated and is not necessarily related to urination or defecation by the stressed animal (Rottman and Snowden, 1972).

A third general function of chemical signals is that of physiological priming. In this case, chemical signals serve to 'prime' or sensitize animals to specific stimuli (i.e. 'primer effect') by triggering a chain or series of physiological events, which lowers the response threshold to these stimuli. This is in contrast to the 'releaser effect' in which the chemical signal (or any communicatory signal) *directly* triggers a response, as in the examples cited previously in this section.

The best examples of primer effects are seen in reproductive behaviors such as the 'ram effect' in seasonally breeding sheep (*O. aries*), discussed in Chapter 9. The introduction of a mature ram in a flock of anestrous ewes facilitates and synchronizes the induction of estrus and can advance the onset of behavioral estrus by 2 weeks to 2 months. The stimulative effect of the ram is mediated primarily by odor (see p. 128–129). A similar phenomenon exists in domestic goats (*C. hircus*). Iwata *et al.* (2003) found that the primer 'pheromone' that induced LH secretion and estrus cycling in female goats at the start of the breeding season was a mix of chemical substances derived from 4-ethyl octanoic acid.

Chemical signals from males can also influence the rate at which puberty is attained. Kirkwood *et al.* (1981), working with pigs (*S. scrofa*), demonstrated that gilts attained puberty about 25 days earlier (day 208 versus day 234) when exposed to a vasectomized boar for 30 min/day starting on day 160. Acceleration of puberty in response to the presence of a boar did not occur when the olfactory lobes of the brain were removed by surgery. Sham-operated gilts (control for the effects of surgery itself) responded the same as intact gilts, indicating that puberty acceleration was olfactory mediated in this species. Vandenbergh (1969) reported a similar puberty-accelerating phenomenon in female house mice (*M. musculus*) exposed to the odors of adult males. Drickamer (1982) found that group-reared female house mice produce a urinary-based chemical signal, which delays puberty in other young females. Only 4–7 days of brief daily exposure is necessary for the puberty-delaying signal to exert its physiological effect.

12 Agonistic Behavior

Agonistic Behavior Defined

Some of the more frequently observed social behaviors of animals are behaviors related to conflict or contest between two or more individuals. These behaviors typically carry the connotation of physical combat and aggression. Yet, conflict often takes place in more subtle ways. To account for the diverse ways that conflict can be expressed, Scott (1958) coined the word 'agonistic', a Greek word meaning 'to struggle'. This term is still widely used today.

Agonistic behaviors include: (i) displays expressing dominance (threat) or submission; (ii) offensive and defensive fighting (i.e. physical combat); (iii) active avoidance; and (iv) passive avoidance (e.g. remaining immobile). The term 'aggression' is used when animals engage in threatening displays and physical combat. Observations and studies on agonistic behavior reveal some important differences between those forms of conflict that involve bodily contact (e.g. physical combat or fighting) and those interactions where no direct physical contact is made.

Adaptive Significance of Agonistic Behavior

Before we get into the expression of agonistic behavior, I would like to discuss briefly its adaptive significance. Why did agonistic behavior evolve when it sometimes appears so detrimental to individuals and groups of animals? To answer this question, we have to consider that the resources available to most free-living animal populations (e.g. food, shelter, mates, etc.) are more or less limited. Competition for these resources increases as the size of populations increase. Consequently, agonistic behaviors have evolved to provide an efficient mechanism for dealing with competition, namely different forms of social organization such as the dominance hierarchy and territoriality. In essence, agonistic behaviors have evolved to establish and maintain some degree of order in groups of animals. This is achieved through agonistic interactions which serve the function of physically spacing animals in their environment and determining access to resources in short supply. By so doing, the 'costs' of competition for those resources are minimized.

Expression of Agonistic Behavior

Physical combat is a relatively costly way to deal with competition. If individuals had to engage in fighting every time they wanted the same resource, there would be a significant toll on general health, well-being and reproductive success. Energy expenditure is high when individuals engage in physical combat and animals are often wounded or killed. Consequently, various agonistic *displays* have evolved which frequently achieve the same end result as overt aggression but which require relatively little energy to express and are relatively safe.

Dominance (threat) displays

Dominance, or threat, displays serve to intimidate conspecifics. Such displays often incorporate visual components which maximize an animal's size or profile. Animals may threaten one another using postural or whole-body displays. For example, dominant animals of some species exhibit a behavior called 'broadsiding' in which they display the largest side of their profile, as if to impress their opponents with their large size (Fig. 12.1). Domestic goats (*C. hircus*) often rear up while broadsiding, thus enhancing their height (Fig. 12.2). Dominant wolves (*C. lupus*) exert 'control' over the anal parts of subordinates while presenting their own anal region for inspection (Fig. 12.3). Fatjó *et al.* (2007) reported that in wolf packs, the erect tail was the most reliable visual indicator of dominance, an observation relevant to understanding dominance displays by domestic dogs. Dominant Siamese fighting fish (*Betta splendons*) approach their opponent head-on and open their gill covers (opercula), greatly increasing their perceived size.

Animals also use weapon displays to advertise their assumed dominance. Horses (*E. caballus*), baboons (*Papio*) and wolves (*C. lupus*) threaten conspecifics with open-mouth displays which reveal their teeth (Fig. 12.4). Horses threaten to kick conspecifics by backing into an opponent while swishing their tail back and forth. Cattle (*Bos*), sheep (*O. aries*) and members of the deer family (e.g. *Odocoileus*) 'show off' their horns or antlers when trying to intimidate conspecifics. This is accompanied by

Fig. 12.1. Broadsiding display by a bull accentuates his size.

Fig. 12.2. Goat rearing up and broadsiding penmate.

Fig. 12.3. Dominant wolf exerts 'control' over the anal parts of a subordinate wolf while presenting his own anal region for inspection (Schenkel, 1948).

lowering their head with nose drawn toward the body, as if to point their weapons at the opponent (Fig. 12.5).

Dominant animals can often be identified by their uninhibited approach to conspecifics, as if expecting other individuals to 'give way' to them. Their 'confident' demeanor and actions can intimidate subordinates ('bully effect').

Visual threat displays may be accompanied by auditory and olfactory signals, which reinforce an animal's perceived dominance. Dogs (*C. familiaris*) growl, bulls

Fig. 12.4. Threat display of a dog protecting its food (courtesy of Melissa Morris).

Fig. 12.5. Antler display and threatening gesture by a mule deer buck. The subordinate buck turns away in submission (courtesy of the late Lu Ray Parker and the Wyoming Game and Fish Department).

Agonistic Behavior

189

(*Bos*) bellow and horses (*E. caballus*) scream. Dominant animals of many species can be recognized by a distinctive odor, as was discussed previously (p. 180).

Submissive displays

Submissive displays serve to reduce the likelihood of provoking an aggressive encounter with a conspecific. Postural displays expressing submission tend to reduce the animal's profile size and decrease its conspicuousness. As previously discussed, submissive dogs (*C. familiaris*) lower their tail or tuck it between their legs and crouch down or roll over on their side or back (see Fig. 11.4). When attacked, subordinate dogs whine and yelp. Such infantile-like sounds can abruptly defuse the aggressiveness of the attacker.

Subordinate animals frequently employ avoidance behaviors to reduce the frequency and intensity of interactions with dominant conspecifics. Submissive animals generally 'give way' to an approaching dominant animal (active avoidance). An alternative is to 'freeze', often in a crouching posture, in the presence of a dominant animal (passive avoidance). Subordinates may also avoid eye contact with a dominant individual (Fig. 12.6), a phenomenon Chance (1962) called 'visual cut-off'. Subordinate animals do not want to risk provoking a more dominant individual by a direct approach or stare, so they are more likely to orient *away* from the dominant animal. The orientation of two animals when meeting may reveal their relative dominance. Low-ranking chickens (*G. domesticus*) and pigs (*S. scrofa*) tend to orient

Fig. 12.6. 'Visual cut-off' by a subordinate goat in response to the threatening behavior of a conspecific. By looking away, the subordinate animal inhibits the aggressive behavior of its dominant conspecific (Hart, 1985).

Chapter 12

away from others more frequently than high-ranking conspecifics (Mankovich and Banks, 1982; McCort and Graves, 1982). In the dog (*C. familiaris*), avoidance behaviors may involve incomplete escape movements (head or body turning), care-seeking behavior (pawing and licking movements) or various 'displacement' behaviors, actions triggered by conflicting motivational states and seemingly unrelated to the conflict at hand (e.g. self-grooming, urination). Avoidance behaviors often represent a socially neutral opportunity for the opponents to disengage without fighting. By momentarily breaking off sensory contact, anger, fear and stress can be reduced in both parties. Unless subjected to a pain-inflicting attack by a dominant animal, avoidance seldom includes elements of rapid escape since flight is often the stimulus which releases chase and attack.

Subordinate animals will sometimes thwart being the recipient of an act of aggression by changing the context of the interaction. Stimulating a competing behavior in the dominant animal is sometimes used to defuse a tense situation. For example, certain species of non-human primates will use conspecific grooming to avoid aggressive interactions. Instead of moving away, a subordinate will approach a dominant individual, while showing its intention to groom. Occasionally, a subordinate will solicit sexual behavior from a dominant animal rather than risk an aggressive encounter. Interestingly, this technique works both ways. Dominant individuals can sometimes gain physical access to subordinates by soliciting grooming or otherwise showing an intention to engage in non-aggressive interactions.

Use of Weapons in Physical Combat

Because fighting is risky or dangerous for both parties involved, physical combat between conspecifics is often expressed in ways to minimize the possibility of physical harm or injury. These inhibitions are most evident in species which are equipped with effective weapons. Animals with horns or antlers normally engage in head-on attacks where the weapons themselves absorb the 'shock' of the encounter. Rather than trying to gore or inflict injury on one's opponent, the two parties engage in pushing/ shoving or 'fencing' matches to determine dominance (Fig. 12.7). Injury is more incidental to these encounters than purposeful. Of course, different 'rules' apply when these weapons are used in defense against predators.

Stimulus Control of Agonistic Behavior

Given that agonistic behavior has an important function in controlling competition for needed resources, we can ask the question, 'Is there a spontaneous drive or motivation to exhibit agonistic behavior?', or we can rephrase the question to ask, 'Is there a physiological "need" for animals to engage in agonistic interactions?'. To answer this question, we can compare the biological factors initiating agonistic behaviors with the factors initiating other behaviors known to be controlled by endogenous stimulation. Consider the hunger drive, for example. Hunger is a time-dependent phenomenon associated with a host of physiological processes. The intensity of hunger is correlated with time since the last meal or feeding. The motivation to obtain a meal increases over time, regardless of whether or not the animal perceives

Fig. 12.7. Oryx antelope possess formidable horns but do not attempt to gore their opponent. Two bulls begin combat with a display (a) and then fence with the tips of their horns (b). After a pause (c), they clash forehead to forehead (d) and engage in a pushing and shoving match using the horns only to maintain contact (Walther, 1974).

suitable food items in its environment (exogenous stimuli). Spontaneous drives, like hunger, are based on measurable and relatively predictable changes in the physiology of an organism over time.

In contrast to hunger, agonistic behaviors do not appear to be influenced by such short-term, time-related physiological phenomena. The probability that an animal will engage in an agonistic behavior is not related to the variable 'time since the *last* agonistic interaction'. Animals do not normally seek out those external stimuli (i.e. conspecifics) releasing agonistic behavior on a predictable, recurring basis. Rather, aggressive behaviors are typically exhibited by animals when conspecifics are encountered in contexts which threaten their social status or their ability to gain access to desired resources. Natural selection does not favor animals that continually initiate aggressive encounters with conspecifics, just for its own sake. Aggression and related behaviors are too risky to the well-being of individuals to engage in spontaneously in contexts which typically offer little or no benefit to the parties involved.

Some aggression is motivated by the desire to protect possessions (e.g. food, nest sites, mates, territory) and some is motivated by fear. Fear-induced aggression is often misinterpreted as protective aggression. For example, I have heard people talk about how their dog (*C. familiaris*) becomes aggressive when strangers come to their house, as if the dog is protecting them (the owners) or their property. While this may be the case, in many instances the dog may be more interested in protecting itself. The presence of a strange human or unfamiliar animal in a familiar place can evoke fear in an animal and trigger self-protecting threatening or aggressive behaviors.

Intact males tend to be more aggressive than intact females in most species. This is due to the presence of much larger concentrations of androgens (e.g. testosterone) in males and enhanced development of brain centers for aggressive behavior. Castration reduces the frequency and intensity of aggressive behavior in most of our domesticated species but, interestingly, not in dogs (*C. familiaris*) (Le Boeuf, 1970). The aggressiveness of beagles castrated at 40 days of age (prepuberally) was similar to that of intact littermates when competing for estrous females or food (bones) as adults.

Active immunization against gonadotropin-releasing hormone (GnRH) is an alternative to physical castration, does not require surgery and is pain free. Testicular growth and function in intact males are dependent on hormonal activity initiated by the secretion of GnRH in the hypothalamus of the brain. My colleagues at UC Davis (Adams *et al.*, 1996) immunized male beef calves (*B. taurus*) against GnRH by injections of a GnRH preparation at 1, 4 and 6 months of age (i.e. prior to sexual maturity). At approximately 16 months of age, we compared the aggressive behaviors of immunized males to that of surgically castrated steers and intact bulls (Price *et al.*, 2003a). As hypothesized, the immunized males exhibited levels of aggressive behavior roughly comparable to steers and significantly lower than intact bulls (Table 12.1). In addition, the intensity of agonistic behavior, when it occurred, was lower in the steers and immunized males. Hormone assays showed that immunization was only partially achieved in a few of the experimental males. Plasma testosterone levels were higher in these animals than in steers, but still much lower than in intact bulls. Interestingly, these partially immunized males exhibited levels of aggressive behavior comparable to intact males, suggesting that only relatively small, threshold amounts of androgen were needed for male cattle to show 'normal' levels of aggressive behavior.

Table 12.1. Mean total number of butts initiated and sparring bouts observed in 12 groups of eight male beef cattle either surgically castrated, actively immunized against GnRH or left intact (five 20-min observation periods per group).

Treatment	Butts initiated[a]	Sparring bouts[a]
Surgically castrated steers	30	9
Immunocastrated males	41	27
Intact bulls	85	51

[a]Per 20-min observation period.

Factors Which Affect Thresholds for Agonistic Behavior

Even if animals do not spontaneously 'seek out' opportunities to engage in agonistic behavior on a regular, recurring basis, we know that there are important individual and species differences in thresholds for responding once eliciting stimuli are encountered. As we discussed in Chapter 1, differences in response thresholds influence the frequency with which certain behaviors are exhibited.

Genetic (evolutionary) considerations

Genetic influences on thresholds for agonistic behavior can be demonstrated by artificial selection. Guhl *et al.* (1960) artificially selected male and female domestic chickens (*G. domesticus*) for high and low levels of aggressiveness and found significant differences by the first (females) and second (both sexes) generations of selection. The experiment was terminated after the fourth generation since the line differences were so great. In the fourth generation, males from the high aggressiveness line dominated males from the low aggressiveness line in 75 of 80 initial paired encounters. High line females dominated low line females in 75 of 76 initial paired encounters. Heritability estimates obtained for bird aggressiveness averaged about 0.20.

Humans may consciously or inadvertently select for reduced agonistic behavior in captive animals by eliminating animals from the breeding population which are particularly aggressive toward conspecifics and (or) humans. The effectiveness of such selection can be significant if the Guhl *et al.* study described above is any indication of how rapidly artificial selection can bring about phenotypic changes for this trait.

In captivity, humans provide most of the resources that animals need for survival and most captive-breeding programs are designed to minimize competition for mates. This facilitates natural and artificial selection (in captivity) for higher optimal thresholds of agonistic behavior (i.e. less frequent agonistic behavior). However, captive animals tend to be housed at higher densities than in nature, thus increasing the opportunities for agonistic behavior to occur.

Natural selection for thresholds of agonistic behavior occurs in both nature and captivity. Thresholds for agonistic behavior in free-living populations are selected for some optimal level. Individuals can be too aggressive (relatively low threshold) for achieving their maximum fitness. Overly aggressive animals place themselves at greater risk of injury and depleting energy reserves needed for reproduction. Overly aggressive individuals may be less attractive as mates. Guhl *et al.* (1945) showed that dominant female White Leghorn chickens (*G. domesticus*) had greater access to food and roosting sites than subordinates, but they were courted less and received fewer matings from males. On the other hand, too little aggressiveness (relatively high threshold) compromises the ability of animals to obtain resources needed for survival and reproduction.

In a few instances, humans have purposefully increased aggressiveness in selected breeds of domestic animals through artificial selection. Certain breeds of dogs (*C. familiaris*), chickens (*G. domesticus*) and cattle (*Bos*) have been selected for their aggressiveness by genetically lowering thresholds for aggressive behavior. Various breeds of dogs have been bred to protect livestock from predators ('guard dogs') and to provide protection for humans and their property. Dog breeds bred for fighting (e.g. pit bull types), as horrible as this practice is, appear to have a relatively high tolerance of pain and a tendency to continue fighting when injured or fatigued. They also have been selected to suppress or eliminate displays which communicate aggressive motivation or intent; it is to their advantage to attack when least expected. This may, at least partially, explain why so many people have been attacked by these dogs without warning. In addition, it is sometimes impossible to visibly distinguish fighting stock from genetic lines of the same breed which are less aggressive.

Intensive breeding of dogs for the mass pet trade (i.e. 'puppy mills') and the show circuit with only minor concern for behavior and physical soundness has, in some cases, resulted in the proliferation of dogs with undesirable temperament and painful congenital physical defects, which can lower thresholds for aggression toward other dogs and people.

Experience

Past experience in terms of wins or defeats in agonistic encounters can have an important effect on response thresholds. Animals who are accustomed to winning sometimes develop a lower threshold for agonistic behavior because of their greater confidence. An animal's memory of wins and defeats with specific individuals affects their response thresholds in subsequent encounters with the same (familiar) individuals. Animals tend to be more cautious (higher thresholds) when interacting with strangers.

Thresholds for agonistic behavior in dogs (*C. familiaris*) can be influenced by physical and mental abuse, such as painful beatings by their owners and lack of socialization to other dogs and people, especially children. Fear-induced aggression in animals is often exacerbated by a physically abusive caretaker or conspecific. Similarly, thresholds for agonistic behavior in animal–human interactions can be changed through training. Humans can suppress the aggressive tendencies of an animal by asserting the right amount of control or dominance. The technique used and the frequency and intensity of use must be adjusted for the specific animal involved, taking species or breed, temperament, size, sex and other factors into consideration.

In Chapter 4 (p. 44–45), we discussed how bulls (*B. taurus*) reared in social isolation were more aggressive toward humans than animals reared with conspecifics (Price and Wallach, 1990a). The explanation given was that the isolate-reared animals never learned that an act of aggression might incite the victim to retaliate. Consequently, these animals did not develop the normal repertoire of social inhibitions shown by the group-reared animals when interacting with humans. People who regularly handle potentially dangerous animals understand that they must maintain some degree of dominance over the animals under their control. Still, there is no guarantee that their dominance will not be challenged eventually.

By play-fighting, young rats (*R. norvegicus*) learn how to behave in agonistic interactions. Van den Berg *et al.* (1999) found that just 2 weeks of social isolation during adolescence prevented young domestic rats from developing the social skills needed to resolve conflict situations with their peers. Pellis (cited by Pennisi, 2000) explained that rats lacking play experience tended to be uninhibited when approaching conspecifics for the first time. Their boldness is viewed as aggressiveness by a rat which has had a normal 'rough-and-tumble' upbringing and the miscommunication often leads to escalated aggression.

Physical aspects of the rearing environment can play a surprisingly important role in setting thresholds for agonistic behavior. Researchers at the University of Wales in the UK (Prayitno *et al.*, 1997) reported that broiler chickens (*G. domesticus*) were more aggressive when reared in red light than under white, blue or green lighting. Domestic rats (*R. norvegicus*) reared in laboratory cages are considered relatively docile and non-aggressive toward one another. If these same animals are reared in an outdoor pen or a laboratory environment where they can dig and live in burrows, they become much more aggressive toward penmates and strange conspecifics (Nikoletseas and Lore, 1981; Boice and Adams, 1983). The consistent practice of rearing laboratory rats in open laboratory cages is no doubt partly responsible for their docile reputation. Interestingly, the aggressiveness of wild Norway rats (*R. norvegicus*) is also suppressed by being reared in laboratory cages (Price, 1978), again suggesting that the laboratory environment either directly or indirectly plays an important role in setting thresholds for agonistic behavior in this species.

Context

We are all familiar with the concept of 'home field advantage' in sporting events. Like humans, animals tend to be more confident and less inhibited when they are in a familiar environment. As we will see in the next chapter, social dominance and accompanying agonistic behaviors are more likely expressed in an animal's own 'backyard' than in their neighbor's space.

The presence of non-participating conspecifics (i.e. bystanders) can also influence response thresholds. Ylander and Craig (1980) demonstrated that the frequency of agonistic behavior in female chickens (*G. domesticus*) was reduced in the presence of a dominant male. Grandin and Bruning (1992) reported that the presence of sexually mature boars (*S. scrofa*) significantly reduced the frequency and intensity of aggressive behavior when mixing young slaughter-weight pigs (Table 12.2). They also noted that the odor of a dominant boar, as on the floor of an enclosure, could inhibit aggression among younger males (personal communication). Similar studies designed to determine the effect of boar presence on the aggressiveness of adult sows (Barnett *et al.*, 1993; Séguin *et al.*, 2006) have provided conflicting results. Interestingly, application of aerosolized boar androstenone (5-α-androst-16-en-3-one) to the snout and head of newly regrouped prepuberal pigs significantly reduced the duration of their aggressive behaviors (McGlone and Morrow, 1988). In another study, McGlone and Anderson

Table 12.2. Incidence of aggression and visible wounding from mixing 50 slaughter-weight pigs in the presence or absence of three sexually mature boars (from Grandin and Bruning, 1992.)

	Mixed with mature boars[a]	Mixed without mature boars[a]
Total number of fights	34	82
Total number pigs wounded	5	27

[a]Two different groups of 50 pigs were used for each treatment. Each group was observed for 50 min. Scores are the sum total for both groups in each treatment.

Chapter 12

(2002) reported that aggressive behavior between newly weaned pigs was reduced by painting a synthetic 'maternal pheromone' on their feed bin at the time of weaning.

Response thresholds can be influenced by the distribution of resources in the animals' environment. Animals frequently exhibit lower thresholds for agonistic behavior when resources are clumped, rather than distributed uniformly. For example, farm animals are more likely to exhibit aggression toward one another when feeding from a small trough than when feeding on grass in a pasture. Violations of 'personal space' (see Chapter 14) are more frequent in trough feeding because of the close quarters, and competition is more intense because the animals are highly motivated to access the feed due to its limited availability.

Some animals may become more aggressive when certain expectations are not met (i.e. they become frustrated). For example, frustration-induced aggression is sometimes seen when animals are accustomed to being fed at a certain time of day and the caretaker shows up without their feed, or the person takes the animals' feed away once it is delivered. In those circumstances, aggression may be directed at the caretaker, rather than other animals in the group. The animals' aggressiveness may escalate if the caretaker responds with aggression.

The physical context in which social interactions take place can also affect response thresholds for agonistic behavior. Barnett *et al.* (1996) found that newly mixed unfamiliar pigs (*S. scrofa*) were significantly less aggressive when introduced to one another during the hours of darkness. Physical barriers to visual communication can reduce the incidence of agonistic behavior in confined populations. However, that may have more to do with reducing the likelihood that conspecifics will encounter one another than with reducing response thresholds.

Physiological influences

The effect of androgens (e.g. testosterone) on lowering the threshold for agonistic behavior is well known (Hart, 1985). It is best illustrated by the decline in frequency of agonistic behavior following castration and its return to normal levels after testosterone injections ('replacement therapy'). Line *et al.* (1985) studied the effect of pre- and postpubertal castration on the aggressiveness of horses (*E. caballus*) toward conspecifics and humans. Castration significantly reduced the frequency of aggressive behavior in both cases and there was little difference in animals castrated before and after puberty (Fig. 12.8). A similar result was found for pre- and postpubertal castration in the domestic cat (Hart and Barrett, 1973; Hart and Cooper, 1984).

Seasonal changes in the frequency of agonistic behavior correspond to cycles of gonadal activity (i.e. hormone production) related to breeding and reproduction. The seasonal relationship of testosterone production and aggressive behavior in Soay rams (*O. aries*) is illustrated in Fig. 12.9.

Since thresholds for agonistic behavior are under hormonal influence and the brain centers mediating agonistic behavior are sensitive to androgens, it is not surprising that males are generally considered more aggressive than females. As an example, most dog bites on humans are by males, 70% in one survey and 87% in another. In the latter survey, it was found that dog breeds most associated with attacks on humans had the highest proportion of bites by males (German shepherds, 86%; pit bulls, 90%; chow chows, 92% and Rottweilers, 98%) (cited by Lockwood, 1995).

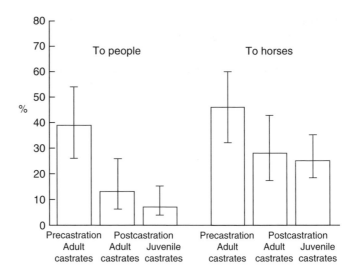

Fig. 12.8. Percentage of horses exhibiting undesirable aggression toward people and other horses both before and after prepubertal or postpubertal castration. Castration reduced aggressiveness toward both humans and conspecifics, regardless of when the castration was performed (Line *et al.*, 1985).

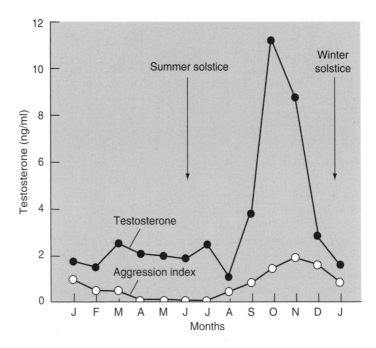

Fig. 12.9. Monthly aggression index and blood plasma concentrations of testosterone in adult Soay rams. Note the marked increase in both variables during the autumn breeding season (Lincoln and Davidson, 1977; figure from Hart, 1985).

Social interactions can also modulate androgen production and secretion, thus affecting thresholds for agonistic behavior. Winning a fight usually increases androgen levels. Interestingly, androgen levels can increase just by watching conspecifics fight. Oliveira *et al.* (2001) demonstrated that male cichlid fish (*Oreochromis mossambicus*) showed increases in testosterone and 11-ketotestosterone, the principal fish androgens, after observing (through a one-way mirror) two neighboring males engaged in a dominance struggle.

Familiarity and mixing of individuals

Animals are less likely to engage in agonistic behaviors with familiar conspecifics than with strangers. Familiar individuals typically have established their dominant–subordinate relationship and thus are less likely than strangers to engage in status-determining behaviors. The presence of an unfamiliar individual is often perceived as a threat to an animal's social status or right to gain access to desired resources (e.g. food, shelter, mates, etc.), thus invoking acts of agonistic behavior.

A circumstance in animal management systems that increases the likelihood of agonistic behavior is on mixing or housing unfamiliar individuals. This often occurs when animals are weaned, bred, regrouped for 'finishing' (placed on a pre-slaughter diet), transported, sold to a new owner, or taken to an auction yard or slaughter facility. Aggression is relatively high during the hours after mixing and then gradually tapers off. Arey (1999) reported that aggression in his pigs (sows) (*S. scrofa*) became stable 7 days after mixing unfamiliar individuals. Mounier *et al.* (2006) found that beef bulls (*B. taurus*) mixed at the beginning of fattening were not only more aggressive during food competition tests than unmixed bulls but also more fearful when physically separated and more stressed during pre-slaughter handling.

It is not uncommon for newly mixed animals to show a sharp reduction in food intake. Stookey and Gonyou (1994) studied the effects of mixing on behavior and weight gain in pigs (*S. scrofa*) during the 'finishing' period prior to slaughter. Animals placed with strangers in unfamiliar enclosures engaged in fighting seven times longer on day 1 (i.e. day of mixing) than control pigs placed in unfamiliar enclosures with familiar penmates. Mixed pigs also gained significantly less weight during the 2 weeks following mixing than control groups (0.77 and 0.87 kg/day/pig, respectively). Other experimental pigs mixed for 24 h and then placed with familiar penmates also gained less weight than the controls (0.80 and 0.87 kg/day/pig, respectively), suggesting that the social stress experienced from being housed with unfamiliar conspecifics during a 24 h period was sufficient to reduce weight gains.

Size of the group and social experience may influence the intensity of aggression shown toward newly introduced conspecifics. Animals maintained in relatively small, stable groups where individuals are clearly recognized are often more reactive than animals maintained in large groups when meeting unfamiliar conspecifics for the first time. Turner *et al.* (2001) found that mixing unacquainted pigs (*S. scrofa*) from groups of 80 individuals resulted in less aggression per animal than mixing unacquainted pigs from groups of 20. Buchwalder and Huber-Eicher (2005) found that individual male turkeys (*M. galopavo*) introduced into small groups ($N = 6$) of similar age birds were the recipient of more aggressive pecks and fights than individuals introduced into larger groups ($N = 30$). The presence and unfamiliarity of strangers introduced into small groups is immediately recognized, whereas strangers introduced into large groups may go unnoticed for a time. Also, animals reared in relatively large groups where individual recognition is seldom attained may become conditioned to ignore or avoid conspecifics on chance meeting as a means of adaptation. They may also develop better social skills relevant to meeting strangers, such as the recognition and sensitivity to physical and behavioral cues associated with aggressive potential or social status (D'Eath and Keeling, 2003).

The reaction of animals to newcomers may be influenced by observing the agonistic interactions of other animals with the stranger. Hogue *et al.* (1996) reported that chickens (*G. domesticus*) were more likely to initiate aggression with an unfamiliar bird if they had seen it defeated by an animal dominant to themselves than if they observed the stranger defeating their dominant penmate.

Temporary mixing of unfamiliar companion animals can occur at horse, dog and cat shows. Horses (*E. caballus*) meet strange conspecifics on trail rides or at other recreational events. Dogs (*C. familiaris*) encounter unfamiliar individuals when taken on walks or to 'dog parks'. Fortunately, humans are often present to control the intensity and duration of such chance agonistic interactions.

The frequency and intensity of aggressive behaviors among strangers can be reduced by simultaneously introducing them (for the first time) to an area unfamiliar to all parties involved. If one of two individuals is familiar with the meeting place, the resident may perceive the stranger as an intruder and a threat to its status and resources.

Dairy cows (*B. taurus*) are frequently exposed to unfamiliar conspecifics after giving birth to their first calf and placed in the milking string, or when moved from one group to another based on stage of lactation. When the heifers on our farm gave birth to their first calf, they were placed in a pasture with the older milk cows. Many of the older cows would gather around the heifer, make threatening displays and engage in sparring. The younger, lighter-weight heifers would invariably lose these battles but, after a day or so, appeared to be accepted into the group and not be given special scrutiny.

A number of techniques have been used to facilitate the acceptance of new individuals into an established group or to facilitate reintroduction of animals that have been away from the group for a period of time. One technique, called 'mutual toweling', has been used when introducing or reintroducing domestic cats (*F. domestica*) to one another. Odors are important to cats in revealing personal and group identity. Towels are used to rub the scent-producing areas of the face, feet, tail and perineum of each cat in the group, effectively transferring individual odors from one cat to another, so that each cat is carrying at least some of the odors of the other individuals. Toweling is conducted once or twice a day and the towels are left in the 'home' areas of the cats until the next 'rub-down'.

Relative size of individuals

Animals similar in size and (or) weaponry are more likely to engage in physical combat with one another than animals very different in size and (or) weaponry. When the differences between individuals are great, prolonged physical combat is seldom necessary to establish a dominant–subordinate relationship. Threat displays of visual, auditory

Table 12.3. Mean (± SD) duration of fighting and frequency of biting in pairs of unfamiliar pigs (*S. scrofa*) of similar or different body weights (from Rushen, 1987).[a]

	Similar weight	Different weight
Fighting duration (s)	591.8 ± 543.9	163.4 ± 118.2
Biting frequency	158.9 ± 97.1	64.1 ± 52.3

[a]Tests were 1 hr in length.

and (or) olfactory cues usually suffice. When animals are evenly matched, physical combat may be the only way to decide who is dominant. Rushen (1987) recorded the aggressive behaviors in pairs of unfamiliar pigs of similar or different body weights (Table 12.3). The mean duration of fighting for similar size pairs in the first hour following introduction was more than 3.5 times greater than for different size pairs. The frequency of biting in the two groups followed a similar pattern (nearly 2.5 times greater in the similar size pairs).

Population size and density

One might assume that the rate of agonistic interactions (per animal and per unit of time) will increase as group size or the density of individuals in a population increases. While this relationship between density and agonistic behavior is sometimes seen in behavioral studies of captive animals, particularly those housed in small enclosures such as cages, it is by no means a predictable pattern. Animals housed in relatively large groups are often more tolerant of one another than animals housed in relatively small groups (Hughes *et al.*, 1997). Because most of our domestic animal species are very social, they tend to stay in relatively close contact with one another, no matter how much space is available. Group size or density may affect how much time certain animals spend being active or when, during the day or night, activity occurs. Subordinate animals may change their active time to avoid interacting with dominant conspecifics. Third-party effects that we talked about earlier (p. 196) may become more common as group size or density increases, thus reducing the frequency of agonistic interactions. Changes in group size or density can affect the type of social organization seen in populations, which can influence thresholds for agonistic behavior. As group size increases, hens (*G. domesticus*) often change their social system from a relatively strict pecking order (dominance hierarchy) to one where unfamiliar conspecifics are tolerated more readily and dominance is determined through direct assessment of aggressive potential and 'status signalling', rather than a remembered dominant–subordinate relationship based on individual recognition (D'Eath and Keeling, 2003). This topic will be revisited in Chapter 14.

13 Social Organization

There are two basic types of social organization in vertebrate species, the dominance hierarchy and territoriality. The common denominator of both of these forms of social organization is 'dominance'. Dominance is an attribute of a relationship between two or more individuals and is achieved and maintained by agonistic behaviors. Dominance is expressed by the ability of an animal to displace another (in space) in competitive and spontaneous interactions. Dominance may involve displacement of another animal from a fixed geographical area (i.e. 'territory'), from non-fixed objects or conspecifics (e.g. food, mates, offspring), or from the area surrounding itself (i.e. 'personal space').

In the dominance hierarchy form of social organization, dominant–subordinate relationships between individuals do not depend on where (in space) the interactions occur. In territoriality, dominance is site specific (i.e. dependent on where in space the interaction takes place). For example, let us say animal A and animal B possess defended territories sharing a common boundary. Animal A is subordinate to animal B when it intrudes into B's territory and animal B is subordinate to animal A when it intrudes into A's territory. If animal A fails to defend its territory to B, the latter will likely take it over and A will be forced to set up a territory elsewhere, or not have one at all.

Dominance Hierarchy

The dominance hierarchy is common among domesticated animals because of their tendency to live in groups or social units and *share* space. It is sometimes referred to as the 'peck order' because some of the first studies on the dominance hierarchy were conducted on chickens (*G. domesticus*) by Schjelderup-Ebbe (1922) and Guhl (1953). The dominance hierarchy is a characteristic of groups, not individuals, even though it is established and maintained by the social behavior of individuals. Some scientists refer to it as a 'relative hierarchy' since it is based on the dominance of individuals relative to one another.

The frequency and intensity of agonistic interactions, particularly aggressive behaviors, are greatest when unfamiliar individuals are first housed together (i.e. during the establishment of dominant–subordinate relationships). Aggressive behavior declines quickly once these relationships become established and its occurrence is relatively infrequent in a stable dominance hierarchy (Fig. 13.1). Stable hierarchies and territories are advantageous because they reduce the incidence of harmful aggression and the time devoted to potentially stressful social interactions, thus freeing up more time and energy for other behaviors important to survival and reproduction.

Variations of the dominance hierarchy

The simplest form of the dominance hierarchy is despotism, in which one individual is dominant over all other members of the group, with no rank distinctions made

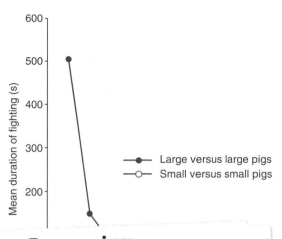

arge and two small,
more and for a
rst 2 h (Rushen,

n artifact of rearing
:annot escape from
d among chickens
atively small cages.
social order among
n male house mice
naining individuals

s in a more or less
'alpha', individual
ninates all but the
roup, the 'omega',
nmon in relatively
re typically stable
ange. Sometimes,
ationships are so
air combinations.
1 the animals are

ther relationships
in Fig. 13.2), no
:s, and when two

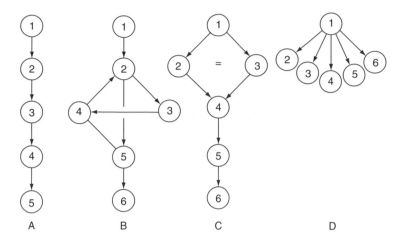

Fig. 13.2. Schematic illustration of types of social hierarchies. A, linear or unidirectional hierarchy; B, linear hierarchy with a triangular relationship; C, linear hierarchy with individuals of equal social status (animals 2 and 3); and D, a despotic system in which one animal dominates all others (Hart, 1985).

individuals appear to be equal in social status (C), the onus is on the observer to determine that these associations are real and not the observer's failure to document adequately the relationships of the animals involved.

Some alpha animals are constantly asserting their 'authority' over members of the group by initiating agonistic displays and aggressive interactions (Fig. 13.3). Other high-ranking animals do not assert their dominance on a frequent basis. Their body language may convey a 'confidence' and 'aloofness' that is, in itself, intimidating, thus inhibiting other group members from challenging their social status. The alpha animal in the group is not necessarily the 'leader' when the group moves from one place to another. This type of 'leadership' is often assumed by the oldest or most experienced animal.

Not all species of domestic animals have a strict dominance hierarchy, even when living in relatively small groups. Although there have been reports of a stable dominance hierarchy among small groups of domestic cats (*F. domestica*) in a laboratory setting (Cole and Shafer, 1966), groups of cats are generally organized in a loose fashion or do not have any recognizable social hierarchy. Dominance is most commonly exhibited by large males. The living conditions of cats (e.g. homes versus farms) and the manner in which they obtain their food (e.g. fed at one location versus hunt for themselves) can influence the nature of the hierarchy. Domestic cats are naturally less social than species with more or less strict linear hierarchies (e.g. chickens and dogs), which may explain this difference.

Captive wolf packs typically form dominance hierarchies with alpha, beta, omega animals, etc. Such packs often consist of unrelated and previously unacquainted animals. In nature, the wolf pack is not organized by a relatively rigid peck order but is usually a family including a breeding pair and their offspring of the previous 1–3 years, or sometimes two or three such families (Mech, 1999). The alpha male and female are merely the breeding animals, the parents of the pack, and dominance

Fig. 13.3. The posture of the dominant wolf (left) expresses challenge and threat. The subordinate wolf shows inhibition and defensive threat (Schenkel, 1967).

contests are very rare, except during competition for food. Parents monopolize food and allocate it to their youngest offspring.

Complete versus partial dominance

In complete or unidirectional dominance, one animal is clearly dominant over another. Subordinates almost never exhibit attacks or threats toward higher-ranking individuals. In partial or bidirectional dominance, less dominant animals sometimes threaten or attack more dominant conspecifics. Young (immature) animals often exhibit partial dominance, particularly when they are in groups of similar sized individuals. Sexually mature animals are more likely to exhibit complete dominance. Beef bulls (*B. taurus*) typically do not show complete dominance toward other similar age individuals until they are 3 years old. Animals raised for meat are commonly maintained in groups of similar size individuals and are sold and slaughtered at a relatively young age. Consequently, animals in these populations are more likely to display partial, rather than complete, dominance. Animals maintained longer, such as companion animals, dairy cattle (*Bos*), laying hens (*G. domesticus*) and the breeding stock of meat-animals, are more likely to display complete dominance.

The establishment of complete dominance leads to a reduction in the frequency and intensity of agonistic behavior, particularly physical combat. Partial dominance does not have this effect. Partial dominance creates the impression that the social status of conspecifics is not fully acknowledged and 'respected'.

Unisexual versus heterosexual hierarchies

In sexually dimorphic species, it is common for adult males and females to establish separate hierarchies. For example, in free-ranging herds of cattle (*Bos*), all adult males dominate all females and seldom show aggression toward them. Mixed-age populations of sexually dimorphic species (like cattle) often exhibit transitional heterosexual hierarchies in which young males are initially dominated by mature females but become dominant to these same females at some point during the maturation process. Adolescent bulls start working their way up through the adult cow hierarchy starting at 1–2 years of age and dominate most cows by about 2–3 years of age. At that time, they occupy the lowest positions in the adult bull hierarchy. Stable heterosexual hierarchies among adult members of the population are most common in species lacking sexual dimorphism.

'Spontaneous' versus competitive interactions

Dominance hierarchies are evident both when animals interact 'spontaneously' in the absence of competition for tangible resources and when they compete for food, mates, shelter, etc. Spontaneous interactions may be initiated simply by a desire for status or to protect one's personal space, a topic discussed in the next chapter. There is generally good correspondence between the hierarchies observed in spontaneous interactions and hierarchies observed when individuals are competing for tangible resources, particularly when the individuals in the population vary in size and aggressive potential. However, variation in rank-based access to *different* resources can be due to individual differences in motivation for those resources or for the same resource at different times. Expression of the hierarchy may even be abandoned in some competitive situations. Banks *et al.* (1979) studied groups of chickens (*G. domesticus*) deprived of food and water for 18 h and then given access to a feeder and a waterer permitting only one bird to feed and drink at a time. The investigators found a significant relationship between social rank and the frequency and duration of feeding. However, the birds did not compete for water. Rather, they drank to their fill on a first-come, first-served basis, regardless of their dominance rank.

Prerequisites for a stable dominance hierarchy

Stability of a dominance hierarchy is dependent on several factors. First, individuals must be able to recognize one another. Individual recognition is necessary for dominant–subordinate relationships to be remembered from one interaction to the next. In most cases, recognition is based on visual, auditory and (or) olfactory cues. White Leghorn chickens (*G. domesticus*) recognize one another primarily by features of the head and neck. The size and shape of the comb is one of the more important cues (Fig. 13.4). Domestic house mice (*M. musculus*) and many other non-primate mammals recognize one another primarily by odor. Sheep make use of sophisticated visual cues from the face and body and olfactory cues from the body and wool of other sheep to recognize individuals (Kendrick, 1994). Stookey and Gonyou (1998) found that familiarity in young pigs (*S. scrofa*) was based on being housed (i.e.

Fig. 13.4. White Leghorn chickens typically recognize one another by features of the head and neck. The size and shape of the comb is particularly important. For example, compare the combs and wattles of these nine red junglefowl cocks, all drawn from actual photos (Collias and Collias, 1996).

reared) together and not on genetic relatedness. Of course, familiarity and individual recognition rapidly decline once group size reaches a certain point. The fact that individual recognition is more readily achieved in smaller populations of animals supports the observation that hierarchies in smaller groups are more stable.

Second, the stability of hierarchies is dependent on the continued reinforcement of dominant–subordinate relationships by agonistic displays. Displays are energetically 'cheap' and less stressful compared to physical combat and thus are a more efficient way to remind subordinates of their social status. Older, high-status individuals can sometimes use agonistic displays to retain their high rank in the group well beyond their prime physical age. In some species, older 'prestigious' individuals will sometimes direct intense dominance displays toward low-ranking individuals in the presence of more able contenders for their high-rank position. They are also reluctant to get involved in fights. Eventually, they are tested physically by more vigorous individuals and their high-rank status is lost. Older alpha individuals can be demoted to the omega position in a short period of time, often with disastrous consequences, including death. I have heard numerous anecdotal accounts of high-ranking individuals ruthlessly attacked by subordinates when handicapped or incapacitated by injury, sickness and seizures, and even immediately after being killed by a hunter.

A third factor contributing to the stability of the dominance hierarchy is group integrity. Introduction of one or more unfamiliar individuals into the group increases the frequency and intensity of agonistic interactions until the social status of the newcomer(s) is decided. Guhl (1953) found that when several new hens (*G. domesticus*) were added to an established group of chickens, the newcomers initially isolated themselves from the resident birds and did not assimilate into the social order for several days or weeks. Moore *et al.* (1993) reported that when 10 new sows and gilts (*S. scrofa*) were added to an established group of 30 sows and gilts, the newcomers socially isolated themselves from the residents at one end of their enclosure. Complete social integration of the newcomers and residents (random intermingling of the two groups) was observed in only one of three pens after 3 weeks had passed. Arnold and Pahl (1974) found that it took a similar amount of time for two unfamiliar flocks of sheep (*O. aries*) to integrate after being placed in the same pasture. Of course, spatial integration (voluntary intermingling between groups) and social assimilation into the dominance hierarchy may not necessarily occur at the same time. Assimilation into the social order usually follows the spatial integration of newcomers and residents. Furthermore, the time required for newcomers to become part of the social order will vary greatly, based on such factors as the size and age structure of the two populations, the amount of space available to the animals, the degree of competition for resources and so on.

Stability of the hierarchy and levels of aggressiveness are usually not affected by removing individuals from the group. However, reinstating individuals who have been temporarily removed from the group may result in increased aggression and a temporary loss in stability, as in the case of newcomer introductions. Time away from the group is an important variable, in this case, since it can determine whether or not the identity of previous group members and their social rank is remembered by the residents. Ewbank and Meese (1971) found that dominant pigs (*S. scrofa*) could be absent from groups of eight for up to 25 days without being the target of aggression and disrupting the stability of the hierarchy on their return. Arey (1999) reported that for groups of six pigs, individuals could be removed for 6 weeks without disruption of the social hierarchy on their return. Schjelderup-Ebbe (1935) found that chickens (*G. domesticus*) failed to recognize penmates after 2 weeks of separation. Species, degree of familiarity prior to separation and prior social rank, as well as other factors, may determine whether an individual assumes its previous rank on reinstatement without incident.

Benefits of the dominance hierarchy

The primary benefit of the dominance hierarchy is that it stabilizes the social relationships of group-living animals by minimizing psychological and physical stress related to competitive interactions. Individuals in groups with a stable dominance hierarchy acknowledge and 'respect' the 'right of access' of animals with higher social status than themselves.

Stability of the hierarchy is reflected in the productivity of the group. Stability not only minimizes stress but also improves survival, growth and reproduction. Guhl and Allee (1944) compared the productivity of small flocks of domestic chickens (*G. domesticus*) in which individuals were either systematically exchanged at frequent intervals (i.e. hierarchies never stabilized) or membership was not changed (i.e. stable hierarchies). The unstabilized groups exhibited more pecks per unit time, ate less feed, gained less weight and produced fewer eggs than the stable groups. Brakel and Leis (1976) added 4 unfamiliar dairy cows (*B. taurus*) to groups of 24 cows to study the effects of group transfer on agonistic behavior and milk production. During the 1h period following transfer on day 1, each of the newly introduced cows was involved in approximately twice as many agonistic interactions as each of the resident cows. The number of agonistic interactions involving transferred cows was reduced sharply on day 2, but remained higher through the first week. Milk yield of the transferred cows dropped 3% on day 1, while the non-transferred animals showed no change in milk production. Milk yield of the transferred cows on day 2 and thereafter was not significantly different than pre-transfer levels.

Factors which influence the attainment of dominance

We saw in the last chapter how factors such as breed, body size/weight, age, sex, experience and weapons (e.g. horns, teeth, etc.) can affect the frequency and intensity of agonistic interactions with conspecifics. Not surprisingly, these same factors also influence the attainment of dominance rank. As noted previously, dominant–subordinate relationships are quickly established in animals with substantial differences in body size and weight. One factor that can compensate for smaller size is the presence (or absence) of weapons such as horns. There are breeds within certain species of livestock (e.g. cattle, sheep) that possess horns, while other breeds have been genetically selected to be hornless ('polled'). Bouissou (1972) studied how the presence or absence of horns in dairy heifers (*B. taurus*) interacted with body weight in the establishment and expression of dominant–subordinate relationships. There were four groups of animals in her study. Two of the four groups consisted of individuals averaging 50 kg heavier than individuals in the other two groups. All of the individuals in one of the 'heavy' and one of the 'light' groups were dehorned. Her findings were as follows:

- When heifers of similar weight were paired, the one having horns was dominant 85% of the time.
- When both subjects had horns but one weighed 25 kg or more than the other, the heavier animal was dominant 90% of the time.
- When both subjects did not have horns, weight was not significant in determining dominance.
- Heavier animals with horns dominated lighter, hornless animals 98% of the time.
- Lighter animals with horns dominated heavier, hornless animals 75% of the time.

Bouissou concluded that horns are of major importance in the *establishment* of dominant–subordinate relationships among dairy heifers and that differences in body weight (of the magnitude in her study) are of secondary importance.

In another study, Bouissou (1964) reported that dehorning some heifers in an established group had no effect on their dominance hierarchy. The dominant heifer in each of two groups of seven individuals was dehorned and, 1 month later, all the remaining heifers except the most subordinate were dehorned. Only 1 of 41 possible interindividual relationships was changed. Bouissou concluded that once dominant–subordinate relations were established, the social experience of individual animals could supersede the advantages conferred by the presence of horns. The lack of aggressiveness selected for in dairy breeds of cattle (*B. taurus*) may have elevated the relative importance of social experience in preserving social stability. It is important to note that Bouissou's study was conducted with relatively small groups of heifers in which dehorning would not prevent individual recognition.

Advantages of dominance rank

High-ranking individuals have greater access to resources in short supply. When food, shelter, mates, space, etc. are limited, dominant animals generally have a distinct advantage. They are often the first in the group to feed and breed and prime space is theirs for the taking. In any given breeding season, dominant animals leave more offspring than subordinates. In many highly social species (e.g. marmosets, wolves), only the dominant female produces offspring. Table 13.1 illustrates the greater mating success of dominant male domestic turkeys (*M. gallopavo*). Dominant males attained more successful matings with fewer mounts and fewer of their mounts were interfered with by lower-ranking males. Osterhoff (personal communication) studied the reproductive success of four beef bulls (*B. taurus*) of different ages maintained together in the same breeding pasture over a period of 5 years. Paternity of the calves born in each year was determined by blood typing. In years 1 and 2, bull A was dominant and sired approximately 75% of the calves (Table 13.2). Bull A lost his dominant status to bull B in year 3 of the study and in that year sired only 16% of the calves. Bull B was dominant in years 3 and 4 and sired 64% and 73% of the progeny in those years, respectively. In year 5, bull C (now 8 years of age) became dominant and sired 63% of the progeny, while bull B's success was reduced to 25%.

You may wonder why the dominant bulls in the study cited above did not sire more than about two-thirds to three-quarters of the calves in each year. In group-living domestic animals, it is not uncommon for two or more females to be in estrus at any given time. Under these circumstances, subordinate males will sometimes find the opportunity to mate with unattended females while the dominant male is actively engaged in mating or driving away other males. In a study on multiple-male harem

Table 13.1. Social rank and mating activity in male domestic turkeys (from Hale, 1953).

Social rank	Total number of mountings	Total number of matings	Percentage of mountings followed by interference from other birds
1 (top)	310	171	8
2	335	114	47
3	334	73	68

Table 13.2. Relation of social dominance to reproductive success among four free-ranging beef bulls over a 5-year period (from Osterhoff, personal communication).

	Bull A	Bull B	Bull C	Bull D
Age at start (years)	10	4	3	2
Calves sired (%)				
Year 1	72[a]	16	7	5
Year 2	76[a]	18	6	Absent
Year 3	12	64[a]	12	12
Year 4	Absent	73[a]	12	15
Year 5	Absent	25	63[a]	12

[a]Dominant bull in that year.

bands of feral horses (*E. caballus*), it was found that only half of the bands had foals sired by one male only (Bowling and Touchberry, 1990).

Dominance does not always ensure greater productivity. Dominant males may not leave the most offspring if they have poor quality semen or if they possess a genital abnormality. Dominance rank is not correlated with milk production in dairy cattle (*B. taurus*) (nine different studies). Dairy cows are typically housed and managed to minimize competition for feed and water. This allows animal breeders to select cattle artificially for milk production without modifying thresholds for aggressive behavior. High-producing dairy cows consume a great deal of energy in milk production. It is not adaptive for them to expend large amounts of energy in dominance rank struggles.

Reducing the impact of dominant animals in captive populations

We have seen how dominant animals will use agonistic behavior and their social status to gain access to resources in short supply. What can we do to lessen the negative impact of dominant individuals on the population as a whole? This is a particularly important question when resources are limited and the health and well-being of subordinate animals could be jeopardized. Removal of dominant individuals from the group is only a temporary fix and individual housing can deny animals normal social interactions, not to mention the added cost of facilities.

Managing captive animals to reduce competition is one way to lessen the impact of dominant animals on their subordinate conspecifics. The relative distribution of resources over available space can be critical, particularly when animals are housed in relatively small areas or pens. For example, feeding time is often a period of intense competition. Hungry subordinates prevented from accessing food can become frustrated and show aggression toward more dominant neighbors. King (1965) studied the interactions of hungry domestic cockerels (male chickens) (*G. domesticus*) when: (i) feed was distributed widely over the floor of their pens; (ii) feed was presented in circular feeders so that all birds could feed at one time if they crowded together; and (iii) feed was available to only one bird at a time. Floor feeding resulted in infrequent aggressive behavior. Aggressive acts increased 36 times when the birds were fed in the circular feeders and 5% of the subordinates attacked or threatened dominant penmates. When only one cockerel could feed at a time, aggression was rampant and peck-order violations (subordinates attacked animals dominant to themselves) were observed in about 50% of the pairs with a known dominant–subordinate relationship. Interestingly, King found that stable dominance relationships were quickly reinstated after termination of the experiment.

Improving the quality of housing by constructing physical barriers to visual communication can sometimes reduce the frequency of agonistic interactions and, thus, social stress. Subordinate animals quickly learn to use these visual barriers to avoid dominant individuals. The physical design of artificial feeding devices can provide subordinate animals with a measure of protection from dominant penmates. Bouissou (1970) found that partitions (i.e. barriers) protecting the head of subordinate dairy heifers (*B. taurus*) while feeding were of particular importance. In 3-min tests with dominant and subordinate cows feeding side-by-side, subordinates fed for about 2 min, on average, when head partitions were in place on the feed trough and less than 45 s in their absence (Fig. 13.5). In fact, a single bar across the top of the trough increased the feeding time of subordinates to 84 s. Although the subordinate animals could clearly see their dominant penmate, they sensed a degree of protection offered

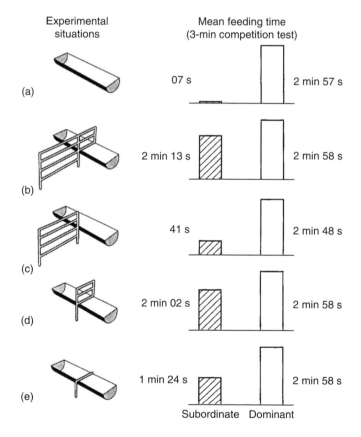

Fig. 13.5. Effect of different types of physical barriers on side-by-side feeding times of hungry cattle classified by dominant status. Barriers providing complete protection of the head resulted in the highest feeding times for subordinates (Bouissou, 1970).

by the bars. Herlin and Frank (2007) reported a similar finding with dairy cows (*B. taurus*). Feeding stations with protective gates which closed when a cow was being fed reduced by 65% the number of aggressive acts and displacements received, compared to animals in stations with the protective gates left open. As a result, cows obtained their daily feed rations with fewer visits to the feeding area. Similar results were noted in horses (*E. caballus*) by one of my former students, Lisa Holmes (Holmes *et al.*, 1987). Either solid or wire-mesh partitions afixed to a feed trough significantly increased the feeding time of hungry subordinate mares (Fig. 13.6). We also found increased feeding times of subordinate mares when a fly mask with

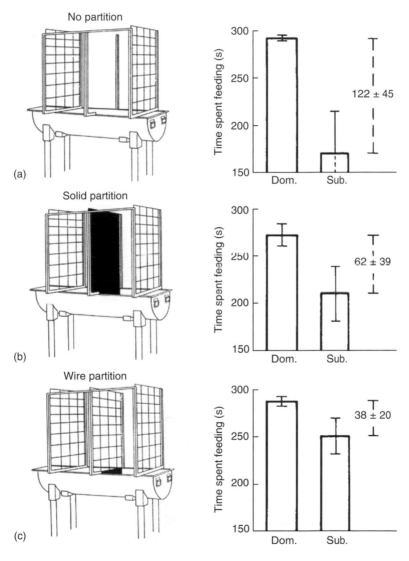

Fig. 13.6. Mean (± SE) time spent feeding and mean (± SE) difference in time spent feeding by dominant and subordinate mares (tested in pairs) under three treatment conditions (Holmes *et al.*, 1987).

black cloth covering the eye region was placed on their dominant penmates (unpublished study). Changes in the body language of dominant mares and the absence of direct eye contact between dominant and subordinate individuals appeared to increase the 'confidence' of subordinate mares when feeding in close proximity to their dominant penmates.

Territoriality

Territorial species are not as likely candidates for domestication as group-living species organized by the dominance hierarchy, because of their greater requirements for space. Consequently, humans have not domesticated very many territorial species, the domestic cat (*F. domestica*) being the primary exception. However, it is important to understand territoriality as a concept or principle of animal behavior because of its importance in the animal world.

Definition and characteristics

Territoriality can be defined as 'the process by which a geographical area is delineated and maintained by the behavior of animals acting individually or in groups as an area of exclusive use of one or more resources'. Let me explain why I have chosen this definition but, before doing so, I would like to note that animals predisposed for territoriality do not necessarily defend all of the space they use in their daily activities. The total area used by an animal is referred to as its 'home range'.

One normally thinks of a territory as a single, fixed geographical area that is defended by the territory holder. While this is the general rule, the geographical area defended can be broken up into one or more parcels and the area defended may change over time. Some animals defend the same area year after year. Oftentimes, song sparrows (*Melospiza melodia*) use the same territory for mating, nesting and offspring care for several years. Weatherhead (1995) found that 55% of territorial male red-winged blackbirds (*A. phoeniceus*) returned, on their annual spring migration, to the same general area over several years and 81% of those reoccupied the exact same site. In addition, 40% of the breeding females returned to the same general area and 76% of those returned to the same marsh. In contrast, transient or temporary territories may be established around a variable or changing resource. For example, the feeding territories of nectar-feeding birds (e.g hummingbirds, sunbirds, etc.) are discontinuous in that sources of nectar are not distributed uniformly over the animal's environment but are patchy. Furthermore, flowers from which nectar is obtained are relatively short-lived and, thus, territories set up around these food resources are but temporary and short-lived.

We generally think of individuals holding territories. In nature, wild female Norway rats (*R. norvegicus*) maintain nesting territories. Breeding male ring-necked pheasants (*P. colchicus*) maintain a territory from which all other males are excluded. Territories of the domestic house cat (*F. domestica*) are held by individuals, not groups. However, there are species which exhibit group territoriality. Mated pairs of birds will defend nest sites and nestlings. Intruders are readily detected and driven away. Wolf (*C. lupus*) packs defend their territories as a group. Unfamiliar wolves are driven off or killed by the pack.

The 'exclusive use' clause in my definition of territoriality implies that the geographical area involved (i.e. the territory) is defended against all intruders. For this reason, many people have chosen to define a territory as 'a defended area'. However, territory holders do not necessarily defend their territories against all intruders. The sex and (or) age of the intruder may be important in determining when and how the area is defended. The intensity of the defense often changes over time, reflecting seasonal changes in thresholds for agonistic behavior. Male song sparrows (*M. melodia*) defend the area in which they live only during the breeding season. The same area may be occupied at other times but not defended.

'Exclusive use' implies that territories are spatially separated (i.e. do not overlap). This is not always the case; some territories are separated in time. Two or more unrelated individuals may share the same territory by being active on the site at different times of the day or night. Leyhausen (1979) reported this phenomenon in feral house cats (*F. domestica*). Nectar-feeding birds may share feeding territories by default, accessing and defending the same flowering plants at different times of the day.

Defense may be active and may involve overt aggression between a territory holder and an intruder. Responses to intruders and neighboring territory holders often vary, depending on the threat posed to the defender. If the relative aggressiveness of a neighboring territory holder is known, the defender will respond accordingly. Male song sparrows (*M. melodia*) engage territory neighbors in countersinging bouts, which contain graded signals of aggression in the form of switching song types or song-type variants (Hyman and Hughes, 2006).

Defense may also occur indirectly and not involve a direct confrontation. Chemical cues (e.g. scent marking), with slower fade-out times than other communicatory signals, may be used to intimidate potential intruders in the absence of the territory holder. Because wolf (*C. lupus*) pack territories are relatively large, it is easy for intruders to enter a territory without being detected. Trespassing is reduced when the resident pack deposits urine on trees, rocks and other objects along the boundary of its territory. Such urine marks serve as 'no trespassing signs' to neighboring wolf packs and has an intimidating effect on would-be intruders. Studies conducted in winter with snow on the ground have shown that wolves urine mark the perimeter of their territory more than twice as much as the territory's center (Peters and Mech, 1975; Fig. 13.7).

Types of territories

Territories may be categorized by function since territories established and maintained by different species often serve different functions.

Mating territories

Mating territories are established and maintained for the purpose of courtship and mating. The best example is the 'lek', traditional male-defended mating areas used by such species as the Uganda kob antelope (*Kobus kob thomasi*), fallow deer (*D. dama*), sage grouse (*Centrocercus urophasianus*), ruff (*Philomachus pugnax*), etc. (Wiley,

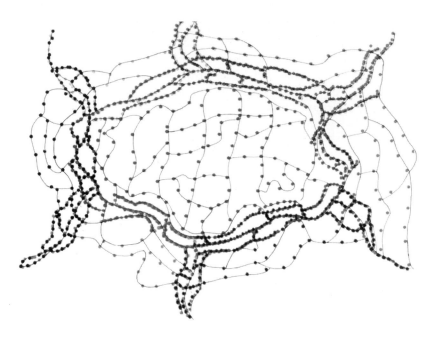

Fig. 13.7. Distribution of urine marks (in the snow) throughout the territory of one wolf pack and the perimeter of the territories of six neighboring wolf packs. Scent marking is more frequent at the edges of territories, as if the wolves were posting 'no trespassing signs'. Travel routes are simulated (Peters and Mech, 1975).

1991; Clutton-Brock *et al.*, 1993). Lekking occurs in species where large individual territories would be less effective in attracting and holding mates. Groups of male Uganda kob and fallow deer defend individual, contiguous mating territories often no more than 10 m in diameter (Fig. 13.8). Estrous females leave their resident herds and travel to a nearby lek area to mate. Female fallow deer prefer to enter the territories of males with harems of females since they are attracted to female groups on the lek (Clutton-Brock and McComb, 1993). Subordinate males unable to establish territories on the lek seldom have the opportunity to breed or, if they do, they mate at a much lower frequency.

Reproductive territories

Reproductive territories serve the function of mating, nesting and offspring care. Feeding occurs outside this area. Colonial nesting species such as gulls and other seabirds frequently defend reproductive territories. One member of the mated pair guards the nest and offspring while its mate is feeding.

Feeding–reproductive territories

Feeding–reproductive territories serve the functions of reproductive territories plus feeding. They are typically much larger than either mating or reproductive territories.

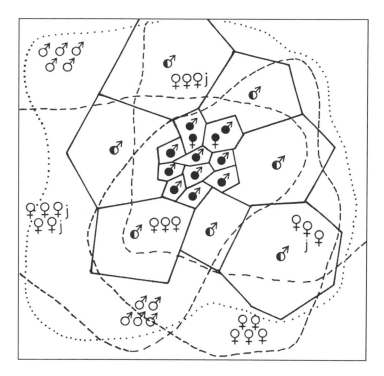

Fig. 13.8. Lek mating territories of the Uganda kob are very small and locally aggregated in dense clusters. Females in estrus leave their groups and enter the lek, where they mate with one or several males. Solid ♂ = male on the lek; partially solid ♂ = male on a 'single' territory; solid ♀ = estrous female on the lek; open ♀♀♀ – female group (Leuthold, 1977).

Many songbird species hold feeding–reproductive territories during the breeding season. All of their basic needs are met in these territories.

Feeding territories

Feeding territories serve the function of feeding only. Nectar-feeding bird species defend specific flowering plants when nectar is available. Young piglets defend their preferred teat on the sow as if it was a feeding territory.

Prerequisites for a stable territorial system

As in the establishment of a dominance hierarchy, the establishment of territories is usually accompanied by frequent, and sometimes intense, agonistic behavior. If space is limited, only the most dominant animals will be successful. The frequency and intensity of agonistic interactions decline sharply once territories have been established in a given area. Members of the population unsuccessful in establishing territories (usually the young or less vigorous individuals) stand ready to fill the void if a territory holder is eliminated through death or other causes.

Stability of the territorial system is dependent on three factors. First, animals must become intimately familiar with the area they are defending. More specifically, they must be able to discriminate between their own territory and the territories of their neighbors. Site recognition is achieved through visual and chemical cues. Physical characteristics of the territory are learned.

Second, stability of the territorial system is dependent on continual advertisement of ownership through visual, auditory or chemical displays. Male songbirds advertise their territories by singing. The male ring-necked pheasant (*P. colchicus*) not only crows to advertise his territory but periodically stands tall and rapidly flaps his wings to provide both visual and auditory evidence of his presence. Domestic cats (*F. domestica*) urine mark. These displays reinforce an animal's rights to its territory and are less stressful, more energy efficient and safe to exhibit than direct confrontations with conspecifics.

A third factor contributing to stability is the integrity of the population. Frequent trespassing by conspecifics and challenges to one's territorial rights create an increase in agonistic behavior and jeopardizes the stability of the territorial system. The loss of a territorial holder creates short-term instability as neighboring territory holders or surplus individuals attempt to take over the voided space. This phenomenon was clearly demonstrated in a study on red grouse (*Lagopus lagopus scoticus*), in which territorial males were artificially removed from the population (Watson and Jenkins, 1968).

Evolution of territoriality

You may ask why the territorial system of social organization has evolved? Why is the dominance hierarchy not sufficient to reduce competition-driven agonistic behavior and to provide social stability in animal populations? It is clear that competition for resources in short supply is a prerequisite for the evolution of both the dominance hierarchy and territoriality. The establishment and maintenance of a territory requires a considerable investment of time and energy, not to mention the risks to health and well-being. If nothing was gained by holding a territory, there would be no selective advantage of territory holders and they would leave no more reproducing offspring for the next generation than animals without territories. So, what are the potential gains from territoriality that cannot be attained through dominance hierarchies?

The obvious difference between territoriality and the dominance hierarchy is that territoriality spaces out individuals over their available habitat. There are three commonly recognized benefits from spacing out individuals relative to sharing space:

- Spacing reduces the overexploitation of resources (e.g. food, shelter) in localized areas.
- Spacing reduces the likelihood of conspecifics interfering with the reproductive behaviors (e.g. mating, parental care) of the territory holder. It offers more privacy.
- Spacing reduces losses due to disease, parasites and predators.

These benefits to survival and reproduction are more likely to be obtained in the territorial system of social organization than under dominance hierarchies. If these benefits outweigh the 'costs' of establishing and maintaining a territory, there will be selective pressure for territoriality to evolve.

In some cases, the cost to the territorial animal will be too great to justify having a territory. The *defendability* of a territory from a cost-benefit standpoint must have a net positive effect on fitness. For example, gulls, seals and other ocean-dwelling species feed over a large area. Their preferred foods can be here today and somewhere else tomorrow. It would not make 'economic' sense for such species to attempt to establish and maintain feeding territories when the food is so transitory and the required area is so large. The costs would be too great relative to the gains. Wild house mice (*M. musculus*) are typically territorial and males will defend their territories vigorously. Butler (1980) found that as the density of house mice populations increases, there comes a point where the animals transition to a dominance hierarchy form of social organization. It is as if the 'cost' of defending a territory becomes too 'expensive' (i.e. interactions with intruders become too frequent) so the animals shift to a less expensive form of social organization in which they engage in sharing space. These examples point out that the propensity of certain species to defend territories is not *necessarily* fixed or unmodifiable. The form of social organization traditionally adopted can be modified, depending on the circumstances under which the animals are living, a concept of special importance to persons housing territorial species in captivity.

14 Personal Space and Social Dynamics

Definition and Measurement of Personal Space

Personal space can be defined as 'the area surrounding an individual which, if encroached upon by another individual, results in an agonistic response (e.g. aggression, avoidance) by the former'. It can be thought of as the area within a balloon surrounding an individual. The personal space balloon can shrink, or even collapse, under certain circumstances, such as when animals rest, engage in mating, care for their offspring, or when they are herded. The balloon can expand in other situations, as when animals are competing for scarce resources, defending a territory, or meeting an unfamiliar conspecific. Because circumstances and an animal's internal state are changeable, personal space is a dynamic variable which helps us visualize and discuss the ever-changing spatial relationships of animals in their environment.

The concept of personal space also helps us understand the dynamics of social interactions in social groups of animals (McBride, 1971). Subordinate individuals typically avoid the personal space of dominant conspecifics, and dominant animals experience the freedom of invading the personal space of subordinates. Ideally, personal space requirements should be considered in estimating the maximum density of animal populations in captivity. Unfortunately, this is difficult to put into practice since the size of the personal space of an individual at any given moment depends on so many changeable factors.

A related term which is much used in the literature, 'interindividual distance' (or 'individual distance'), refers to how close one animal can get to another before the latter responds with some form of agonistic behavior (Hediger, 1950). The response may be an act of aggression or simply avoidance (active or passive). Interindividual distance can be highly variable, for the reasons we discussed above in reference to the size of an animal's personal space. It provides a measure of the size of an animal's personal space at any given time and can be likened to the distance from an individual to the outer edge of its personal space (i.e. the radius of the personal space balloon). Complicating the picture is the fact that the radius is not equal on all sides of the animal but tends to be greater in the direction the animal is facing. The concept of interindividual distance in gregarious animals reflects a balance between attraction to conspecifics and repulsion, as when personal space is invaded. Interindividual distance also reflects an animal's threshold for agonistic behavior. Most studies on personal space use interindividual distance as their unit of size because it is more practical to measure than the defended area around an individual.

Some Factors Influencing the Size of Personal Space

In general, there is an inverse relationship between thresholds for agonistic behavior and personal space requirements (i.e. the lower the threshold, the larger the personal space). Factors influencing the size of the personal space of animals are similar to those affecting thresholds for agonistic behavior (e.g. species, sex, age, hormonal state, social status and the specific circumstances surrounding each social interaction). Immature individuals tend to have smaller personal spaces than adults. Females tend to have smaller personal spaces than males. Size of the personal space is influenced by stage of the reproductive cycle or time of the year in seasonally breeding species. Personal space requirements rise with testosterone levels. The context of the interaction is also important. Adult males tend to have larger personal spaces when interacting with other adult males than when interacting with females or immature males. Marler (1956) found that male chaffinches (*Fringilla coelebs*) (a sparrow-like bird) in non-reproductive condition have an average interindividual distance of about 20 cm when interacting with male conspecifics and about 10 cm when interacting with females. Domestic ungulates tend to relax or ignore their personal space requirements when they are being herded or moved as a group.

The nature of the agonistic response exhibited by an animal can determine the size of its personal space. When two familiar individuals meet, their dominant–subordinate relationship dictates who will likely be the aggressor and who will likely exhibit avoidance behaviors (either active or passive). In such circumstances, the more aggressive (i.e. dominant) individual tends to have a smaller personal space than the avoiding (i.e. submissive) animal. The avoiding animal initiates avoidance behavior at a greater distance than the dominant animal initiates aggressive behavior. McBride (1964) recorded the interindividual distances of unfamiliar female chickens (*G. domesticus*) when meeting for the first time in pairs. He found that the average interindividual distance of hens initiating avoidance behavior (e.g. avoiding direct visual contact by turning the head and (or) body away from the other party) was about 76 cm. In contrast, offensive or defensive aggressive behaviors were not initiated until the birds averaged about 38 cm from one another. Keeling and Duncan (1989) recorded the use of space and interindividual distances of multiple groups of three chickens (*G. domesticus*) at two densities (0.65 m^2 and 1.3 m^2). They found that interindividual distances were greatest when one of the animals was the most dominant bird. It seems reasonable that subordinates are more likely to keep a 'safe' distance from the alpha animal than from other conspecifics.

Enclosure size can also influence interindividual distances. Keeling and Duncan (1989) reported that in 1.3 m^2 pens, chickens (*G. domesticus*) tended to orientate away from each other at distances greater than 35 cm and toward each other at distances less than this, whereas in smaller pens (0.65 m^2) this change in orientation occurred at about 25 cm. It appeared that the birds were attracted to one another in the larger pens (i.e. ample space had been provided), while a degree of repulsion was experienced in the smaller pens (i.e. available space was insufficient). They suggested that this might reflect a degree of flexibility in the interindividual spacing mechanism. The idea that interindividual distance may decrease with increasing density seems plausible, considering that interindividual distances can be nearly non-existent at different times, such as when animals are herded, during mating, etc.

This discussion highlights the complexities associated with the concepts of personal space and interindividual distance. In spite of these complexities, knowledge of personal space requirements is important in designing facilities for captive animals and developing management practices that minimize artificially forced interactions between animals under our care.

Social Dynamics and Food-Animal Well-Being

The intensification of animal agriculture has had a dramatic effect on husbandry techniques and animal productivity but has raised a multitude of questions regarding the effects on the physical and psychological well-being of confined animals. For example, should density recommendations be based on the space requirements of the dominant, intermediate-rank or lowest ranking individuals in a population? If subordinate animals have the greatest space requirements in any given environment, should their needs dictate density recommendations?

Density-dependent effects

Studies have been conducted on the effects of density on animal productivity, reproduction and social stress by systematically manipulating the number of animals in populations, the space available to them, or both. Absolute density refers to the number of individuals in the population (group size) without taking space per individual into consideration. Relative density refers to the number of animals per unit of space. The effect of changes in absolute and relative density can be confounded by the social behavior of animals and the quality of the available space. Unless placed in small cages or pens, animals do not typically distribute themselves uniformly over the space available. Animals with strong social affinities (e.g. sheep, goats, chickens, etc.) are sometimes found in close proximity to one another, even when ample space is provided. As a case in point, Febrer *et al.* (2006) examined the behavior of broiler chickens (*G. domesticus*) reared at five different maximum (relative) densities currently represented in commercial broiler chicken houses in the UK, Denmark and other European countries. In particular, they examined the spatial distribution of chickens when neither feeding nor drinking to determine if they met certain criteria for: (i) being attracted to one another or (ii) showing avoidance behaviors. The results showed that the chickens found the close proximity of conspecifics more attractive than aversive at all of the densities studied. The birds would consistently cluster together, even though additional space was provided. With few exceptions, the behavior of the chickens did not change as density increased. This challenges the position that housing birds at the maximum densities currently allowed is unacceptably aversive and that the birds would prefer to have more space available to them.

It is important to note that broiler chickens are normally slaughtered as juveniles at about 6 weeks of age, when they weigh 2–3 kg. Studies have shown that broiler chickens do not normally exhibit avoidance behaviors toward one another until they are 6–10 weeks of age, when they become reproductively mature and start to establish dominance hierarchies. It follows that a different result might have been obtained in Febrer's study (cited above) had it been conducted on more mature birds. However, considering the fact that broiler chickens, like many meat animals, are routinely slaughtered before reaching maturity, we cannot assume that conclusions reached in studies on adult animals are necessarily valid when setting guidelines for the rearing and management of their younger counterparts.

Even if young domestic food animals prefer to cluster rather than distribute themselves evenly over available space, we cannot assume that their well-being will not be compromised in some way. Febrer *et al.* (2006) also addressed that point in their study on broiler chickens and reported that mortality, number of birds culled and leg problems did not increase with increasing density. They reminded the reader, however, that in the broiler industry, environmental issues affecting health and welfare were more common at high-density stocking rates. Levels of ammonia in broiler houses and high humidity can readily build up if the housing units are not managed properly. One can argue that density-dependent *environmental* issues may currently be more important than density-dependent effects on behavior when assessing broiler chicken productivity and welfare.

Of course, if recommended maximum densities are ignored, populations can reach a point where the animals will experience negative effects on productivity and welfare. Stress-inducing agonistic interactions may become more important at such high densities and subordinate individuals will have greater difficulty avoiding aggressive penmates.

Since density and productivity are not *necessarily* correlated at reasonable density levels, could it be that the frequency and intensity of agonistic interactions experienced by individuals within a population serve as a better predictor of stress and reduced productivity than density, per se? In a study with female pigs (*S. scrofa*), Mendl *et al.* (1992) found that there were more adverse physiological effects (denoting stress) of being persistently aggressive and often displaced than either being aggressive and usually winning or being unaggressive and involved in few interactions. In addition, the aggressive, unsuccessful pigs gave birth to young with low birth weights. Al-Rawi *et al.* (1976) investigated the frequency of agonistic interactions and three measures of productivity (age at sexual maturity, egg production and mortality) in five commercial strains of White Leghorn chickens (*G. domesticus*) housed in cages at three different absolute densities (4, 8 and 14 birds per cage). Cage and feeder space per bird (relative density) were held constant, while absolute density (group size) was varied. The results indicated that the frequency of agonistic interactions and mortality increased with increasing group size, while egg production declined (Table 14.1). Age at sexual maturity was not affected. However, strain differences in frequency of agonistic interactions did not correlate with productivity for each of the three group sizes (Table 14.2). The less aggressive strains were not necessarily the most productive and the most aggressive strains were not necessarily the least productive. In fact, the least aggressive and most aggressive strains (1 and 5, respectively) showed almost

Table 14.1. Mean frequencies of agonistic behavior, egg production and mortality for three absolute densities of caged chickens (*G. domesticus*) (from Al-Rawi *et al.*, 1976).[a]

| No. birds per cage | Agonistic acts per hen per h | | Hen-housed egg production (%)[b] | Mortality (%) |
	During all activities	During ingestive and active periods		
4	8.9	11.1	67.0	5.8
8	14.6	19.6	64.2	7.8
14	12.5	18.7	57.2	12.3

[a]Space per chicken (relative density) remained constant for the three absolute densities.
[b]Percentage of birds (introduced) producing an egg each day (does not account for mortality).

equal egg production. The relatively poor correlation between strain differences in agonistic behavior and productivity begs us to ask whether strains (and individuals) can differ in their relative *tolerance* to social stress. It appears that the effects of absolute density (group size) on the frequency of agonistic behavior and productivity are best predicted within, rather than between, strains or breeds of animals.

Determinants of stress and well-being

Volumes have been written on the various physical and psychological conditions which influence animal stress and well-being. A thorough review of this topic is outside the scope of this text. However, it is important to consider some of the factors which influence how animals react to these stressors. The species, breed or strain (within species or breed) of the animal, age, sex, context of the interaction, social status and prior experience can influence individual responses to potential stressors. You will note that these are some of the same factors which influence thresholds for agonistic behavior and the size of an animal's personal space.

Individual responses to potential stressors are highly variable. The last section suggested that some strains of chickens (*G. domesticus*) were more tolerant of social stress than others. Responses to stressors will change with physical maturation, social experience/status and the context in which the stressor is presented. Arnone and Dantzer (1980) measured the pituitary–adrenal response (a common physiological measure of stress) of unacquainted pairs of domestic pigs (*S. scrofa*) on meeting for the first time and related these measurements to the intensity of their aggressive behaviors and the degree of dominance exhibited by each pig. Increases in plasma corticosteroid concentrations (over baseline) were significantly greater when the two animals engaged in intense fighting (Fig. 14.1). In addition, subordinate animals (of the pairs) exhibited greater increases in corticosteroid concentrations than the dominant individuals. Subordinate animals are particularly vulnerable to social stress under captive conditions where space is limited and escape from dominant conspecifics is not an option.

Table 14.2. Mean frequencies of agonistic acts and egg production for three densities (absolute) of five strains of caged chickens (*G. domesticus*) (from Al-Rawi *et al.*, 1976).[a]

	Strain				
	1	2	3	4	5
Agonistic acts/hen/h	5.1	9.8	20.4	21.0	26.0
Hen-housed egg production (%)[b]					
No. birds per cage					
4	68.6	69.9	61.9	66.3	68.6
8	64.6	69.1	57.1	66.3	64.1
14	58.2	60.2	44.9	63.5	59.3

[a]Space per chicken (relative density) remained constant for the three absolute densities.
[b]Percentage of birds (introduced) producing an egg each day (does not account for mortality).

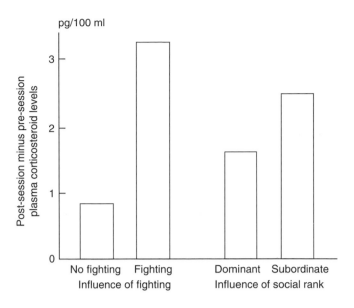

Fig. 14.1. Mean increase in plasma corticosteroid levels (pg/100 ml) when pairs of unacquainted pigs are placed together for 20 min. Corticosteroid levels increased more when pairs fought and increases were somewhat greater for subordinate than dominant animals (Arnone and Dantzer, 1980).

Protection of animals from social conflict and disturbance

Housing systems for captive animals tend to be relatively open and devoid of the kind of physical objects which, in nature, preclude visual communication and provide subordinate animals a degree of protection from dominant conspecifics. In spite of this handicap, subordinate animals living in captivity learn various coping strategies to minimize social interactions and conflict. One adaptation is to become less active and less involved in social interactions (Mendl *et al.*, 1992). Inactivity reduces the likelihood of an interaction with a feared penmate. Another adaptation is to signal submission frequently when in the company of dominant individuals. In addition, submissive animals learn to avoid direct eye contact with dominant conspecifics ('visual cut-off'). McBride *et al.* (1963) photographed the movements of female chickens (*G. domesticus*) under relatively high-density conditions and found that each hen continually adjusted its orientation and spacing to avoid directly facing dominant penmates. As distance between them decreased, a point was reached where they faced each other more directly. Subordinates also learn to hide behind penmates to avoid direct eye contact with certain individuals. Whenever possible, animal caretakers should provide partitions or shelters in otherwise open enclosures or cages to facilitate visual cut-off and avoidance of aggressive penmates. Relatively simple and inexpensive improvements in the quality of captive environments can significantly reduce the social stress experienced by animals. For example, installing vertical panels in otherwise open chicken pens distributed birds more uniformly (Cornetto and Estevez, 2001) and dramatically decreased the frequency that resting birds were disturbed by penmates (Cornetto *et al.*, 2002).

Consequences of social strife

The decline in productivity seen in stressed animal populations is measured not only in the loss of commercially important animal products such as meat, milk, eggs, etc., but also reduced growth rates, poorer reproduction, lowered resistance to disease and parasitism and increases in mortality. Negative effects on reproduction can be manifested in a number of ways, including slower rates of sexual maturation, prolongation of estrous cycles, reduction in the incidence of ovulation and implantation, intrauterine mortality and abnormal maternal behavior.

This book has repeatedly emphasized that an individual's social status can have important effects on its physiology and health. While subordinate animals tend to bear the brunt of social stress in captive populations, it exacts a price on dominant individuals as well (Sapolsky, 2004). The price of dominance is perhaps best revealed in free-living populations of animals, where food and other needed resources are often less abundant than in captivity. Dominant animals in nature are sometimes required to assert and reinforce their rank continually to retain their high status. Dominance in wild female olive baboons (*P. cynocephalus anubis*) bears the price of a higher miscarriage rate (Packer *et al.*, 1995). Creel *et al.* (1996) found higher stress-related glucocorticoid levels in alpha African wild dogs (*Lycaon pictus*) and dwarf mongooses (*Helogale parvula*) than in subordinate members of their study populations, a likely consequence of their frequent use of aggressive behavior to maintain their superior status. Beletsky and Orians (1992) summarize work showing that dominant male red-winged blackbirds (*A. phoeniceus*) with large territories have higher levels of corticosterone and testosterone than males without territories. Interestingly, the higher levels of stress hormones found in the dominant animals of these latter three studies did not appear to affect their reproductive success negatively, at least in the short term. The price of dominance in these species may be a shorter reproductive lifespan.

Changes in social status can have a greater effect on the physiology and health of dominant individuals in the group than on their subordinate counterparts. The stress associated with loss of social status is believed greater than the stress associated with gaining status (e.g. Sapolsky, 1992). Defeats in agonistic interactions serve as important triggers for changes in immunocompetence (ability to mount an immune response). Hawley (2006) demonstrated that dominant male house finches (*Carpodacus mexicanus*) experienced a significant drop in immune response simply by losing one place in the dominance hierarchy (dominating one fewer individuals in the group). In contrast, comparable *increases* in social status of subordinate individuals had no detectable effects on immunocompetence. This difference may reflect relative changes in access to resources accompanying the loss or gain in social status.

The higher incidence and transmission of disease-producing pathogens and parasites among populations of captive animals living under relatively high-density conditions makes it imperative that our management decisions do not compromise the ability of our animals to mount an appropriate immune response when needed. Veterinarians have long recognized the importance of preventative medicine for maintaining the health of our captive wild and domestic animals. Managing the behavior of our animals to minimize social strife can be just as important as managing the physical environment to protect their health and well-being.

Chapter 14

15 Human–Animal Interactions

Animals are an integral part of our lives. We have domesticated animals for food and other products, companionship and recreation, all of which are important to our quality of life. Animals still plow fields in many developing countries, herd livestock, assist law enforcement officers, rescue people and guide the blind. Animal research enhances our knowledge and understanding of human biology and medicine. Many wild animals are housed in zoos and wildlife parks simply for our enjoyment and edification. Humans are naturally fascinated with animal behavior and enjoy learning about the adaptations of animals to their environments. It is hard to imagine what our lives would be like if we humans were the only animals on our planet.

Basic Responses of Captive Animals to Humans

One of the most basic responses of captive animals is a tendency to fear humans and novel or startling stimuli associated with humans. Jones and Gosling (2005) found that in 51 publications on temperament in domestic dogs (*C. familiaris*), fearfulness was the most frequently studied trait. Captive animals are born without knowledge or fear of humans or other objects in their environment. This is important to the newborn else it would be afraid of its parent. However, a generalized fear response to unfamiliar stimuli typically unfolds within hours, days or months of birth, depending on the species, and typically remains with the animal for the rest of its life. Depending on their experience, animals will continue to fear certain stimuli, while their fear of other stimuli is reduced or lost. The animal's personal experience with humans can have a profound effect on its relationship to people and the extent to which humans are feared.

Animals that are fearful of people will exhibit avoidance behaviors in their presence. If a fearful animal cannot escape from the presence of humans, it may show submissive or cowering behaviors or it may threaten and attack (defensive aggression). When the animal can escape from humans, its degree of fearfulness can be measured by its 'flight initiation distance' (FID). FID is how close a human (or other animal) can get to an animal before it flees. FID is a commonly used name for 'interindividual distance' (see preceding chapter) in situations where the animal's response to humans is avoidance. The animal's 'flight zone' is that area (i.e. 'personal space') surrounding an individual which, if encroached upon, initiates flight. The FID can be thought of as the radius of an animal's flight zone. Animals which have large FIDs to people are very sensitive to the presence or approach of a person and will flee at great distances. Captive animals that are fearful of people and exhibit relatively large FIDs are generally more difficult to handle and care for, experience poorer welfare and poorer reproductive success.

Tameness, Taming and Tameability

Tameness is the extent to which an animal is willing to accept the presence of humans. It is measured on a continuum. Untamed animals are highly motivated to avoid humans and almost never approach them. In contrast, highly tame animals often approach humans and seldom show avoidance behaviors. Tameness is an important behavioral trait of captive animals since it usually facilitates animal handling and management and adaptation to the captive environment. *Taming* is an experiential learning process occurring during the lifetime of an individual in which an animal's avoidance of people is reduced and willingness to approach people is increased. *Tameability* is the capacity to be tamed. Tameability, like tameness, is a desirable trait of captive wild and domestic animals. Some species, breeds and individuals are more tameable than others.

Genetic Influences on Tameness and Tameability

In nature, it is adaptive for animals to be fearful of unfamiliar and startling stimuli. Such stimuli could be harmful (e.g. predators, poisonous animals) to the animal and avoiding them is adaptive. In captivity, most harmful stimuli have been eliminated and adaptation to man and the environment he provides is facilitated by low levels of emotional reactivity (i.e. emotional responsiveness to changes in one's environment). Natural selection for emotional reactivity is relaxed when wild animals are brought into captivity and selection during ensuing generations favors animals that are less reactive. Animals that are *naturally* less fearful of unfamiliar and startling stimuli are generally more tameable. Thus, it is not surprising that an animal's general emotional reactivity and tameability are linked and both are responsive to genetic selection. Studies dealing with major gene effects, heritability estimates, artificial selection for tameability and twin studies have demonstrated an important genetic component to an animal's basic fear of humans.

Major gene effects on ease of handling

We saw in Chapter 3 (p. 23–24) that wild Norway rats (*R. norvegicus*) and deermice (*P. maniculatus*) homozygous for the 'black' pelage-color gene are easier to touch, stroke and handle than their agouti-colored counterparts (Cottle and Price, 1987; Hayssen, 1997). Furthermore, it was reported that 70–80% of the inbred rat and mouse (*M. musculus*) strains examined were primarily homozygous for the black allele, even though its phenotypic expression might be masked by genes at the albino locus. These and other reports have strongly suggested that the black coat-color allele has a major effect on the tameness and ease of handling when present in the homozygous recessive condition.

Heritability estimates

Breed differences in tameness have been reported for most domesticated companion and farm animal species. Le Neindre *et al.* (1995) found a significant sire effect

(34 sires) and a heritability estimate of 0.22 for docility scores obtained in handling tests administered to 906 cattle (*B. taurus*) of the Limousin breed. The test consisted of leading individual animals to a corner, retaining them there for 30 s and then stroking them. (The 0.22 heritability estimate means that 22% of the variability in docility scores can be attributable to genetic differences between the individuals tested.) Hemsworth *et al.* (1990), working with pigs (*S. scrofa*), found that the behavioral variable 'time to physically interact with humans' was also a moderately heritable characteristic (0.38 heritability estimate).

Artificial selection

Animal caretakers purposefully, or unconsciously, select tame over untame animals for breeding stock because they are easier to handle and manage. Artificial selection for tameability (capacity to be tamed) has been highly successful, again highlighting the role of genetics in the expression of these traits. One of the most celebrated studies is a 40-year project selecting silver foxes (*Vulpes vulpes*) for tameability, conducted in Siberia by Dmitry Belyaev (see summary by Trut, 1999). Belyaev's research created a population of tame foxes fundamentally different in temperament and behavior from unselected farmed foxes (control population) and which exhibited a number of physiological and morphological traits similar to other domesticated animals. In each generation, fox pups were monthly offered food from the experimenter's hand, while the experimenter attempted to stroke and handle the pup. At 7–8 months of age, they were assigned to one of four tameness categories. Foxes in the least-tame category would flee from the handler or bite when stroked or handled. The most-tame foxes were eager to establish human contact, whimpering to attract attention and sniffing and licking the researchers like domestic dogs (Fig. 15.1). Eighteen percent of the foxes in the tenth generation of selection qualified for this most-tame category. By the 20th generation, 35% qualified and, after 40 years of selection, the figure reached 70–80%. Interestingly, an animal's relative degree of tameness is first noticed before it is 1 month old. It was determined that about 35% of the variation in the foxes' tameness could be accounted for by genetics.

A study in Denmark successfully selected mink (*Mustela vison*) for 'fearful' and 'confident' behaviors toward humans (Malmkvist and Hansen, 2001). In each generation, mink were subjected to a handling test (Fig. 15.2) and a 'stick' test (Fig. 15.3) in which a wooden tongue spatula was inserted through the wire-mesh wall of the animal's cage. Fig. 15.4 shows the response to selection of the fearful, 'confident' and non-selected control lines over the first 11 generations. In addition, fearful mink exhibited higher plasma cortisol levels after handling than mink from the 'confident' line. Furthermore, cross-fostering of offspring of 'confident' mothers on fearful mothers and vice versa had no effect on the subsequent timidity scores of the mink, essentially ruling out postnatal maternal effects as a significant contributor to the strain differences obtained. Malmkvist and Hansen (2001) also reviewed selection experiments demonstrating that mink selected for reduced fear of humans also showed less fear of novel objects, less fear of unfamiliar mink placed in their cage and increased willingness to mate in captivity, supporting the hypothesis that tameness is closely linked to an animal's general emotional reactivity.

Fig. 15.1. Silver foxes selected for non-aggressive behavior toward humans are offered food (courtesy of the Institute of Cytology and Genetics, Novosibirsk, Siberia).

Lankin (1997) studied the behavior of 11 breeds of sheep (*O. aries*) and concluded that breeds subjected to intensive selection for commercial purposes were tamer toward humans than breeds which had not been subjected to intense artificial selection (Fig. 15.5). The East Friesian breed has been intensively selected for meat and milk production and is particularly tame toward humans. Animal breeders may unconsciously select tamer animals for breeding because they are easier to manage and handle (Price, 2002).

Twin studies

Twin studies can also provide valuable information on the development of tameness. Lyons *et al.* (1988a) examined the tameness of dairy goats (*C. hircus*) toward humans, both within and between twin sets. One individual of each twin set was reared by its natural mother and the other individual was hand reared apart from its mother.

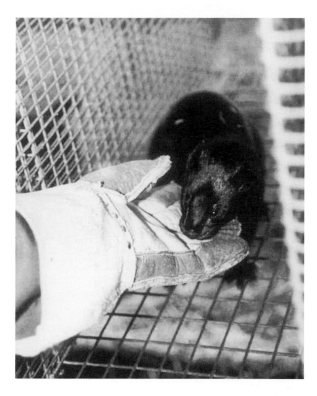

Fig. 15.2. Handling test for mink used in selecting mink for 'fearful' and 'confident' behavior toward humans (courtesy of the Danish Institute of Agricultural Sciences).

The hand-reared and mother-reared goats were ranked separately for tameness, based on their scores in a standard interaction test with humans. In all cases, the tameness scores of the hand-reared goats were better than for their dam-reared twin. Moreover, the tameness ranking of individuals within the hand-reared treatment closely correlated with the tameness rankings of their co-twins in the dam-reared treatment (Fig. 15.6). Although there was no control for prenatal maternal effects in this study, the results suggested a significant genetic contribution to the development of behaviors characterizing tameness in this species.

Experiential Effects on the Taming Process

Some domestic animals reared in the wild (feral condition) can be just as fearful of people as wild animals and some wild animals reared in close contact with people can be just as tame as domestic animals. For example, feral dogs (*C. familiaris*) reared in the wild with little or no contact with humans exhibit flight initiation distances (from humans) very much like that of other free-living, wild canids. Conversely, wild rats (*R. norvegicus*) reared in laboratory cages and subjected to daily handling when young are nearly as tame as domestic rats. These observations suggest that tameness can have a significant experiential or learning component.

Fig. 15.3. 'Stick' (spatula) test for mink used in selecting mink for 'fearful' and 'confident' behavior toward humans (courtesy of the Danish Institute of Agricultural Sciences).

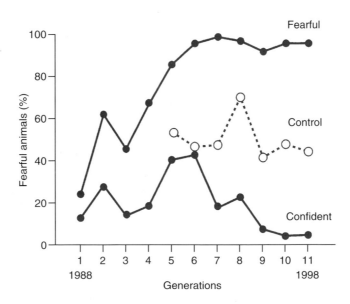

Fig. 15.4. Percentage of mink reacting fearfully (i.e. showing avoidance in the 'stick' test) over 11 generations of selection for 'confident' and 'fearful' behavior (Malmkvist and Hansen, 2001).

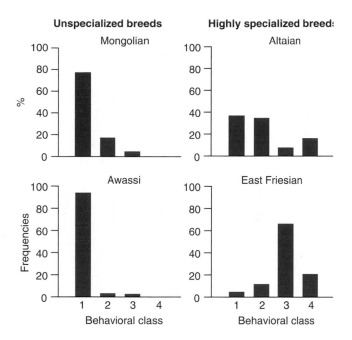

Fig. 15.5. Percentage distribution in avoidance of humans between two relatively unspecialized breeds of sheep not selected for productivity and two commercial breeds of sheep selected for productivity. Animals in behavioral class 1 showed the greatest avoidance of humans and those in behavioral class 4 exhibited the least avoidance of humans (Lankin, 1997).

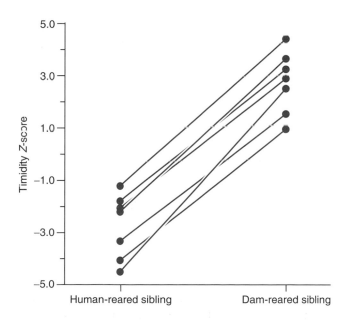

Fig. 15.6. Timidity *Z*-score for seven sets of twin dairy goats. In each twin set, one was human reared and the other dam reared (Lyons *et al.*, 1988a).

Two experiential phenomena have been identified as contributing heavily to the taming process: (i) habituation to humans and (ii) positive associative conditioning. Acquired tameness is often achieved by some combination of these two factors.

Habituation

A degree of tameness can be achieved without any deliberate effort on the part of the animal caretaker. In habituation, the animal's fear of humans is gradually reduced by repeated exposures in a neutral context, when a person's presence has no obvious reinforcement properties. It is noted in the chapter on learning (p. 52) that habituation is situation specific. Thus, habituation to people is facilitated by repeated exposures that are spatially and temporally predictable; that is, they occur in the same place and at a similar time of day. Jones (1993) found that domestic chicks (*G. domesticus*) showed decreased avoidance of humans following twice daily exposures to either: (i) placing the experimenter's hand in the chick's cage with no attempt to initiate physical contact; (ii) allowing chicks to observe their neighbors being picked up and handled regularly; or (iii) allowing visual contact with a person who simply stood in front of and touched the wire-mesh wall of the chick's home cage. Domestic horse (*E. caballus*) foals showed less fear of humans at least up to 1 year in age if they had contact with humans during their first 5 days after birth in the form of observing their mothers being fed by hand and brushed (Henry *et al.*, 2005). Wolves (*C. lupus*) maintained in enclosures with minimal exposure to people were more fearful of humans than wolves reared in enclosures permitting frequent exposure to humans (Ginsburg and Hiestand, 1992). Malmkvist and Hansen (2001) noted that mink (*M. vison*) selected for 'fearfulness' toward people exhibited increasingly more 'confident' behaviors over repeated exposures to people. Of course, their captive mink may have also learned to associate humans with positive reinforcers such as food.

Positive associative conditioning

Taming may also be achieved by humans becoming associated with such primary reinforcers as food, water, shelter and various forms of attention, such as petting and grooming. Tsutsumi *et al.* (2001) reported that minipigs (*S. scrofa*) subjected to routine care (feeding and cage cleaning) gradually became tamer with age and met established criteria for tameness at about 10 months of age. Comparable animals given 'positive contact' (brushing, patting and talking to the animals) on opening their cage door reached the same degree of tameness in only 4 weeks after the initiation of contact. Lyons *et al.* (1988a) reported that hand-reared goats (*C. hircus*) exhibited much shorter flight initiation distances than mother-reared dairy goats. The hand-reared goats had considerable contact with humans at feeding times, whereas their mother-reared counterparts had little contact with people up to 14 weeks of age. At 14, 22 and 30 weeks of age, the two groups of goats were scored for their response to being captured and restrained. The majority of the hand-reared goats did not avoid the captor and did not struggle when first restrained (Fig. 15.7). The majority of the mother-reared goats showed significant avoidance of the captor and considerable struggling when captured. Differences in timidity persisted beyond the first year of age, even in the milking parlor during their first lactation.

Fig. 15.7. Hand-reared does in the 'capture test' showing zero flight initiation distances (Lyons *et al.*, 1988a).

Positive human contact and gentle handling in the days immediately following birth is an effective tool for taming wild animals. Galef (1970) tested the effects of several rearing experiences on the ease with which wild Norway rats (*R. norvegicus*) could be handled by humans. Second and third generation laboratory-reared wild rats were: (i) reared by either wild or domestic foster mothers; (ii) reared with either wild or domestic littermates; (iii) given either minimal or maximal exposure to humans in a laboratory environment; and (iv) either not handled at all or handled for 2 min daily from days 10 to 23. At weaning (about 23 days of age), each rat was subjected to a handling test which scored such behaviors as difficulty of capture, escape behaviors, vocalizations and bites directed toward the investigator's hand. Galef found that only direct handling experience (2 min/day) improved the ease of capture and handling. Markowitz *et al.* (1998) demonstrated that mother-reared lambs (*O. aries*) handled during the first 3 days following birth subsequently exhibited shorter flight initiation distances toward people than lambs handled on days 4–9. Aengus and Millam (1999) showed that parent-reared orange-winged Amazon parrot chicks (*Amazona amazonica*) could be tamed by occasional handling without risk of imprinting on humans. Chicks incubated and hatched by wild-caught parents were handled periodically during the first 2 months after hatching and then evaluated for tameness as fledglings by their willingness to approach the handler, perch on a finger, be touched on the head and by their respiratory rate in the presence of the handler. Handled birds were more tame for all measures (Fig. 15.8), thus supporting the position that neonatal handling of parent-reared parrots provides a minimal labor and low-technology alternative to hand rearing for enhancing their tameness toward humans.

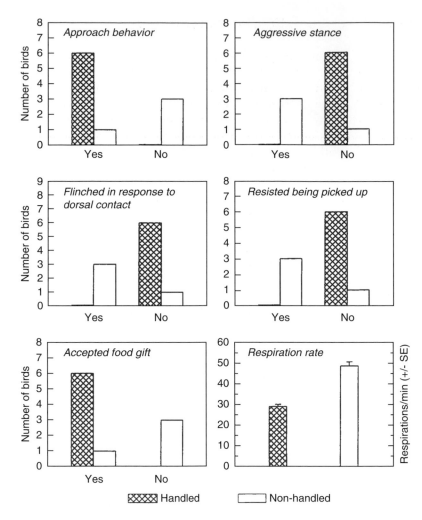

Fig. 15.8. Differences between handled and non-handled parrot chicks in behavioral and physiological measures of tameness (Aengus and Millam, 1999).

For some species such as the domestic dog (*C. familiaris*), humans can take on the role of a social object (i.e. 'companion') whose presence itself is rewarding. In these contexts, people become an important reinforcer. Topál *et al.* (2005) provide evidence that the strong social attachment of domestic dogs to their owners may be an effect of domestication since hand-reared wolf pups (*C. lupus*) fail to show a marked preference for their owner over strange humans at 4 months of age. This difference in owner attachment may be a result of selective breeding during domestication for dependency and attachment to humans. Domestic dogs have evolved a capacity for social bonding to humans that is functionally analogous to that observed in human infants.

Parental and Conspecific Effects on Tameness

The component of tameness acquired by contact with humans cannot be transmitted from dam to offspring. In the Lyons *et al.* (1988a, 1988b) study with dairy goats (*C. hircus*), flight initiation distances (to humans) were the same for mother-reared young, regardless of whether their mothers were timid (mother reared) or tame (hand reared), providing the young were tested alone. However, these same relatively timid, mother-reared young exhibited less fear of people in the presence of tame conspecifics, especially their mothers and familiar penmates (Lyons *et al.*, 1988b; Fig. 15.9). In addition, corticosteroid levels (physiological measure of stress) of timid goats were significantly lower when exposed to humans in the presence of their mothers or familiar penmates than when tested alone or with strange goats (Fig. 15.10). In contrast, the social context at testing had little effect on the behavioral or physiological responses of *tame* (hand-reared) young. It was concluded that social companions, particularly mothers, could modulate or buffer the behavioral and physiological responses of timid animals to humans.

Cattle ranchers will tell you that the flight initiation distances of untamed beef cattle (*Bos*) to people in a pasture or rangeland setting are much shorter when a person is on horseback than on foot. Several ideas have been proposed to explain this phenomenon. One is that the body language of the horse does not communicate alarm; therefore, the cattle are not alarmed. Another is that the human profile is masked by being on a horse. However, the rider can dismount and move several meters away from the horse without the cattle showing concern. Perhaps the general stimulus configuration and gait of an approaching horse and rider is similar enough to that of any ungulate that it does not signal alarm. A person approaching cattle on foot, however, provides a different unfamiliar profile and gait for animals not habituated to seeing people in a pasture or rangeland context. Stimulus novelty may be the critical factor.

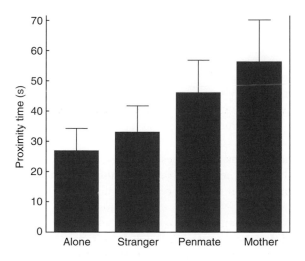

Fig. 15.9. Mean (± SE) time (s) young goats spent in proximity to humans in four social settings including the presence of tame strangers, familiar adult penmates and their tame mothers (Lyons *et al.*, 1988b).

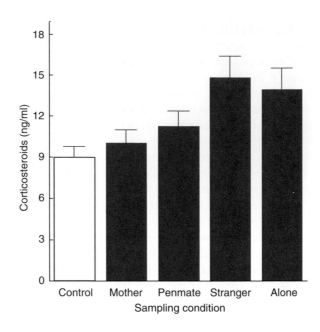

Fig. 15.10. Mean (± SE) control and post-test serum corticosteroid concentrations in young goats following encounters with humans conducted in four social settings (Lyons *et al.*, 1988b).

Effect of the Physical Environment on Tameness

Environmentally induced developmental events reoccurring in each generation can play an important role in the taming process of domestic and captive-reared wild animals. Clark and Galef (1977) demonstrated that the tameness and docility of captive gerbils (*M. unguiculatus*) could be influenced by the physical design of their cages. Laboratory-reared gerbils given access to an enclosed hiding place (e.g. burrow system or shelter) early in life exhibited greater avoidance of a human-like stimulus than gerbils reared in open laboratory cages. Elicitation of escape responses (flight) into a concealed chamber was found to be the 'critical' experience in the development of this avoidance response. Once flight and concealment responses were established, experience in an open-cage environment had little influence on their behavior. Harri *et al.* (1998) reported that farmed blue fox cubs (*Alopex lagopus*) provided with a nest box in their cage were more fearful toward humans than foxes without a nest box. If nest boxes are offered, foxes take advantage of the opportunity to hide and, as a result, do not habituate as well to humans and to the everyday activities of farm staff and visitors.

Human Behavior and Posture Affect Tameness

The way in which humans approach and interact with animals can affect the degree of tameness exhibited. Hemsworth *et al.* (1986) compared the behavior of pigs (*S. scrofa*) in response to a passive, stationary person versus a person who actively approached them. He also noted the pigs' reaction to different human body postures. Pigs approached the experimenter significantly more when he stood passively in the test arena rather than when the pigs were actively approached. Squatting elicited more approach behavior than when the experimenter stood erect. Lyons *et al.* (1988a) found that young dairy goats (*C. hircus*) spent more time within a body length of a stationary person than a moving or pursuing person. Conversely, time spent avoiding (moving away from) humans was greatest when they were pursued. It is not surprising that a prominent, rapidly approaching human provides a more threatening stimulus to animals than a passive stationary person.

Aside from their presence or posture, animal handlers can minimize flight initiation distances by exhibiting behaviors that are perceived as non-threatening. In a study reported by Vas *et al.* (2005), 30 dogs (*C. familiaris*) representing 19 breeds were repeatedly approached by the same person, alternately exhibiting friendly or threatening gestures. Friendly gestures consisted of walking toward the dog in a normal manner, talking to it and petting it when contact was made. Threatening gestures included approaching the dog slowly and haltingly without speaking, while staring at the dog's eyes. The dogs' reactions to the person were different and appropriate for the person's behavior. A second experiment was conducted with 20 Labrador retrievers, 20 'sled dogs' and 20 Belgian sheepdogs. The person first approached the dogs in a friendly way, as described above, and later approached the dogs in a threatening manner. The Belgian sheepdogs were most responsive to the person's changed behavior, while the effects on the retrievers and sled dogs were minimal.

I enjoy photographing wildlife and learned during my summers as a ranger-naturalist in Alaska that I could get relatively close to the wild ungulates in the Park, namely moose (*Alces americana*), Dall mountain sheep (*O. dalli*) and caribou (*Rangifer arcticus*), if I approached the animals at an angle, as if I was walking past them. At the same time, I also responded to their body language. If the animal I wanted to photograph appeared alarmed (head and ears upright and looking directly at me, body tense, etc.), I stopped moving and looked the other way or at the ground until the animal gave me cues that it was less concerned about my presence (e.g. chewed its 'cud', started feeding, looked away from me). I then resumed my angular approach until the animal again appeared alarmed. I would repeat this scenario until I got close enough for a photograph.

One day I spotted a band of 16 male Dall sheep (*O. dalli*) bedded down on a steep mountainside. I used rocks and terrain to approach within 100 m of the animals without being seen. At that point, there was nothing to hide behind to help me get close enough for a photograph with my 100 mm lens. I did the only thing I could think of. In a crouched position, I slowly moved into plain sight of the animals while looking in a different direction. I could see in my peripheral vision that the younger animals (smaller horns) immediately got up and appeared alarmed. Two or three of the older rams appeared alarmed but did not stand up. I remained immobile until the older rams started chewing their 'cuds' (ruminating) and then moved a bit closer. When the rams stopped chewing their cuds, I stopped. When cud chewing resumed, I moved closer. Using this technique, I got about 50 m from the rams in about 30 min and took their photograph (Fig. 15.11). The ram with the largest horns (full curl) was still lying down at that point. I noticed that each time I moved forward, the younger rams would look at me and then look at this older ram. They appeared ready to flee but did not do so because, I suspect, their patriarch was still lying down.

Fig. 15.11. Dall mountain sheep in Denali National Park, Alaska. The older ram lying down was the animal I watched closely as I approached. The younger rams (uphill) appeared to be taking their cues from the older males.

Recognition of Individual Humans

Tameness may vary depending on the specific person encountered, the attributes of the person at the time and the location of the interaction. Although individual animals do not always behave differently toward different people, it has been demonstrated that many species of animals (cattle, sheep, pigs, rabbits, seals, rats and chickens) can recognize individual humans (see reviews by Rushen *et al.*, 1999; Davis and Taylor, 2001). It appears that visual cues (e.g. color of clothing, human behavior) are particularly important in identifying people, but other sensory modalities (e.g. auditory or olfactory cues) may also be involved, depending on the circumstances. Cattle (*B. taurus*) can use the color of clothing alone to distinguish between different humans, and their response to individuals may change as clothing color is changed. Cattle can also use relative body height or facial features in recognizing individual humans (Rybarczyk *et al.*, 2001).

The context of the interaction (e.g. place) may influence the animal's response to people. The location of prior positive or negative handling experiences can determine the behavioral response of cattle toward humans. They may choose to approach or avoid the same person in two different places if that person treated them differently (e.g. pleasantly or harshly) in the two places (Rushen *et al.*, 1999).

Effect of Human–Animal Interactions on Animal Productivity

The animal handler's basic attitude toward animals and technical skill level can affect how he or she responds to animals and the body language he or she displays in their presence. These factors, in turn, influence how animals respond to their handler and their general fearfulness of people. Rough or unpleasant handling of animals consistently increases their fear of people and fear is negatively correlated with economically important measures of productivity, such as growth, milk and egg production and reproductive success (Fig. 15.12). Hemsworth *et al.* (1981a) studied the effects of quality of handling on growth rate in juvenile pigs (*S. scrofa*). The experimenter entered each pig's pen three times per week for 2 min when the pig was between 11 and 22 weeks of age (11-week period). Pleasantly treated pigs were stroked whenever they approached the experimenter. Unpleasantly treated pigs were slapped, restrained with a 'snout noose' or given a slight electric shock with an electric prodder. At the end of the 11-week treatment period, the pleasantly treated pigs had gained 3.5 kg more than the unpleasantly treated animals. Not surprisingly, corticosteroid levels (measure of stress) of the unpleasantly treated pigs rose threefold after a 2-min exposure to humans, whereas no change was observed in the pleasantly treated animals (Fig. 15.13). In addition, at 25 weeks the unpleasantly treated pigs were significantly more likely to avoid the experimenter in a 2 × 3 m pen than the pleasantly treated animals.

In another study, Hemsworth *et al.* (1981b) looked at the reproductive performance of pigs (*S. scrofa*) on 12 farms under the same ownership and management but with different caretakers. Reproductive success of the sows on some of these farms was better than others and the investigators sought to determine if this variation in reproductive

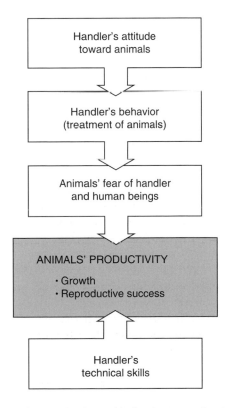

Fig. 15.12. Effect of the handler's attitude and behavior toward animals on the animal's fear of humans and their productivity (Hemsworth *et al.*, 1989).

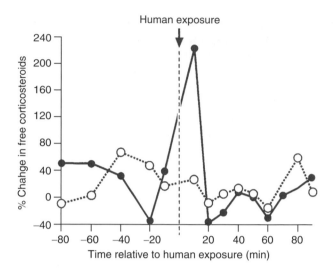

Fig. 15.13. Percentage change in free corticosteroid concentrations in negatively (solid circles) and positively (open circles) handled gilts before and after a 2-min exposure to a human (Hemsworth *et al.*, 1981a).

success was associated with the animals' behavioral response to people. The animals' reaction to people was measured in two tests. In one test, sows were given a small amount of feed in the feed trough at the front of their stalls. Once the sow started feeding, the experimenter gradually lowered his hand into the trough to touch the sow on the side of its snout. Sows that remained at the trough got a score of zero. Those that withdrew from the trough were given 10 s to return to the trough with the experimenter's hand present. A score of one to ten was recorded based on the number of seconds it took them to return. The second test measured the behavioral responses of sows to a person standing on one side of a relatively small pen. The results demonstrated that the reproductive performance of the sows on the 12 farms was significantly associated with their response toward people. Farms where the sows showed the strongest avoidance responses toward people also had the poorest farrowing percentages (percentage of sows that gave birth) and average total number of piglets born per year. The experimenters concluded that differences in the reproductive performance of the sows on the 12 farms were related to differences in the relationship that had developed between the caretaker and his animals at each farm. The manner in which the caretaker handled his animals was likely a critical factor in the development of this relationship.

Similar studies (Breuer *et al.*, 2000; Hemsworth *et al.*, 2002) have shown that rough handling of dairy cows (*B. taurus*) and fear of humans result in lowered milk production. Likewise, fear of humans is inversely related to egg production by laying hens (*G. domesticus*) (Barnett *et al.*, 1992) and feed conversion (to meat) of broiler hens (Hemsworth *et al.*, 1994). Müller and von Keyserlingk (2006) showed that the speed of beef cattle (*B. taurus*), as they traveled down a raceway after being weighed, was inversely related to average daily weight gain. Even the quality of animal products can be related to the treatment animals receive from the farmer. Abusive handling of cattle prior to slaughter can result in meat with a shorter shelf life and darker than normal color caused by a depletion of muscle glycogen. Physical stress (strenuous exertion) and psychological stress (adrenalin secretion from excitement) are the main causes of muscle glycogen loss. Lensink *et al.* (2001) found that the quality of veal calf (*B. taurus*) meat (color) was higher on farms where the animals were cared for with 'positive' behaviors during rearing. Studies indicate that one-fifth to one-third of the variation in productivity seen across farms in the pig, dairy and poultry industries is related in some way to human–animal interactions.

Seabrook (1972) reported that dairymen most likely to exhibit positive relationships with their cows (*B. taurus*) were confident, patient, persevering, independent and socially introverted. They tend to provide more positive stimulation such as touching and petting, are more consistent in how they react to their animals and are more likely to exhibit firm and sure movements. Herdsmen who are insensitive, rigid and overconfident often limit the extent to which they learn from their handling experiences and are less likely to develop improved handling techniques.

Hemsworth *et al.* (2002) designed a 'behavior intervention' program for workers on dairy farms to improve their attitudes and behavior toward cows (*B. taurus*). The program involves instruction and multimedia presentations on how worker attitudes and behavior when handling cows affects cow behavior and productivity (milk production). Video clips demonstrate positive and negative handling techniques and discuss how rough handling increases animals' fear of humans, thus creating stress and lowering productivity. The authors reported that workers on 'intervention' farms developed more positive attitudes about the amount of physical and verbal effort required to move and handle cows, and subsequently exhibited more positive responses toward cows (e.g. pats, strokes, resting hand on animals, talking to animals and slow deliberate movements) and fewer negative responses (e.g. hits, kicks, slaps) than workers on control farms not exposed to the behavioral intervention program. Furthermore, controlled experiments showed that cows on the intervention farms exhibited shorter flight initiation distances than cows on the control farms, suggesting they were less fearful of the workers. Not surprisingly, milk production on the intervention farms was more than 5% greater than on control farms during the peak milk production period. Intervention farms where cow fear levels declined showed 10% higher levels of milk production during the first 3 months than intervention farms in which cow fear levels were not reduced. More than 75% of the workers on the intervention farms reported improvements in their behavior toward cows and greater ease in handling their cows than before the program was initiated.

Studies like those reviewed in this section (see also review by Hemsworth, 2003) demonstrate the importance of exposing farm workers to structured tutorial programs aimed at improving human attitudes and behavior toward livestock. Employers cannot assume that people in general, and even those experienced in handling farm animals, understand how worker attitudes about animals and their own behavior can have a significant effect on the behavior, productivity and welfare of the animals under their care.

Effect of Early Handling and Hand Rearing on the Reproductive Success of Captive Wild Animals

Early handling of captive wild animals not only increases tameness but also can improve reproductive success. Dalsgaard and Pedersen (1999, cited in Mononen *et al.*, 2001) reported that blue foxes (*A. lagopus*) repeatedly handled by humans at 7–10 weeks of age were less fearful of people and reproduced more reliably than non-handled foxes which had only been subjected to normal farm activities. One of my graduate students, Brian Clark, and I studied the effects of handling on the reproductive success of captive-reared wild Norway rats (*R. norvegicus*) (Price, 1980; Clark and Price, 1981). Handling was for 1–2 min daily during the post-weaning period from day 26 to day 32. At sexual maturity, 23 of 24 (96%) handled male rats successfully copulated (and sired young) when left overnight with a hormone-induced domestic estrous female. In contrast, only 27 of 61 (44%) non-handled male wild rats mated with estrous females. In addition, 24 of 25 (96%) handled wild females conceived and 21 (88%) of those gave birth. This is an extraordinarily high reproductive success rate for first-generation, captive-reared wild female rats. These results suggest that early handling may be one of the better ways for zoos, wildlife parks and animal research laboratories to improve the reproductive success of species that are typically poor breeders in captivity.

While *early handling* can increase the reproductive success of wild animals in captivity, *hand rearing* may have the opposite effect. Beck and Power (1988) found that female lowland gorillas (*Gorilla gorilla gorilla*) hand reared by humans from birth were less likely to reproduce in captivity than mother-reared female gorillas (35% versus 71%, respectively). Lack of early social experience with other gorillas was believed to be the determining factor. Mellen (1992) observed a similar result from hand rearing domestic cats (*F. domestica*). Twenty-one female kittens were divided equally into three treatment groups: (i) human reared alone; (ii) human reared with a single sibling; and (iii) mother reared with a single sibling. Only two of seven females hand reared without siblings (i.e. reared alone) and three of seven females hand reared with siblings successfully copulated as adults, compared to six of seven subjects mother reared with siblings. In addition, many of the cats in the human-reared-alone group were very aggressive toward human handlers and other cats, whereas the cats in the mother-reared treatment were calm and friendly toward human caretakers and appeared playful and calm around other cats. The human-reared-alone cats could be described as lacking 'trust' in both their human caretakers and the males with whom they were paired. Through interactions with their mother, kittens may acquire a degree of 'trust' of cats larger than themselves, such as male sexual partners.

Undesirable Aspects of Tameness

A high degree of tameness can be undesirable. Highly tamed animals may direct behaviors toward humans (e.g. aggression, sexual behaviors) which they would be reluctant to exhibit if they were more fearful of people, as my research assistant and I found out with our hand-reared beef bulls (*B. taurus*) (p. 44–45). Many people, particularly children, are injured each year by domestic dogs (*C. familiaris*) in unprovoked attacks. Wolf–dog hybrids (*C. lupus* × *familiaris*) inherit the capacity to become bonded to people but retain a stronger predatory drive than most dog breeds (Polsky, 1995). The combination of tameness toward humans and a relatively strong predatory drive sets the stage for potentially dangerous aggressive interactions directed toward people, particularly children and strangers, who may inadvertently

Fig. 15.14. Hand-reared crane chicks were occasionally led afield by a costumed caretaker and given opportunities to forage for insects and to use natural marsh habitat (Ellis *et al.*, 2000).

trigger an attack. It is estimated that only 5% of wolf–dog hybrids make truly satisfactory pets.

Tameness is also undesirable when rearing animals in captivity for release in the wild. Human caretakers will often disguise themselves to prevent their subjects from becoming socialized to humans (Fig. 15.14). It is particularly important to prevent animals from associating people with food else it will be more difficult for them to adapt to their wild existence.

Links Between Tameness and Brain Biochemistry

Strains of animals artificially selected for and against aggressiveness toward humans have been used to investigate the physiological basis of tameness. Popova *et al.* (1991a) demonstrated that tameness in silver foxes (*V. vulpes*) could be directly linked to the brain serotonergic system. Foxes selected for tameness had higher levels of the neurotransmitter, serotonin, in the midbrain and hypothalamus of their brains than their unselected wild counterparts bred in captivity. Serotonin is believed to inhibit fear-induced defensive aggression. Interestingly, similar changes in brain chemistry have been found in mink (*M. vison*) (see review in Nikulina, 1990) and Norway rats (*R. norvegicus*) (Popova *et al.*, 1991b) selected for reduced aggressiveness toward humans.

Selection for tameness in foxes (*Vulpes*) and rats (*Rattus*) has also influenced the catecholamine system of the brain (Cuomo-Benzo *et al.*, 1977; Nikulina, 1990). Catecholamines are hormones or neurotransmitters in the sympathetic nervous system (e.g. adrenaline, noradrenaline, dopamine) that prepare the body to meet emergencies such as cold, fatigue, social stress, etc. In both species mentioned above, tame animals possessed higher levels of noradrenaline in the anterior hypothalamus of the brain than was found in control animals that exhibited aggression toward humans. In addition, 12 generations of selection for tameness in silver foxes (*V. vulpes*) halved basal levels of plasma corticosteroids (measure of stress) relative to their unselected controls. Levels were halved again after 30 generations of selection (see summary by Trut, 1999).

Other studies have shown that early handling of domestic rat pups (*R. norvegicus*) can affect the development of their bodies' defenses against stress, particularly those associated with the hypothalamic–pituitary–adrenal (HPA) system (see review by Liu *et al.*, 1997). Liu *et al.* (1997) demonstrated an inverse relationship between the frequency with which rat mothers groomed and licked their offspring and their offsprings' HPA-mediated behavioral and endocrine responses to stress during adulthood. Rat pups handled by humans receive copious amounts of grooming and licking by their mothers when the young are returned to the nest and subsequently demonstrate improved HPA responsiveness to stress as adults. Perhaps the greater tameness and reproductive success of captive animals subjected to early handling by humans is due to reduced physiological responsiveness to stressors typically associated with the captive environment.

16 Animal Handling and Movement

Management Procedures Which Require Some Degree of Handling

Most captive animals require some degree of handling to keep them in good health. Both companion and food animals receive regular health checks, vaccinations and treatment for parasites. Young livestock are handled when tagged or branded for identification and when males are castrated. Animals with injuries, disease or infections must be handled to administer treatment. Weaning or regrouping animals is often accompanied by handling. Handling is sometimes necessary when animals are moved from one location to another. When the distance is relatively short, the animals may be carried, led (by a tether) or herded. Transport equipment (cars, trucks, train) is often necessary for long-distance translocation.

Some animals are small and docile enough to be picked up physically and restrained by their owner or veterinarian. Larger farm animals are often restrained for treatment using mechanical devices such as head gates or squeeze chutes. These devices are designed for the safety of both the animal and its handler(s). Larger wild animals are sometimes tranquilized or temporarily anesthetized using dart guns prior to tagging, health treatments or movement. Tranquilization minimizes the risk of injury to animals and handler(s).

Most companion animals are docile enough that many routine treatments can be administered with a minimum of restraint. Domestic dogs (*C. domesticus*) typically learn to accept routine grooming practices. Horses (*E. caballus*) and certain species of zoo animals (e.g. elephants) can be trained to receive foot treatments (e.g. shoeing, nail trimming) without objectionable behaviors. Monkeys have been trained to present an arm through the wall of their enclosure to facilitate injections without physical restraint. Handling is more difficult for those animals which have had painful or frightening experiences when restrained or handled. There are dogs which must be tranquilized to get their toenails clipped. Mistreated animals can be aggressive toward their human handlers. Wild and domestic animals with little exposure to humans can show 'panic-type' behaviors when approached by people. 'Wild' cattle are known to jump fences, crash through barriers, attack humans and struggle violently in restraining devices, while tame cattle typically exhibit little or no concern when herded or handled. Unruly cattle constitute a safety hazard for their handlers, cost more to own and are less profitable to sell. They are more susceptible to stress and some diseases and do not convert feed (to meat) as efficiently as tranquil cattle. Their 'agitated' behavior tends to persist over time and over successive handling sessions (Grandin, 1993).

Handling procedures that are highly stressful can reduce the animal's immune function, causing lowered resistance to disease and reduced reproductive success. Reproductive performance can be affected at several levels (e.g. failure to ovulate, sexual behavior problems, failure to conceive, miscarriages, premature births, maternal behavior problems). Stressful handling can result in lowered productivity (e.g.

growth rate, milk and egg production, etc.), as we saw in the last chapter (p. 240–243). Because the results of stressful handling experiences are seldom immediate, handlers often fail to make a connection between the handling event and the animal's health problem, poorer reproductive performance or reduced productivity. Compromised immune systems may not lead to illness for many days (or weeks) after a stressful handling event. Poor reproduction and slower growth rates are frequently attributed to genetics, nutritional deficiencies, etc., without considering the potential impact of handling stress. Good animal managers view rough handling as unacceptable.

Low-stress handling benefits animals and their handlers. In addition to minimizing handling impact on health, reproduction and productivity, animals that are handled carefully have less risk of injury to themselves, other animals and their human handlers. Low-stress handling reduces animal damage to enclosures and equipment. Handling is more efficient and enjoyable for the people involved. Routine handling procedures can sometimes be accomplished in less time and with fewer people, which, in many cases, represents significant monetary savings.

Facility Design for Handling Large Domestic Animals

Some basic guidelines should be followed when designing facilities to handle large domestic animals. These guidelines are based on the natural behavioral tendencies of livestock and the experience of animal handlers and behavioral researchers (Grandin, 2007a). Some of the more important ones are listed below.

- Facilities should be designed so that animal traffic is in one direction only. Cattle (*Bos*), sheep (*O. aries*), etc., tend to be followers. When they see animals going in a different direction, they want to turn and follow them.
- Forward movement in alleyways and chutes is facilitated by curved as opposed to straight pathways (Fig. 16.1). Livestock are motivated to keep in visual contact

Fig. 16.1. Curved alleys and solid sidewalls facilitate animal movement through work areas (courtesy of Temple Grandin).

with conspecifics and must stay in closer contact with the animal in front of it when pathways are curved. Although visual contact can be maintained at greater distances in straight alleyways, bigger gaps are often left between animals. Right-angle corners in an alleyway create the appearance of a 'dead end', thus discouraging forward movement.

- Movement is facilitated by solid sidewalls (Fig. 16.1). Solid sidewalls prevent animals from being distracted by events going on outside the alleyway or chute (e.g. people, other animals or vehicles moving about).
- Animals should be allowed to see beyond (i.e. through) temporary barriers (Fig. 16.2). Doors or guillotine panels are sometimes placed along alleyways and chutes to separate animals or temporarily prevent them from moving forward. If these barriers are constructed of solid wood or metal, they will appear as a 'dead end' to the animals and will impede forward movement up to the barrier. Doors and panels constructed of materials that animals can see through encourages movement up to the barrier, particularly if the lead animal can see conspecifics on the other side.
- Projections on sidewalls should be avoided to prevent bruising and injury to the animals. Door hinges, bolt ends and other metal materials used in construction should be recessed or hidden in the walls of alleyways, chutes and pens.

Fig. 16.2. Temporary barriers in working chutes should be constructed of materials that animals can see through; otherwise, they will stop short of the barrier, as if confronting a 'blind alley' (courtesy of Temple Grandin).

- Inclines (i.e. ramps) should be gradual (not greater than 20 degrees) and provide good footing for the animals. Steep, smooth slopes become very slippery when wet or covered with manure. Slippery substrates inhibit animal movement and can cause injuries if the animal falls. Sheep (*O. aries*) and other livestock move up inclines more readily than down.
- Entry points to alleyways and chutes should *gradually* 'funnel in' animals and exit points should *abruptly* funnel them out. If the funnel leading up to a chute is too narrow, or if too many animals are driven into the funnel at a time, they often become so packed or wedged together that movement into the chute is slowed, or stopped entirely. Exit points should be wide enough to allow animals to exit slowly and not be pushed by animals to their rear.
- Handling facilities should be constructed to provide uniform lighting. Shadows and 'zebra stripes' (Fig. 16.3a,b) on the sides and floor of an alleyway or chute tend to be avoided by livestock, perhaps because of associating shadows with unsure footing.
- Alleyways, chutes and pens should be reasonably well lit. Animals sometimes balk at moving from a well-lit area to one that is dark. This is particularly important when moving animals on a sunny day into an unfamiliar building that is dark inside. Animals move best when they can clearly see the area ahead.

Techniques for Moving Large Domestic Animals

Positioning and movements of handlers

Grandin (1998) describes one of the better techniques for moving large domestic animals through chutes or alleyways (Fig. 16.4). Animals tend to move forward in a chute when a handler within the animals' flight zone walks past them in the opposite direction to which they are facing (i.e. direction of desired movement). Chutes designed for cattle are usually equipped with 'catwalks' along the side of the chute to put the handler's torso in full view of the animals. Animals are inclined to move forward when the handler passes the animal's shoulder, the so-called 'point of balance' between stimulating forward or backward movement. The handler's speed of movement along the catwalk depends on the rate at which the animals move forward as the handler passes by them. When the handler reaches the rear of the chute, he or she should leave the catwalk and move to the front of the chute, while out of sight of the animals, to begin the next pass.

Grandin uses a similar technique when moving animals out of a pen or corral. The handler moves gradually into the flight zone of the lead animals while walking in the opposite direction of desired movement (away from the gate). After the handler passes a number of animals, he or she moves outside the flight zone of the closest animals, while returning to the front of the group to begin another pass, as described. This keeps the group moving forward, aided by the animals' tendency to follow those in front. This cycle is repeated as many times as necessary until all of the animals are removed from the pen. A fence or wall of the corral can be used to contain the animals on the side opposite the handler. Preferably, a large herd or flock of animals should be broken up into smaller groups to move them most efficiently and in a low-stress manner. It is important to avoid getting directly behind the animals. Livestock have a 'blind' spot

Fig. 16.3. Livestock tend to avoid zebra stripes (a) and shadows (b), thus inhibiting movement of animals along alleys and corridors (courtesy of Temple Grandin).

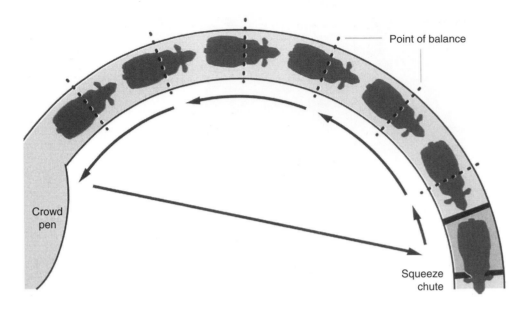

Fig. 16.4. Cattle will move forward when a handler, moving in the opposite direction along a single-file race, passes the shoulder ('point of balance') of each animal. The handler moves out of the flight zone of the animals while returning to the head of the alley (Grandin, 1998).

directly behind them and will turn their heads to locate the position of a handler in the rear. Since animals tend to move in the direction their head is facing, this response will sometimes cause the animals to veer away from the direction of group movement.

Sorting animals at a gate (i.e. retaining some in the corral and allowing others to exit) is more difficult to master since the handler must move in and out of the flight zone of animals very quickly with accurate timing (Grandin, 2007b). In this case, one or more handlers slowly move a group of 8–10 animals toward the gate, preferably in single or double file, while the sorting handler assumes a position at the gate opposite the fence or wall containing the animals on the other side (Fig. 16.5a). The distance from the sorting handler to the opposite side of the gate should be great enough not to discourage approaching animals from passing by the handler and through the gate. When an approaching animal is to be retained in the corral, the handler moves at an angle toward the animal and the opposite fence (Fig. 16.5b) until the animal turns back into the corral (Fig. 16.5c,d). If the next animal approaching the gate is to exit, the handler immediately moves back to his or her original position outside the animal's flight zone (Fig. 16.5e), letting the animal move through. If an approaching animal is to exit but starts to turn back into the corral, the handler immediately steps away from the gate and in the direction the animal is turning, leaving more space for the animal to pass by and turning it back toward the gate. Mastery of this technique is largely dependent on the sorting handler's ability to read the body language and intentions of the approaching animal. A good sorting handler focuses on the head of the approaching animal to determine whether its inclination is to move forward and exit or to turn back into the corral. At the instant the animal moves its head in an undesired direction, the handler steps to the left or right, whichever is appropriate, to discourage movement in that direction. Immediacy in the handler's response is very important.

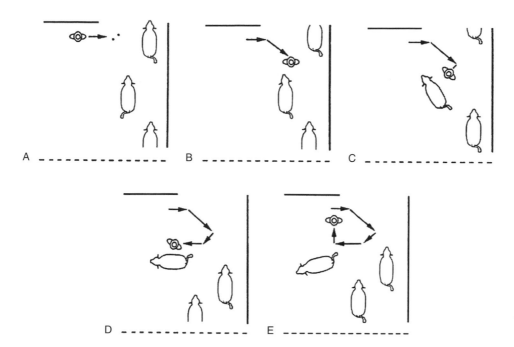

Fig. 16.5. Handler movement patterns for turning an animal back when sorting at a gate (Grandin, 2007b, citing Smith, 1998).

The techniques used in moving animals from one pasture or field to another are similar to those used in moving animals out of a pen or corral (Grandin, 2007b). First, the animals are gathered into some semblance of a group. If the pasture is very large, the handlers may choose to be on horseback or on an all-terrain vehicle and (or) use herding dogs. The animals will move in the desired direction more quickly and efficiently if a following stimulus is provided, preferably one or more riders on horseback (Fig. 16.6). Livestock are moved more efficiently and with less stress on the animals and handlers if a following stimulus is provided. Fewer handlers are required when moving animals along a fenceline. Once movement is started, the animals may simply follow along behind those in front of them. Stress is minimized when movement is at a relaxed pace (i.e. walk). Animals that are 'pushed' or rushed are more likely to split

Fig. 16.6. Cattle following a rider on horseback along a fenceline. Another rider with dogs is in the rear.

off from the group. Ordinarily, handlers should not pursue single individuals that split from the group. This takes the handler out of position relative to the herd and other handlers. Furthermore, the gregarious nature of most livestock species will likely motivate these wayward individuals to rejoin the group rather than be left alone.

Use of social facilitation in handling large animals

It is fortuitous that many of our domestic animal species are gregarious and readily follow conspecifics. I was attending a livestock auction at a county fair one summer when a steer (*B. taurus*), which had just been brought into the auction ring, panicked and ran out of the ring, out of the fairground and across the road into a farmer's wheat field. I expected some men on horseback to go out in the field to rope the animal. Instead, the steer's owner tethered another steer familiar to the escaped animal to the back of his truck and slowly led this animal out into the field toward the escaped steer. The panicked steer saw the animal tied to the truck and followed it back on to the fairground, where it was easily placed in a pen without incident. Needless to say, I was impressed with the owner's understanding of animal behavior and the astuteness with which he handled this difficult situation.

Horses (*E. caballus*) not willing to enter a barn or trailer or cross a stream may do so willingly if they can follow another horse into those areas. Social facilitation can make an otherwise difficult handling task easy for both animals and their handlers.

Mothers are distressed by separation from their young and are strongly motivated to follow wherever their young are taken, as long as the young remain in sight. I have used this technique many times to move ewes (*O. aries*) and cows (*B. taurus*) that have recently given birth from a paddock into a barn. The newborn is placed on a vehicle or carried in plain view of the mother and then moved at a slow pace to wherever you want the mother to go. She will follow close behind (Fig. 16.7). I must add that a small percentage of cows, especially of Brahman (*B. indicus*) descent, can be aggressive toward humans when their newborn calves are approached and captured. Children are particularly vulnerable in this situation.

In parts of Europe and Asia, shepherds use trained sheep (*O. aries*) and goats (*C. hircus*) in their flocks to facilitate movement of the herd. These animals are often hand reared and trained to come to the shepherd's calls. Several of these trained animals in a flock can entice the rest of the animals to follow the shepherd's lead.

Social facilitation can also be used to facilitate movement of livestock in chutes and alleyways. Lead sheep in chutes often balk if they become fearful. Such animals are a problem because they refuse to move forward and cannot move backward (because of animals behind them). Some sheep producers keep a trained 'Judas' sheep or goat on hand to serve as a 'follower' stimulus. The trained Judas animal is motivated to move through the chute for a food reward at the other end and, in the process, serves as an escort for the other animals.

Temporary social isolation is one of the most potent stressors for livestock and other group-living animals (Moberg *et al.*, 1980; Rushen, 1986). Some animals exhibit panic-type behaviors when they find themselves separated from conspecifics and in an unfamiliar environment. Many serious injuries to handlers and animals have occurred under these conditions. Single animals requiring treatment should be accompanied by a familiar conspecific to prevent isolation-induced panic or stress.

Fig. 16.7. Ewe following a person holding her newly born lamb.

Fig. 16.8. Cattleguard at the entrance to the UC Sierra Foothill Research and Extension Center.

Fear-producing stimuli which interfere with animal handling and movement

Animal handlers should be constantly aware of stimuli which can interfere with animal movement and increase the level of stress experienced by their animals (Grandin, 2007a). Inappropriate encroachment into the flight zone of animals by people and dogs (*C. familiaris*) is one of the primary reasons animals are difficult to handle or move. Other fear-producing stimuli which can interfere with animal movement include noises from people shouting, metal-on-metal clanging noises, high-pitched or hissing sounds, objects moving in the breeze (e.g. flags, pieces of paper, clothing on a fence), air drafts blowing in the animals' faces and reflections off puddles of water. Livestock are reluctant to step on anything that has an unusual feel or sound (e.g. drains, wooden floor). 'Cattleguards', used in place of gates on roadways (Fig. 16.8), are avoided by livestock because of this principle.

Training Animals to Facilitate Handling and Movement

Habituation or desensitization of domestic animals to humans, handling procedures and restraint are best accomplished at an early age. Many horse (*E. caballus*) breeders begin to expose foals to the sights, sounds and tactile stimulation associated with handling within hours or days after birth. Foals are physically restrained, their torso, ears and feet are rubbed, they are groomed, walked through trailers and chutes and halters are placed on their heads in an effort to reduce or eliminate the fear animals often exhibit when first introduced to these stimuli and experiences at an older age. However, neonatal handling does not always eliminate handling problems at older ages. Lansade *et al.* (2005) found that systematic handling of foals for 2 weeks following birth improved ease of handling (e.g. time to fit a halter, time to pick up feet, defensive reactions and percentage time walking when being led) up to 6 months of age, but at 1 year there were no differences between their handled and unhandled subjects. Handled and unhandled yearlings were also similar in their responses to isolation from conspecifics, presence of a human, presence of a novel object and a startle stimulus. Some professional horse trainers maintain that extensive handling during the first few days after birth often results in 2-year-old horses that are more difficult to saddle train, in spite of being easier to catch and lead. A rule of thumb is to give foals just enough unobtrusive early handling to make them relatively calm when in contact with people and accepting of handling techniques but not so much early handling that they become 'disrespectful' of people and unresponsive to training procedures later on. Recognizing and understanding the animal's 'body language' and anticipating its behavior are critically important when handling and training horses and other large domestic animals. Also, handling neonates within hours of birth can be hazardous to the handler if the mother becomes protective.

An animal's first experiences with a handling technique can set the stage for how they subsequently react to the same technique. An animal is more likely to show a negative or fearful response to a specific handling procedure if its first experience with the technique is aversive. Conversely, pleasurable first-handling experiences tend to facilitate subsequent handling. Cattle (*Bos*) producers have noted that their animals develop less fear of alleys and squeeze chutes if, on first exposure, they are allowed to walk straight through them without being detained or restrained.

Animals can learn to associate a place or location with certain handling techniques. Rushen *et al.* (1998) conducted a study in which a single handler treated the same dairy cattle (*B. taurus*) adversely in one location and gently in another. The cattle subsequently avoided the handler in the location where they had been treated harshly and showed little fear of the handler in the location where they had been treated gently.

Animals often increase their tolerance of certain handling procedures through repetition. Abbott *et al.* (1997) found that 'difficult to handle' pigs (*S. scrofa*) periodically taken out of their pens, through a weighing scale and back to their home pens became more willing over time to leave their pens. On the day of pre-slaughter transport, the pigs were easier to move, required less time for loading and were less likely to stress handlers and be subjected to harsh treatment during loading. Pigs stressed during pre-slaughter handling and transport often yield poorer quality meat resulting from 'porcine stress syndrome'.

Target training

Target training with animals has become popular in recent years because of its ability to motivate animals to perform desired behaviors in a humane manner, especially those behaviors involved in movement and handling. Target training has been used with a wide variety of species such as dogs, horses, pigs, seals, dolphins, whales, chickens, pigeons and even frogs. It has been used to train horses (*E. caballus*) to lead (i.e. move with their handler), stand quietly for veterinary exams, grooming, hoof trimming, shoeing or being sexually mounted and for entering stalls and trailers. It can be used to train dogs (*C. familiaris*) to heel, retrieve, stand on a scale to be weighed, go to their beds or to learn tricks such as jumping up to touch a high target in the same manner as dolphins and whales in aquarium shows. Target training is appealing because it uses only positive reinforcement and animals frequently appear eager to learn the task at hand. It can be used as a primary training technique or integrated into an existing training program.

In target training, the animal is first conditioned to respond to a bridging stimulus, as described on p. 56 in the chapter on learning. The bridging stimulus (e.g. clicker, whistle, voice command) activated by the handler communicates to the animal that it has just performed a desired response and it will soon receive a reward from the handler. The use of a bridging stimulus distinguishes target training from simple operant conditioning where the animal is rewarded immediately on performing a certain response. Clickers or whistles are often used as bridging stimuli because they produce sounds *unique* to the animal's living environment and can be easily heard. Voice commands are less effective because captive animals frequently hear human voices and become somewhat desensitized to them.

In target training, the animal is conditioned to approach and contact either a moving (mobile) or stationary object. The target in mobile target training can be the handler's hand or an extension of the arm and hand, such as a ball on the end of a pole. The animal is initially conditioned to receive a food reward if it touches the ball with its nose (Fig. 16.9). Since the ball on the pole can move with the handler, the target can be used to entice the animal to move anywhere the handler wishes. At first, the animal is rewarded every few steps that it takes in pursuit of the target. Reinforcement is offered less frequently as the animal more predictably follows the target. Mobile target training is especially useful with large, physically strong animals such as horses (*E. caballus*) and large breeds of dogs (*C. familiaris*) that must be under control at all times.

The target in stationary target training can be different than the one used in mobile target training. It can be a dog's bed or an object fastened to a spot where the handler wants the animal to go (e.g. the front of a stall or trailer, a feeding station). The handler himself is the target when he or she instructs an animal to

Fig. 16.9. Target training a young horse (courtesy of Sophia Yin).

'come'. Again, the animal's task is to make contact with the object and remain in contact with it. At first, the process is started with the animal very close to the target. Once the animal has learned to approach and touch the target, the next training session is started with the handler and animal a bit farther apart. This is repeated until the animal readily approaches the target from a distance. As in mobile target training, a bridging stimulus is used immediately to tell the animal it has performed a desired response and a reward will be forthcoming. Because the bridging stimulus becomes associated with a food reward, just the sound of the (auditory) stimulus will eventually motivate the animal to seek out a known stationary target.

Both mobile and stationary targets can be used to entice an animal to perform a set of behaviors. For example, a mobile target can be used to lead a horse (*E. caballus*) to a trailer and encourage it to enter. A stationary target, clearly visible in the front of the trailer stall, can then motivate the horse to move to the front and stand in a designated spot.

Once the animal consistently performs a desired behavior without the need for continuous reinforcement (see p. 62 for a discussion of various reinforcement schedules), use of the target can be discontinued and just the bridging stimulus used to reinforce desired behaviors. For example, when training a dog (*C. familiaris*) to 'heel' with target and bridging stimuli, the ball on the end of a stick serves as a definitive marker for how far forward you want the dog to be relative to your body. Once the dog learns that it will be rewarded only for staying behind that point, the ball and stick can be left home and only the bridging stimulus used to reward the dog when it heels properly.

Electronic devices used in training

This chapter would not be complete without mentioning the use of various electronic devices in training animals. Electronic devices capable of delivering a momentary mild electric shock, or 'nip', are available to inhibit dogs (*C. familiaris*) from barking, straying too far from their owners and for containing dogs and cats (*F. domestica*) in yards or out of gardens ('invisible fence®'). In each case, the animal wears a collar capable of emitting a momentary electric shock of an intensity just strong enough to get the animal's attention and be mildly uncomfortable. 'Bark collars' deliver a shock each time the dog barks. Collars designed to prevent animals from straying are capable of emitting an audible tone or vibration in addition to the shock. The tone or vibration is used to warn the animal that a shock is imminent if the animal does not respond appropriately. When training dogs to stay within a certain distance of their owner, the handler remotely activates the tone or vibration, followed by a low-intensity shock. The dog quickly learns that the warning signal precedes the shock and that it must respond by turning back toward the handler whenever it hears the signal if it is to avoid the shock (avoidance conditioning). Punishment (i.e. shock) during the initial training phase is delivered remotely at the exact time the animal is performing the undesired behavior. Dog owners without the capability of remote punishment too often call their dogs back and then reprimand them, causing the animal to think it is being punished for obeying the owner's command to 'come'.

Invisible fences work on the same principle as the remote trainer, except that the signals and shock, if necessary, are delivered automatically when the animal reaches the perimeter of the yard or garden. The perimeter is delineated by an underground wire that carries a radio signal, which activates the receiver on the animal's collar when it is in close proximity. Small flags are used during training to give the dog a visual cue of the fence's location. Wireless invisible fences are also available which have the advantage of being portable. A transmitter emits a radio signal in a circular field, which can be adjusted in size from about 10 to 60 m diameter. When the animal approaches the boundary of the circular signal field, it receives a warning beep. If the animal continues toward the boundary, it receives a momentary mild shock. Animals quickly learn to stay within the protected area.

Persons using these electronic devices with dogs (*C. familiaris*) should begin with conventional training using verbal commands, whistles, etc., to teach the animal what it is that the owner desires. Once the dog has mastered these commands, the electronic device should be used to reinforce what was previously learned in situations where the owner does not have direct control over the animal's behavior. At first, the lowest intensity shock level on the electronic device should be used, which probably will not be felt by the animal. The intensity is then gradually increased until a response is seen. That is the intensity to be used for training. Schalke *et al.* (2007) found that beagle dogs which were properly trained to associate the electric stimulus with their action (touching a dummy rabbit fixed to a motion device), and thus were able to predict and control delivery (i.e. avoidance) of the electric stimulus, showed no considerable or persistent increase in stress indicators (heart rate and salivary cortisol).

Although today's electronic collars are much more animal friendly than earlier models, improper use of electronic collars is unacceptable. Many people still feel that any use of electric devices to train animals is inhumane. I respect that opinion but, before adopting the use of an alternative training method, one should consider the likely impact of the technique on the animal or the possible consequences of insufficient training. The safety and well-being of both dog and owner are paramount. Many dogs off-leash have been injured or killed by vehicles, attacks from wild animals such as poisonous snakes and predators, drowning in rivers and canals, etc., when straying from home or disobeying their owner's commands. I know of a person who has started a business training dogs to avoid poisonous rattlesnakes using electronic collars and muzzled snakes. Pet control and safety is an important issue that deserves an open-minded approach.

Other training techniques and equipment

Conventional venipuncture techniques for caged monkeys require restraint of the animal by two or three technicians using mechanical restraint devices. Reinhardt (1991, 1996) described how rhesus monkeys (*M. mulatta*) could be trained to cooperate for venipuncture and intramuscular injections in their home cages. Training is based on offering food treats for cooperative behavior such as allowing the caretaker to touch and groom them, showing no resistance to having a leg gently pulled out of their partially-opened cage door or voluntary presentation of an arm through the bars of their cage (Fig. 16.10a–c). Uncooperative behavior is never punished and training trials are never terminated when the animal fails to cooperate, else the animal is reinforced for actions that avoid discomfort. Reinhardt compared cortisol levels (physiological measure of stress) of monkeys physically restrained on restraint tables or in restraint apparatuses versus monkeys trained to receive venipuncture in their home cages. Cortisol levels increased 63% and 50% over pre-handling, baseline levels when the monkeys were restrained on tables or in restraint apparatuses, while the increase was only 16% (statistically non-significant) for monkeys trained to accept this procedure in their home cages. In addition, only one person was required to perform the home-cage procedure compared to two or three technicians when using the more conventional forms of restraint. The home-cage technique was also faster.

Pigs (*S. scrofa*) at the University of Maryland (USA) were trained to stand immobile for blood collection using target training techniques. Total training time for each animal was approximately 1 h, accomplished in 5-min sessions (or less) over a period of 2–3 weeks. The only problem encountered was that the pigs competed to be first in line to have their blood drawn and receive their reward.

The 'Gentle Leader®' headcollar is an effective device for discouraging dogs (*C. familiaris*) from pulling and lunging on their leashes. The leash is attached to a ring under the dog's chin, which is attached to a strap over the dog's muzzle and a strap over the back of its neck behind the ears (Fig. 16.11). When the dog pulls on its leash, pressure is applied to both the muzzle and back of its neck and, in addition, the animal's head is pulled to the side. Interestingly, a dominant wolf (*C. lupus*) will sometimes reinforce its rank position by placing its mouth over the muzzle of subordinate pack members. Also, young pups appear relaxed when picked up by the back of their necks. I wonder if there is a connection between either of these two factors and the effectiveness of the 'gentle leader'?

Horse (*E. caballus*) handlers need to keep their animals under control at all times. Most horses used for breeding and recreation are trained to the use of a halter and lead. Halters are placed on young foals at an early age to habituate them to its feel on their head. A rope or leather lead is then attached to the noseband of the halter and the animal is trained to follow the handler.

Unruly animals can put their handler, other animals and themselves in danger of injury. Difficult to handle horses (*E. caballus*) can be controlled by attaching a chain shank to the lead, which extends through the ring on the noseband of the halter and across the animal's nose (Fig. 16.12). When the horse pulls on the lead, the chain applies painful pressure across the nose of the animal, which causes the horse to stop or back up. Discomfort is experienced only when the horse is pulling on the lead

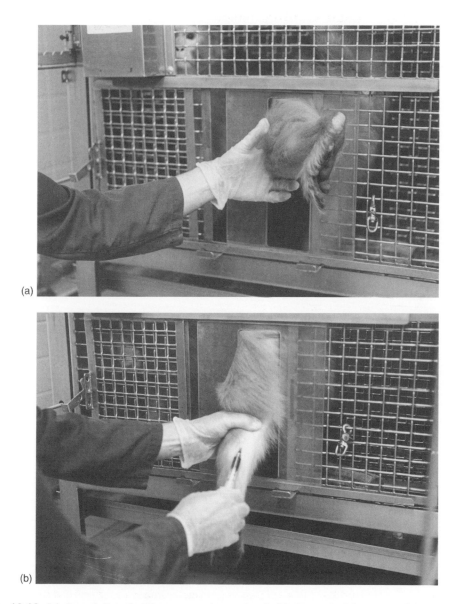

Fig. 16.10. (a) An adult male Rhesus monkey extends his leg out of the cage, (b) blood is collected and (c) he receives a food reward (Reinhardt, 1991, 1996).

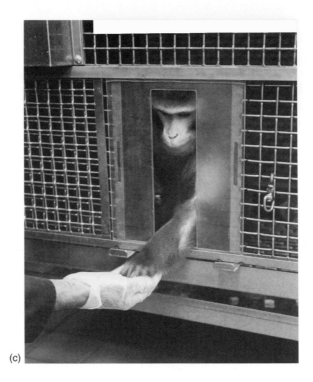

(c)

Fig. 16.10. continued

Fig. 16.11. 'Gentle Leader®' headcollar.

Fig. 16.12. Chain shank for controlling unruly horses.

(positive punishment) and it is experienced immediately, an important factor in the conditioning process. Choke chains on dogs work on this same principle; the discomfort is immediate and only when the animal performs the undesired behavior.

The 'twitch' and 'Stableizer®' are also used to restrain unruly horses (*E. caballus*). The twitch consists of a loop of rope or chain attached to a pole, which is placed over the horse's upper lip and twisted to apply pressure. Most horses will become docile and stand quietly when the twitch is applied. Lagerweij *et al.* (1984) found that application of pressure to the lip in this manner released endogenous opiate-like endorphins, which act as a sedative. The 'Stableizer' fits over the animal's head and applies steady pressure behind the ears and on the upper gum of the mouth between the front teeth and the upper lip. Sustained pressure at these two sites has a sedative effect on horses, mules, llamas and zebras.

There is evidence that horses (*E. caballus*) transported in a rear-facing position maintain their footing and balance better when traveling on rough roads. Also, their broad, fleshy hindquarters make contact with the bulkhead during braking and deceleration rather than their more sensitive head and chest. Waran *et al.* (1996) found that horses transported in the rear-facing position had slower heart rates, vocalized less and were less likely to hold their heads in an abnormally high position. Some horse owners have found it easier to back their horses into trailers, likely because they are not faced with a dark entrance.

Farmers, ranchers and pet owners have developed many simple and ingenious techniques to facilitate handling their animals. Swine producers frequently use easily held solid panels to move pigs (*S. scrofa*) along corridors (Fig. 16.13). The panel appears as a 'dead end' when an animal balks and starts to turn back. It also reduces

Fig. 16.13. Panel used to facilitate movement in pigs.

the risk of leg injuries to the handler. Some techniques are quite unorthodox. This was illustrated to me one day when I observed some people trying to load a pig on to a truck. The animal refused to go up the ramp provided and was too large to lift on to the truck. After many minutes of struggling with the stubborn pig, an experienced pig handler passed by, turned the pig around so that its rump faced the truck and then put its head in a pail. The pig immediately walked backwards up the ramp and into the truck as if it was trying to get its head out of the pail.

Use of Dogs in Handling and Moving Livestock

Dog breeds (*C. familiaris*) used in handling and moving livestock (i.e. conducting dogs) can be roughly divided into 'herders' (sometimes called 'headers' or 'retrievers') and 'heelers'. Herders include those breeds which have a natural tendency to circle around and gather animals. The Border Collie (Fig. 16.14) is the breed most commonly used for herding livestock in the USA. Kelpies (Fig. 16.15) and Australian Shepherds

Fig. 16.14. Border Collie.

Fig. 16.15. Black-and-tan Kelpie.

(Fig. 16.16) are also popular in certain areas of the world. Herder dogs will naturally position themselves on the opposite side of a group of livestock from the handler. They can be taught to keep the stock in a group and to move the stock toward the handler. During training, handlers teach their dogs to respond to vocal commands, whistles and hand signals. They learn such commands as 'fetch' (the animals), 'go closer', 'stop', 'go faster' and 'slow down'. Trained working dogs in the field perform these tasks with a minimum of direction from the handler. A well-trained dog can take the place of several human handlers. When moving livestock, the handler can serve as a following stimulus for the livestock, while the dog takes up the rear and keeps the animals together and moving forward ('pulling'), or the handler can assume a position behind the dog while moving the animals forward ('driving') (Fig. 16.17).

'Heelers' are dogs which chase livestock and sometimes bite at their heels. The Australian Cattle Dog (Fig. 16.18), also known as the Australian Heeler, Blue Heeler or Queensland Heeler, is popular in Australia and western USA. Heelers are typically used with cattle and in rough terrain such as mountainous areas and brush-covered

Fig. 16.16. Australian Shepherd.

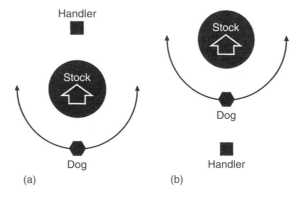

Fig. 16.17. 'Pulling' (a) and 'driving' (b) livestock with dogs.

Fig. 16.18. Australian Cattle Dog.

rangeland. Heelers do not have a natural propensity to herd and tend to chase live-stock, which limits their usefulness. However, they can be taught to herd.

Dogs and Other Animals Used to Protect Livestock

Dogs (*C. familiaris*), llamas (*Lama glama*) and burros (*E. asinus*) are used to protect livestock from predators (Smith *et al.*, 2000). Popular guard dog breeds in the USA include Great Pyrenees (Fig. 16.19), Anatolian Shepherds (Fig. 16.20) and the Italian Maremma (Fig. 16.21). Guard dogs are not 'attack' animals but protect livestock by their defensive behaviors (e.g. barking, urine marking) and their relatively large, intimidating size. The best guard dogs are those which have been reared with the species they are to guard, starting at 4–8 weeks of age (Fig. 16.22), prior to the sensitive period for socialization (see p. 36–37). They are given minimal contact with people during this period and thereafter to minimize the probability of becoming socialized to humans. They must live with livestock for their entire working lives. One-fourth to one-third of the dogs reared for guarding are retired early, either because they are overly aggressive toward unfamiliar humans or livestock or are ineffective in reducing predation.

It is estimated that sheep (*O. aries*) producers in western USA lose 1–2.5% of their adult animals and 4–9% of their lamb crop each year to predators, primarily coyotes

Fig. 16.19. Great Pyrenees.

Fig. 16.20. Anatolian Shepherd.

Fig. 16.21. Italian Maremma shown with Border Collie for size comparison (Coppinger and Coppinger, 2007).

Fig. 16.22. Guard dog being raised with sheep (courtesy of USDA Sheep Experiment Center, Dubois, Idaho).

(*C. latrans*). Lamb losses can reach 20% in some flocks. This badly depletes, if not eliminates, the producer's profits. A 10-year study on the effectiveness of five breeds of guard dogs in protecting sheep from predators revealed a 64% average annual reduction in predation. Fifty-three percent of the producers surveyed reported their losses were reduced to zero. One large rancher in South Dakota with 30,000 sheep suffered a 10% annual loss to coyotes. After he started using Great Pyrenees guard dogs, his losses were reduced to 1% or less (Francis, 1991). A Wyoming rancher lost about 450 sheep/year to coyotes before using guard dogs. With dogs, his losses declined to 10–15 animals. Two or three dogs can protect 1000 ewes and their lambs.

Llamas (*L. glama*) and burros (*E. asinus*) are also effective livestock guardians (Figs. 16.23 and 16.24), particularly against coyote predation. Their large size and

Fig. 16.23. Llama being used to guard sheep (courtesy of William Franklin).

Fig. 16.24. Burro being used to guard sheep.

aggressiveness toward predators is intimidating. A survey of 145 sheep (*O. aries*) producers revealed their average annual losses of sheep to predators declined from 11% to 1% of their flocks after they started using llamas as guard animals. More than half had their losses reduced to zero. One California sheep producer found that a burro cut his losses to coyotes from 200 ewes and lambs/year to almost zero. As with dogs, not all llamas and burros are effective guardians. Age, personality traits, social habits, physical condition and prior training and handling can determine their success.

Table 16.1. Predatory behaviors exhibited by wild canids (e.g. coyote, wolf), conducting dogs and protecting dogs (from Coppinger *et al.*, 1987).

Wild canids:	'Eye' → Stalk → Chase → Bite → Dissect → Ingest
Conducting dogs:	'Eye' → Stalk → Chase
Protecting dogs:	None (only play-chase and play bite)

Two advantages of llamas and burros over dogs are that they do not have to be fed a special diet and they do not have to be reared with the species they will be guarding. They are naturally gregarious and protective. Large, curious and attentive llamas make the best guard animals. Dominance rank and aggressiveness toward conspecifics is not correlated with aggressiveness toward canids. Male llamas make the best guardians because of their greater aggressiveness. A disadvantage is that they will sometimes attempt to mate with estrous ewes during the breeding season. Female burros generally work better than males with sheep since males can be aggressive toward the animals they are guarding.

Coppinger *et al.* (1987) and others have noted that dogs (*C. familiaris*) used to conduct livestock (i.e. headers and heelers) exhibit many of the predatory behaviors of wild canids. They 'eye' their quarry (Fig. 16.14), stalk it and engage in chase. They differ from wild canids in that bite, dissect and ingestive behaviors have been lost (Table 16.1). In contrast, protecting dogs have lost all aspects of their ancestors' predatory behaviors. It is as if their behavioral development is terminated at the juvenile stage, a phenomenon often referred to as neoteny.

17 Atypical Behavior and Behavioral Therapy

Atypical Behavior Defined

Atypical behavior may be operationally defined as those behaviors that are exaggerated in terms of frequency and (or) intensity, disoriented in relation to stimuli or occur in the absence of normal eliciting stimuli. Many of the behaviors we call 'atypical' are actually components of 'normal' behavior directed toward inappropriate stimuli.

Atypical behaviors are generally considered maladaptive or undesirable, particularly when they become obsessive or pathological, or result in injury. However, some behaviors that appear abnormal represent an attempt by the animal to adapt to an atypical or impoverished environment. From that standpoint, some atypical behaviors may be beneficial to the animal under the conditions in which it lives.

Although the expression of atypical behavior is often an indication that something is lacking in the animal's environment, it does not *necessarily* mean an animal's current welfare status is poor.

Causes of Atypical Behavior

Atypical behaviors are seldom seen in wild, free-living animal populations. However, they are commonly observed in captive animals, particularly when individuals are living in relatively small, sterile environments. Atypical behaviors are most likely to develop when animals are prevented from developing and (or) exhibiting their 'normal' behavioral repertoire (Mason and Rushen, 2006). Some causes of atypical behaviors include:

- Locomotor, perceptual and (or) social deprivation associated with a reduction in the quantity and (or) quality of space.
- Unnatural social groupings (i.e. increased or forced exposure to social stimuli).
- Forced exposure to humans and novel objects.
- Nutritional deficiencies and reduced or unpredictable feeding opportunities.
- Hormone imbalance (e.g. cystic ovaries can result in female mammals experiencing continual estrus).
- Pain resulting from injury to the body or structural abnormalities (e.g. arthritis).
- Early weaning.
- Brain damage or psychosomatic disorders resulting in bizarre behaviors.

Stereotyped Behaviors

Behaviors exaggerated in terms of frequency and (or) intensity are often referred to as stereotyped. 'Exaggerated' refers to outside the normal range of variability for the species. Some examples are given below.

Locomotor behaviors

- 'Weaving' in horses (*E. caballus*) – shifting their body weight back and forth from one foreleg to another, often accompanied by swinging the head and neck from side to side.
- 'Whirling' – particularly common in kenneled dogs (*C. familiaris*) when aroused.
- 'Pacing' in large zoo animals.
- 'Backflipping' (backwards somersaults) in caged rodents.

Mouth-based behaviors

- Wood chewing and 'cribbing' in confined horses (*E. caballus*). 'Cribbing' is when the horse places its upper incisors on a solid object such as a fence or stall door and bears down on its teeth while arching its neck and sucking in air (Fig. 17.1).
- Tongue rolling and intersucking (Fig. 17.2), as in cattle (*B. taurus*).
- Bar gnawing in caged rodents.
- Bar biting (Fig. 17.3) and tail biting in confined pigs (*S. scrofa*).
- Wool eating and wool pulling in confined sheep (*O. aries*).
- Feather pecking and cannibalism in chickens (*G. domesticus*).

Stereotyped behaviors are quite common in confined domestic animals. For example, in the UK, it is estimated that over 15% of confined horses (*E. caballus*) exhibit some type of atypical behavior such as stereotyped weaving or cribbing (Waran and Henderson, 1998). The figure may be as high as 20–35% among stabled racehorses.

Some stereotyped behaviors have harmful side effects. For example, wood chewing in horses can result in the ingestion of wood splinters, which can cause severe problems in the stomach and intestines and increase the risk of colic. Splinters can also get wedged between teeth or imbedded in gums. Cribbing in horses can cause

Fig. 17.1. Mare cribbing on a metal pipe (courtesy of Lisa Nash Holmes).

Fig. 17.2. Non-nutritive 'intersucking' by two dairy calves after being fed (courtesy of Janet Baer).

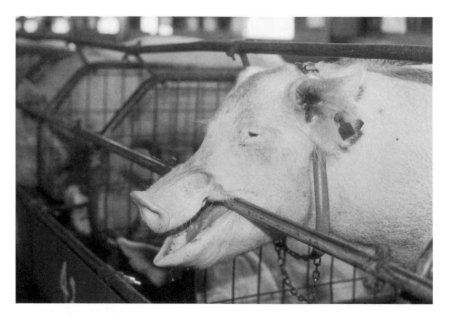

Fig. 17.3. Bar biting by a sow.

the incisors to wear down faster than they can grow out, leading to erosion of the teeth and malocclusions, which sometimes take years to correct. Cribbing horses become ineffective grazers if the incisors do not meet when the mouth is closed. Tooth problems can also lead to colic. Wool eating in sheep (*O. aries*) often causes the formation of compact wool balls in the animal's stomach (abomasum). These balls can cause colic, anemia and general poor condition. The animal may die if the balls obstruct the intestine. Feather pecking in chickens (*G. domesticus*) damages or removes the feathers of their victims. Feather loss is painful, increases the risk of further injury and causes significant heat loss, thus increasing energy requirements.

Treatment of Atypical Behaviors

A number of approaches have been used to treat atypical behaviors in captive animals, including pharmacological techniques, reinforcement of alternative behaviors, punishment, genetic selection and environmental enrichment. Mason *et al.* (2007) have concluded that of these approaches, environmental enrichment provides the best opportunity to improve the behavior and welfare of captive animals exhibiting atypical behaviors, with the fewest undesirable side effects. Environmental enrichment falls into four general categories; foraging opportunities, structural complexity, social companionship and novelty-induced sensory stimulation (Miller and Mench, 2005). The effects of these approaches on the prevention and expression of atypical behaviors are highly variable. Some examples are presented below.

Foraging opportunities

On pasture, horses (*E. caballus*) spend up to 16 h of their day grazing. In confinement, they can eat all that is nutritionally required in 2–3 h. Winskill *et al.* (1996) reported on the use of the 'Equiball®' to increase time spent feeding by confined horses. The Equiball is a round, barrel-like object (Fig. 17.4) that delivers small amounts of food as the horse pushes it around the floor of its stall or pen. The Equiball has been found to increase the general level of activity of confined horses and to reduce certain stereotyped behaviors. I am not necessarily endorsing the use of the Equiball but using it to show how animal caretakers can use their creative skills and ingenuity to provide a more stimulating environment for animals that must be housed under relatively sterile conditions.

Placing hay in the cages of individually housed domestic rabbits (*O. cuniculus*) reduced several behaviors considered atypical, including bar biting, sham chewing and excessive fur licking (Lidfors, 1997).

Modification of the animal's physical environment

Stereotyped and other atypical behaviors are most commonly observed in captive animals living under conditions where the quantity and (or) quality of space is severely restricted. Performance of atypical behaviors in captive animals can often be reduced or eliminated by providing greater opportunity for physical exercise or stimulus exploration. Some examples follow.

Weaving and cribbing in horses (*E. caballus*) confined to stalls or corrals can often be reduced or eliminated by placing them on pasture or providing them with something in their stall to manipulate, like a tetherball hung from the ceiling.

Fig. 17.4. The 'Equiball®' feeding device releases small quantities of feed as the horse pushes it around with its nose (courtesy of Natalie Waran).

Captive animals exhibit fewer atypical behaviors if the complexity of their environment is increased with items to climb or manipulate, substrate to dig or dust bathe in and places to hide or escape from conspecifics, etc. For example, Gebhardt-Henrich *et al.* (2005) demonstrated that caged golden hamsters (*M. auratus*) were less likely to exhibit stereotyped bar chewing if they were given an activity wheel for exercise. Interestingly, they also found that breeding females with activity wheels produced more offspring per litter. Zoos have moved toward more naturalistic exhibits in recent decades to improve the quality of space available to their animals and enhance the aesthetic appeal to human visitors.

There is increasing evidence that an early sensitive period exists for experiences which reduce or prevent the development of atypical repetitive behaviors. Hadley *et al.* (2006) found that fewer deermice (*P. maniculatus*) developed stereotyped behaviors when housed in a relatively large, more complex environment for intervals of time prior to 124 days of age. The effect persisted even when the mice were subsequently housed in smaller standard cages. Older mice did not benefit from this same enrichment. One of my graduate students, Michele Callard, found that simply placing a nest box in the cages of captive-bred wild roof rats (*R. rattus*) resulted in significantly lower rates of backflipping following weaning (Callard *et al.*, 2000). Shelters can insulate or buffer captive animals from fear-evoking stimuli that may otherwise contribute to the development and expression of repetitive stereotyped behaviors.

Modification of the animal's social environment

Most domestic animals are gregarious by nature and seek out the company of con-specifics. Ironically, many farm animals, and a large percentage of our companion animals, are forced to live in social isolation for much of their lives. Providing socially isolated animals with a companion is often effective in reducing the incidence of atypical behaviors. Even allowing an animal to see itself in a mirror has been effective. Enhancing the social environment of captive animals is less subject to habituation than modification of the physical environment.

Weaving in stabled horses (*E. caballus*) has been reduced by providing a social companion (e.g. horse, goat, sheep), a mirror, or even a poster of a horse's face (Mills and Riezebos, 2005). The frequency of spot picking in birds, a stereotyped behavior in which the tip or side of their bill is used to touch a specific place on their body or a cage object such as a perch, can be reduced by providing a view of a bird in an adjacent cage or by placing a small mirror in the bird's cage.

Novelty-induced sensory stimulation

Non-human primates in zoos are stimulated by interactive devices such as tools which scatter food, blasts of air that they can direct at visitors and water sprays that they can activate to cool off on a hot day. However, zookeepers have found that such devices need to be changed periodically to keep the animals entertained. Videotapes and video-game joysticks are also stimulating to captive primates (Platt and Novak, 1997).

In general, the stimulative effect of the introduction of toys and other novel objects into the living environment of captive animals is relatively short-lived due to habituation and thus is of limited usefulness in preventing and treating atypical behaviors. Enrichment devices which challenge the cognitive abilities of captive animals may prove to be most effective, as long as the tasks do not lead to excessive frustration and stress (Meehan and Mench, 2007).

Addressing specific biological 'needs'

When an atypical behavior is related to a specific biological 'need', the best treatment may be to address that requirement.

Targeting nutritional deficiencies

The development of atypical behaviors can sometimes be linked to nutritional deficiencies. For example, the incidence of tail biting in confined fattening pigs (*S. scrofa*) is greater when they lack sodium in their diet (Fraser, 1987).

Wood chewing in horses (*E. caballus*) can be related to a lack of roughage in their diet. Horses fed large amounts of hay are less likely to exhibit wood chewing than horses placed on a high concentrate diet with little roughage. Willard *et al.* (1977) found that horses fed hay spent ten times more time feeding than those fed a concentrate diet, but only spent one-fifth the time chewing wood.

One theory proposed to explain cribbing in confined horses (*E. caballus*) is that it is a response to visceral discomfort in the digestive system caused by excess acid resulting from being fed a cereal-based diet. The theory holds that the additional alkaline saliva secreted in response to cribbing helps to neutralize these acids, thus reducing the discomfort. Hemmings *et al.* (2007) discuss this and other possible explanations of cribbing in horses.

Targeting specific motivational states

Some behaviors become atypical because the animal's motivation to exhibit the behavior cannot be satiated. These behaviors are typically related to basic 'needs' such as eating/chewing/suckling and locomotor activity. For example, non-nutritive sucking, that is, sucking on the ears or other parts of a conspecific's body and sucking inanimate objects such as bars and other objects in the animal's pen, is common in dairy calves (*B. taurus*) and other young ungulates which are taken from their mothers at birth and artificially fed milk from a bucket or artificial teat affixed to a bottle or pail. In either case, milk is drunk quickly, often before the physiological mechanisms of satiation can take effect (about 10 min in calves) and before the animal's motivation to suckle has been satiated (see p. 158). Researchers (e.g. see review by de Passillé, 2001; Veissier *et al.*, 2002) have found that sucking on conspecifics and object sucking can be reduced by allowing the animals to suckle a dry (non-nutritive) or wet (water) artificial teat after the milk meal is consumed to provide more opportunity for suckling. Vieira *et al.* (2008) obtained the same result by feeding larger quantities of milk. In the latter study, restricted-fed calves spent twice as much time on the teat per meal as calves fed milk *ad libitum*. The restricted-fed calves drank their entire meal in one suckling bout and then engaged in many short non-nutritive sucking bouts, whereas the *ad libitum* calves drank their meal in several suckling bouts with little or no non-nutritive sucking after the meal had been consumed.

Non-nutritive sucking after being fed from an artificial teat is less prolonged than when young are fed from a bucket because milk is consumed more rapidly from buckets, resulting in a longer interval before satiation mechanisms reduce the motivation to suckle. Non-nutritive sucking after a meal can also be reduced by slowing the rate at which milk flows from an artificial teat, thus prolonging feeding (suckling) time. Providing hay for the animals to munch on after a meal also reduces non-nutritive sucking.

Some foraging species such as pigs (*S. scrofa*) are predisposed to mouthing and chewing objects in their environment. Pigs kept on barren soil chew soil and stones. Pigs housed in relatively small, sterile pens and fed intermittently are attracted to chewable objects such as pen fixtures and bars and the ears or tails of penmates. These behaviors frequently become habitual and tail and ear biting can result in serious welfare issues for the victims. Severe tail-biting outbreaks in confined pigs can often be avoided by providing copious amounts of straw bedding on the floor of their pen for them to manipulate (Van de Weerd *et al.*, 2006). It was found that intensively-reared pigs readily chew on ropes and chains hung from the ceiling of their pens. Dantzer and Mormède (1981) demonstrated that 'chain pulling/chewing' by pigs reduced plasma concentrations of corticosteroids (Fig. 17.5), suggesting that the opportunity to perform such activities served to reduce stress.

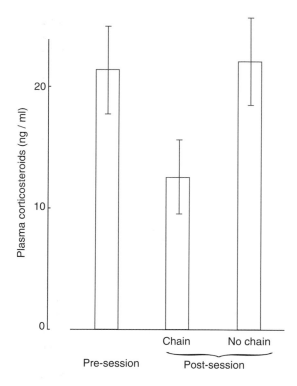

Fig. 17.5. Mean (± SE) circulating corticosteroid concentrations in pigs before and after they developed stereotyped chain pulling. When the chain was taken away, corticosteroid concentrations returned to their pre-treatment levels (Dantzer and Mormède, 1981).

Unpredictable feeding schedules can alter behavior patterns and increase the level of stress in captive animals. For example, Ulyan *et al.* (2006) found that when capuchin monkeys (*Cebus puella*) were fed on an unpredictable schedule for 6 weeks, they were less active and exhibited fewer social interactions than when feeding times were the same each day. They also had higher than normal levels of cortisol ('stress hormone'), which did not decline over the 6-week study period. Carlstead (1986) demonstrated that domestic pigs (*S. scrofa*) given reliable cues that food was forthcoming were less aggressive between feeding bouts than pigs given unreliable feeding cues. She hypothesized that unpredictable feeding cues increased the pigs' level of frustration.

Feather pecking and feather denudation in chicken (*G. domesticus*) production units is seen as an important animal welfare issue. One explanation for feather pecking is that it is redirected ground pecking, a natural foraging behavior of free-living chickens and other gallinaceous birds.

Blokhuis and Wiepkema (1998) noted that the incidence of ground pecking was six times greater for birds on litter compared to birds on barren slatted floors (Fig. 17.6). It is believed that in the absence of litter such as straw, wood shavings and other attractive substrate materials, the motivation of chickens to ground peck is not satisfied and feather pecking becomes a satisfying alternative behavior. In essence, feather pecking becomes a substitute for ground pecking and cagemates become the primary target. McAdie *et al.* (2005) found that feather pecking was significantly

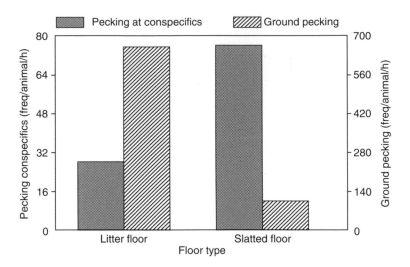

Fig. 17.6. Mean frequencies of pecking at conspecifics and ground pecking by chickens housed on litter or slatted floors (Blokhuis and Wiepkema, 1998).

reduced in a strain of chickens selected for feather pecking by providing a device consisting of strands of white string (eight 16 cm lengths of white propylene baling twine tied together at one end with midpoints at the height of the birds' head) for the birds to peck. Another hypothesis to explain feather pecking is that it is a substitute for dust bathing. Dust bathing is also inhibited by a lack of litter, or the right kind of litter. Dust-bathing hens prefer substrates with a fine structure, such as peat moss or sand (De Jong *et al.*, 2007). Whether feather pecking is encouraged by reduced opportunity for either ground pecking or dust bathing, it appears that lack of substrate is the primary cause. Chickens are highly motivated to seek out and use substrate materials and will work to gain access to it (Harlander-Matauschek *et al.*, 2006).

The finding that early weaning increases the incidence of atypical behaviors in rodents has triggered renewed interest in how breeding stocks of laboratory rodents are managed. Würbel and Stauffacher (1997, 1998) found increased exploratory activity and corticosteroid levels in mice (*M. musculus*) weaned prematurely or at a lower than normal body weight. They proposed that removing young mice from their parents at a time when suckling motivation was high plus handling by humans and being placed in a clean, unfamiliar cage abruptly increased their level of arousal and stimulated an increase in exploration and (or) general activity. The increased locomotor behavior resulting from these sudden changes in their social and physical environments likely became the source behavior for the development of repetitive stereotyped behaviors such as wire gnawing (Fig. 17.7) when other outlets for activity were not available. Callard *et al.* (2000) noted that their young roof rats (*R. rattus*) initiated backflipping during the first night after weaning, an atypical behavior never observed prior to weaning.

Providing activity wheels for caged rodents can prevent, reduce or eliminate certain stereotyped locomotor behaviors such as backflipping. Activity wheels appear to accommodate the behavioral 'need' for locomotor activity in confined rodents. Caged mink (*M. vison*) are also highly motivated to use activity wheels (Hansen and Jensen, 2006).

Chapter 17

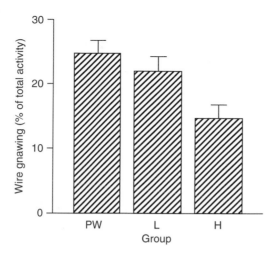

Fig. 17.7. Levels of stereotypic wire gnawing for domestic mice at 80 days of age. Mice weaned 3 days prematurely (PW) at 17 days and mice with low weaning weights (L) at 20 days, relative to their littermates, exhibited significantly more wire gnawing than mice with relatively high weaning weights (H) (Würbel and Stauffacher, 1998).

Fear of novel objects ('neophobia') can sometimes lead to unusual and bizarre behaviors in captive animals maintained in relatively sterile and stable (unchanging) environments. Fox and Millam (2007) found that frequent introduction and rotation of novel objects in the cages of orange-winged Amazon parrots (*A. amazonica*) significantly reduced the neophobic tendencies of most of their birds.

Not all attempts to treat atypical behaviors are successful. These behaviors can become so ingrained that the animals continue to exhibit them even when moved to environments allowing full expression of their behavioral repertoire. Horses (*E. caballus*) which develop cribbing behavior when housed in stalls will sometimes continue this behavior after being placed on pasture. Fox and Millam (see previous paragraph) found that *highly fearful* parrots did not show a reduction in their fear of novel objects, even when repeatedly exposed to them. These examples suggest that it may be easier to prevent than eliminate many atypical behaviors.

Behavior Modification

Behavior modification techniques are designed to reduce or eliminate atypical behaviors or responses to stimuli perceived as undesirable or detrimental to the animal and (or) its owner (Hart *et al.*, 2006). Behavior modification is most frequently used with companion animals because of their close relationship with humans. Several such techniques are described below.

Punishment

Administering aversive stimulation (positive punishment) when an animal exhibits an undesired behavior or taking away something the animal wants (negative punishment) is often sufficient to terminate an undesirable behavior (see p. 50). Punishment is most effective if it is administered while the animal is engaged in the unwanted behavior and if it is administered each time the animal exhibits the behavior. Punishment should be intense enough to get the animal's attention and terminate the behavior but not so intense that it causes undesirable side effects such as aggression directed at the handler. Punishment need not be physical. Sometimes, the command 'no' in an authoritative voice is all that is needed. However, verbal reprimands should not be linked to a positive reinforcer such as giving the animal attention, which sometimes is what the animal is seeking. When verbal reprimands provide the sought-after attention, they shorten the duration of unwanted behaviors but not their frequency. Punishment is most effective in reducing or eliminating learned or conditioned responses. It is not recommended for modifying behaviors based on unconditioned responses or emotional reactivity (i.e. anxiety, fearfulness, etc.). In general, punishment should not be used as a first line or early use treatment for behavior problems and should only be used when animal owners are aware of its possible side effects (see the American Veterinary Society of Animal Behavior 'Position Statement – Guidelines for Punishment'; www.avsabonline.org).

Counterconditioning

Counterconditioning involves training animals to associate the stimulus evoking an atypical or undesired behavior with an unrelated and rewarding event or activity. For example, it is common for house dogs (*C. familiaris*) to become alarmed and bark when someone knocks on the door or rings the doorbell. The sounds of the knocking and bell are associated with strangers, which can evoke fear in many of our four-footed friends. Individuals of some breeds of dogs can become aggressive under these circumstances. Behavior therapists sometimes recommend presenting the dog with a preferred food item when strangers announce their presence at the door. The sound of the doorbell (or knocking) and the appearance of a stranger now become associated with a pleasurable experience or event, which counters or competes with the animal's fearfulness. The initial counterconditioning training may proceed faster by having a neighbor or friend come to the door at frequent intervals for a while.

Some animals become hyperactive at feeding time. Ill-mannered large animals can even be dangerous to the caretaker. When horses (*E. caballus*) show nudging, pushing, pawing or other competitive and begging behaviors at feeding time, it is best to back off and wait until the animal looks away and stands quietly and then quickly offer the feed. The animal must learn that it will be fed only when it stands quietly back from the feed trough, a seemingly unrelated activity. Similarly, well-mannered dogs (*C. familiaris*) are trained to sit while food is being placed in their food bowls and eat only after being given permission.

Flooding

Flooding refers to the systematic forced exposure of an animal to an aversive stimulus until the response is reduced or extinguished. In flooding, the animal is continuously exposed to the aversive stimulus until it demonstrates a complete cessation of its fearful responses, or it appears 'relaxed'. It is important not to terminate the aversive stimulation while the animal is still exhibiting the fearful behavior else the treatment can make the fearfulness worse. This technique should not be used if there is risk of injury or undue stress to the animal. An example of flooding is placing a saddle on a naive horse (*E. caballus*) and letting it buck until it stops and appears 'relaxed'. Flooding can be thought of as forced habituation. Many behavioral therapists do not use flooding because of the stress it can cause their animal patients.

Progressive desensitization

In some ways, desensitization can be viewed as the opposite of flooding. It involves exposure to ever-increasing intensities of an aversive stimulus in incremental steps until the undesired response is reduced or extinguished. It is used when flooding is deemed to be overly aggressive or too stressful on the animal. For example, desensitization has been used to reduce or eliminate fearfulness exhibited to the sounds of electric hair clippers used in grooming. One person restrains the animal while another person turns on the clipper from a distance. Once the animal shows no signs of escape or alarm, the person with the clipper moves a little closer to the animal. When the animal appears 'relaxed' at this distance, the clipper is moved a bit closer. This process is repeated until the clipper is in close proximity to the animal and in plain sight. If the animal shows fearfulness at any point, the clipper is moved away until the 'relaxed' posture is resumed. Eventually, the clipper is placed on the animal's body so it can feel its vibration. Unless the animal's reaction to the vibration is severe, the clippers should remain on the animal's body until signs of fearfulness have dissipated (flooding), else the fear of the vibrating clipper may be enhanced. The handler should then move the clipper along the animal's body as if clipping hair.

When desensitization involves reducing the animal's fear of an aversive auditory stimulus that is out of the handler's control (e.g. thunder and lightning), a tape recording can be made and the fear-provoking sounds played back to the animal at the handler's convenience. The sounds are initially introduced at relatively low amplitude (decibel level). The amplitude is gradually increased each time the animal's body language suggests that it is relaxed. This process is repeated, as many times as necessary, until the animal exhibits no observable response when the sounds are played as loud as they are normally heard. Desensitization *gradually* habituates the animal to the aversive stimulus.

Desensitization may not permanently eliminate a phobia. An animal can become resensitized to the stimulus by a subsequent 'bad' experience (e.g. skin pinched by the clipper). Counterconditioning the animal in addition to desensitization can increase the effectiveness of the treatment.

Animals often associate fear-inducing experiences with the place that the event occurred. In this case, the animal must be desensitized to the place as well. Counterconditioning may again prove useful.

An example: separation anxiety in dogs

Animal behavior therapists and some veterinarians have the training and experience to treat a host of behavior issues in animals. Some common behavior problems in

---

dogs (*C. familiaris*) and cats (*F. domestica*) include dominance-related aggression, fear-induced aggression, excessive barking, destructive behaviors, inappropriate elimination behaviors, sexual behavior problems, ingestive behavior problems, stereotypic or obsessive-compulsive behaviors, fears and phobias and separation anxiety. Companion dogs with separation anxiety typically eliminate, vocalize or engage in destructive behaviors when separated from their owners. Placing the dog in a cage or crate when household members leave eliminates the dog's ability to deface or destroy property but can intensify the dog's anxiety and some dogs have been known to break teeth, pull out toenails and otherwise injure themselves trying to escape from their cages. Separation anxiety is just that; the animal is not acting out of disobedience or spitefulness and it is not a result of being spoiled. The animal is simply fearful of being separated from its 'pack'. A dog can learn cues that departure is imminent (e.g. packing lunch pail, getting car keys, etc.) and exhibit signs of anxiety even before its owner departs.

Treatment of separation anxiety in dogs typically involves progressive desensitization to the owner's departure (Lindell, 1997). The initial steps include graduated timed departures. The object is to teach the dog that in certain situations it is safe and the owner's return is imminent. The dog is placed in its usual location on departure and, after the animal becomes calm, the owner goes out the door and returns after a few seconds. The dog is ignored on leaving and for 10–15 min after returning so that its greetings and excitable behaviors are not reinforced. At no time should the owner or other family members give the dog attention or release it from its crate or room when the dog is whining, barking or pawing for attention. The owner repeats this process several times in a day, gradually and randomly increasing the amount of time away from the house based on previous successes and knowledge of the approximate time it takes the dog to show signs of separation distress. Dogs can also be desensitized to pre-departure cues by the owner purposefully going through his or her usual pre-departure routine from time to time but then staying at home.

Counterconditioning might be employed if the dog fails to remain calm on the owner's departure. In counterconditioning a dog prone to separation anxiety, the animal is asked to perform an alternative behavior such as sit–stay or down–stay. When it complies, the owner walks toward the door and then returns, giving the dog a treat if it has continued to obey the command. This is repeated as many times as necessary to train the dog to comply prior to the owner's departure. Graduated timed departures can then begin. Counterconditioning combined with desensitization provides a more long-lasting solution to separation anxiety than desensitization alone. Counterconditioning creates a more positive emotional state in the animal than desensitization because of the addition of food reinforcement.

Pharmacological treatment

Drugs can be administered to reduce or eliminate some undesirable behaviors. For example, tranquilizers are sometimes used to reduce the incidence of anxiety-based disorders in dogs (*C. familiaris*) and cats (*F. domestica*). Some unwanted behaviors caused by physiological or structural problems can be treated with drugs. Pharmacological treatments should be used judiciously and only with the advice of a veterinarian or certified animal behavior therapist.

'Dog Appeasing Pheromone®' (DAP) has recently come on the market to treat fear-induced and stress-related behavior problems in dogs (*C. familiaris*) such as destructive behavior, barking and whining, house soiling, excessive licking, separation anxiety and phobias such as fear of thunderstorms and fireworks (Sheppard and Mills, 2003). DAP is a synthetic version of a chemical substance produced by lactating female dogs which is believed to have a calming effect on puppies. When applied on a collar or disseminated in a room, it may help puppies and adult dogs cope with stressful situations such as being adopted, moving, temporary isolation, visits to the veterinarian and exposure to novel and unpredictable situations which may lead to the development of phobias. DAP is most effective when used in conjunction with other behavior modification techniques (Levine *et al.*, 2007).

Similarly, 'Equine Appeasing Pheromone®', a synthetic analogue of a chemical produced in lactating horses (*E. caballus*), has been found to reduce stress-related behaviors and lower heart-rate increases in response to fear-evoking stimuli (Falewee *et al.*, 2006).

A synthetic analogue of feline cheek gland 'pheromone', marketed under the name 'Feliway®', has been successfully used in clinical trials to reduce urine marking and vertical scratching in domestic cats (*F. domestica*) (e.g. Frank *et al.*, 1999). It is particularly effective when cats are moved to new places or exposed to unfamiliar conspecifics. Cats typically mark objects in their environment with urine (i.e. urine spraying) and by glandular secretions on their feet and faces. These chemical 'signposts' give the cat a sense of familiarity with its environment and communicate its residency (i.e. territory) to other cats. When Feliway is diffused into an area novel to a cat, it gives the animal an immediate sense of 'home', thus reducing its motivation to mark the area.

While DAP and Feliway cannot be guaranteed to eliminate behavior problems, they represent a more natural approach than psychotropic drugs in managing undesirable behaviors in pet animals. Such approaches to behavioral therapy are much safer and more appealing than the use of drugs and should be encouraged.

References

Abbott, T.A., Hunter, E.J., Guise, H.J. and Penny, R.H.C. (1997) The effect of experience of handling on pigs' willingness to move. *Applied Animal Behaviour Science* 54, 371–375.

Adams, T.E., Daley, C.A., Adams, B.M. and Sakurai, H. (1996) Testes function and feedlot performance of bulls actively immunized against gonadotropin-releasing hormone: effect of age at immunization. *Journal of Animal Science* 74, 950–954.

Aengus, W.L. and Millam, J.R. (1999) Taming parent-reared orange-winged Amazon parrots by neonatal handling. *Zoo Biology* 18, 177–187.

Albone, E.S. (1984) *Mammalian Semiochemistry: Investigation of Chemical Signals Between Mammals.* John Wiley & Sons Ltd., Chichester, UK.

Alcock, J. (1989) *Animal Behavior: An Evolutionary Approach*, 4th edn. Sinauer Associates, Inc., Sunderland, Massachusetts.

Alexander, G. and Shillito, E. (1977) Importance of visual cues from various body regions in maternal recognition of the young in Merino sheep (*Ovis aries*). *Applied Animal Ethology* 3, 137–143.

Alexander, G., Stevens, D., Kilgour, R., de Langen, H., Mottershead, B.E. and Lynch, J.J. (1983) Separation of ewes from twin lambs: incidence in several sheep breeds. *Applied Animal Behaviour Science* 10, 301–317.

Alexander, G., Stevens, D. and Bradley, L.R. (1989) Maternal acceptance of alien lambs in ewes treated and untreated with oestrogen at birth. *Australian Journal of Experimental Agriculture* 29, 173–178.

Alexander, R.D., Hoogland, J.L., Howard, R.D., Noonan, K.M. and Sherman, P.W. (1979) Sexual dimorphisms and breeding systems in pinnipeds, ungulates, primates, and humans. In: Chagnon, N.A. and Irons, W. (eds) *Evolutionary Biology and Human Social Behavior.* Duxbury Press, North Scituate, Massachusetts, pp. 402–435.

Al-Rawi, B., Craig, J.V. and Adams, A.W. (1976) Agonistic behavior and egg production of caged layers: genetic strain and group-size effects. *Poultry Science* 55, 796–807.

Anderson, T.M., Pickett, B.W., Heird, J.C. and Squires, E.L. (1996) Effect of blocking vision and olfaction on sexual responses of haltered or loose stallions. *Journal of Equine Veterinary Science* 16, 254–261.

Arathi, H.S., Ho, G. and Spivak, M. (2006) Inefficient task partitioning among nonhygienic honeybees, *Apis mellifera* L., and implications for disease transmission. *Animal Behaviour* 72, 431–438.

Arey, D.S. (1999) Time course for the formation and disruption of social organisation in group-housed sows. *Applied Animal Behaviour Science* 62, 199–207.

Arnold, G.W. and Pahl, P.J. (1974) Some aspects of social behaviour in domestic sheep. *Animal Behaviour* 22, 592–600.

Arnone, M. and Dantzer, R. (1980) Does frustration induce aggression in pigs? *Applied Animal Ethology* 6, 351–362.

Baker, B.S., Taylor, B.J. and Hall, J.C. (2001) Are complex behaviors specified by dedicated regulatory genes? Reasoning from *Drosophila. Cell* 105, 13–24.

Baldwin, G.A. and Shillito, E.E. (1974) The effects of ablation of the olfactory bulbs on parturition and maternal behaviour in Soay sheep. *Animal Behaviour* 22, 220–223.

Bane, A. (1954) Studies on monozygous cattle twins. XV. Sexual functions of bulls in relation to heredity, rearing intensity and somatic conditions. *Acta Agriculturae Scandinavica* 4, 95–208.

Banks, E.M., Wood-Gush, D.G.M., Hughes, B.O. and Mankovich, N.J. (1979) Social rank and priority of access to resources in domestic fowl. *Behavioural Processes* 4, 197–209.

Baraza, E., Villalba, J.J. and Provenza, F.D. (2005) Nutritional context influences preferences of lambs for foods with plant secondary metabolites. *Applied Animal Behaviour Science* 92, 293–305.

Bardach, J.E. and Todd, J.H. (1970) Chemical communication in fish. In: Johnston, J.W., Jr, Moulton, D.G. and Turk, A. (eds) *Advances in Chemoreception*, Vol. 1, *Communication by Chemical Signals*. Appleton Century Crofts, New York, pp. 205–240.

Barinaga, M. (2003) Newborn neurons search for meaning. *Science* 299, 32–34 (summary of work by many researchers).

Barnett, J.L., Hemsworth, P.H. and Newman, E.A. (1992) Fear of humans and its relationships with productivity in laying hens at commercial farms. *British Poultry Science* 33, 699–710.

Barnett, J.L., Cronin, G.M., McCallum, T.H. and Newman, E.A. (1993) Effects of 'chemical intervention' techniques on aggression and injuries when grouping unfamiliar adult pigs. *Applied Animal Behaviour Science* 36, 135–148.

Barnett, J.L., Cronin, G.M., McCallum, T.H., Newman, E.A. and Hennessy, D.P. (1996) Effects of grouping unfamiliar adult pigs after dark, after treatment with amperozide and by using pens with stalls, on aggression, skin lesions and plasma cortisol concentrations. *Applied Animal Behaviour Science* 50, 121–133.

Barnett, S. (1963) *The Rat: A Study in Behavior*. Aldine, Chicago, Illinois.

Beach, F.A. (1948) *Hormones and Behaviour*. Paul B. Hoeber, New York.

Beach, F.A. (1976) Sexual attractivity, proceptivity, and receptivity in female mammals. *Hormones and Behavior* 7, 105–138.

Beck, B.B. (1980) *Animal Tool Behavior*. Garland STPM Press, New York.

Beck, B.B. and Power, M.L. (1988) Correlates of sexual and maternal competence in captive gorillas. *Zoo Biology* 7, 339–350.

Bednekoff, P. and Balda, R. (1996) Observational spatial memory in Clark's nutcrackers and Mexican jays. *Animal Behaviour* 52, 833–839.

Beletsky, L.D. and Orians, G.H. (1991) Year-to-year patterns of circulating levels of testosterone and corticosterone in relation to breeding density, experience, and reproductive success of the polygynous red-winged blackbird. *Hormones and Behavior* 26, 420–432.

Bench, C.J., Price, E.O., Dally, M.R. and Borgwardt, R.E. (2001) Artificial selection of rams for sexual performance and its effect on the sexual behavior and fecundity of male and female progeny. *Applied Animal Behaviour Science* 72, 41–50.

Bernon, D.B. and Siegel, P.B. (1983) Mating frequency in male chickens: crosses among selected and unselected lines. *Genetics, Selection and Evolution* 15, 445–454.

Blackshaw, J.K., Blackshaw, A.W. and McGlone, J.J. (1997) Buller steer syndrome review. *Applied Animal Behaviour Science* 54, 97–108.

Blockey, M.A. de B. (1976) Serving capacity – a measure of the serving efficiency of bulls during pasture mating. *Theriogenology* 6, 393–401.

Blockey, M.S. de B. (1978) Serving capacity and social dominance of bulls in relation to fertility. *Proceedings of the First World Congress of Ethology Applied to Zootechnics*, Madrid, pp. 523–530 (cited in Price, 1987).

Blockey, M.S. de B. (1979) Observations on group mating of bulls at pasture. *Applied Animal Ethology* 5, 15–34.

Blockey, M.A. de B., Straw, W.M. and Jones, L.P. (1978) Heritability of serving capacity and scrotal circumference in beef bulls. *Journal of Animal Science* 47 (Supplement 1), 254 (Abstract).

Blohowiak, C. (1987) Baffling black ducks. *Zoogoer* 16, 18–19.

Blokhuis, H.J. and Wiepkema, P.R. (1998) Studies of feather pecking in poultry. *The Veterinary Quarterly* 20, 6–9.

Blount, J.D., Metcalfe, N.B., Birkhead, T.R. and Surai, P.F. (2003) Carotenoid modulation of immune function and sexual attractiveness in zebra finches. *Science* 300, 125–127.

Bluhm, C.K. (1988) Temporal patterns of pair formation and reproduction in annual cycles and associated endocrinology in waterfowl. In: Johnston, R.F. (ed) *Current Ornithology*, Vol. 5. Plenum Press, New York, pp. 123–185.

Boice, R. (1968) Conditioned licking in wild F1 and domestic Norway rats. *Journal of Comparative and Physiological Psychology* 66, 796–799.

Boice, R. (1970) Effect of domestication on avoidance learning in the Norway rat. *Psychonomic Science* 18, 13–14.

Boice, R. and Adams, N. (1983) Degrees of captivity and aggressive behavior in domestic Norway rats. *Bulletin of the Psychonomic Society* 21, 149–152.

BonDurant, R.H., Darien, B.J., Munro, C.J., Stabenfeldt, G.H. and Wang, P. (1981) Photoperiod induction of fertile oestrus and changes in LH and progesterone concentrations in yearling dairy goats (*Capra hircus*). *Journal of Reproduction and Fertility* 63, 1–9.

Bouissou, M.-F. (1964) Observations sur la hiérarchie sociale chez les bovines doméstiques. *D.E.S.*, Faculte des Sciences, Paris (cited in Bouissou, 1972).

Bouissou, M.-F. (1970) Role du contact physique dans la manifestation des relations hierarchiques chez les bovines. Conséquences practiques. (The role of physical contact in the manifestation of hierarchical relationships in cattle.) *Annales de Zootechnie* 19, 279–285.

Bouissou, M.-F. (1972) Influence of body weight and presence of horns on social rank in domestic cattle. *Animal Behaviour* 20, 474–477.

Bowers, J.M. and Alexander, B.K. (1967) Mice: individual recognition by olfactory cues. *Science* 158, 1208–1210.

Bowling, A.T. and Touchberry, R.W. (1990) Parentage of Great Basin feral horses. *Journal of Wildlife Management* 54, 424–429.

Boyd, G.W. and Corah, L.R. (1986) Evaluating serving capacity of yearling beef bulls: a field trial. *Journal of Animal Science* 63 (Supplement 1), 207.

Brakel, W.J. and Leis, R.A. (1976) Impact of social disorganization on behavior, milk yield and body weight in dairy cows. *Journal of Dairy Science* 59, 716–721.

Breuer, K., Hemsworth, P.H., Barnett, J.L., Matthews, L.R. and Coleman, G.J. (2000) Behavioural response to humans and productivity of commercial dairy cows. *Applied Animal Behaviour Science* 66, 273–288.

Britt, J.H., Scott, R.G., Armstrong, J.D. and Whitacre, M.D. (1986) What determines the amount of mounting and standing activity for cows in heat? *1986 Report of Department of Animal Science*, North Carolina State University, Raleigh, North Carolina, pp. 44–46.

Brown, J.R., Ye, H., Bronson, R.T., Dikkes, P. and Greenberg, M.E. (1996) A defect in nurturing in mice lacking the immediate early gene fosB. *Cell* 86, 297–309.

Buchwalder, T. and Huber-Eicher, B. (2005) Effects of group size on aggressive reactions to an introduced conspecific in groups of domestic turkeys (*M. gallopavo*). *Applied Animal Behaviour Science* 93, 251–258.

Burman, O.H.P., Ilyat, A., Jones, G. and Mendl, M. (2007) Ultrasonic vocalizations as indicators of welfare for laboratory rats. *Applied Animal Behaviour Science* 104, 116–129.

Burritt, E.A., Mayland, H.F., Provenza, F.D., Miller, R.L. and Burns, J.C. (2005) Effect of added sugar on preference and intake by sheep of hay cut in the morning versus the afternoon. *Applied Animal Behaviour Science* 94, 245–254.

Butler, R.G. (1980) Population size, social behaviour and dispersal in house mice: a quantitative investigation. *Animal Behaviour* 28, 78–85.

Caldji, C., Tannenbaum, B., Sharma, S., Francis, D., Plotsky, P.M. and Meaney, M.J. (1998) Maternal care during infancy regulates the development of neural systems mediating the expression of fearfulness in the rat. *Proceedings of the National Academy of Sciences* 95, 5335–5340.

Callard, M.D., Bursten, S.N. and Price, E.O. (2000) Repetitive backflipping behaviour in captive roof rats (*Rattus rattus*) and the effects of cage enrichment. *Animal Welfare* 9, 139–152.

Capehart, J., Viney, W. and Hulicka, I.M. (1958) The effect of effort upon extinction. *Journal of Comparative and Physiological Psychology* 51, 505–507.

Carlstead, K. (1986) Predictability of feeding: its effect on agonistic behaviour and growth in grower pigs. *Applied Animal Behaviour Science* 16, 25–38.

Carlstead, K., Mellen, J. and Kleiman, D.G. (1999) Black rhinoceros (*Diceros bicornis*) in US zoos: I. Individual behavior profiles and their relationship to breeding success. *Zoo Biology* 18, 17–34.

Carr, W.J., Martorano, R.D. and Krames, L. (1970) Responses of mice to odors associated with stress. *Journal of Comparative and Physiological Psychology* 71, 223–228.

Carroll, J., Murphy, C.J., Neitz, M., Hoeve, J.N. and Neitz, J. (2001) Photopigment basis for dichromatic color vision in the horse. *Journal of Vision* 1, 80–87.

Carter, C.S. and Getz, L.L. (1993) Monogamy and the prairie vole. *Scientific American* 268, 100–106.

Carter, D.S. and Goldman, B.D. (1983) Antigonadal effects of timed melatonin infusion in pinealectomized male Djungarian hamsters (*Phodopus sungorus sungorus*): duration is the critical parameter. *Endocrinology* 113, 1261–1267.

Chance, M.R.A. (1962) An interpretation of some agonistic postures: the role of 'cut-off' acts and postures. *Symposia of the Zoological Society of London* 8, 71–89.

Cheng, M.Y., Bullock, C.M., Li, C., Lee, A.G., Bermak, J.C., Belluzzi, J., Weaver, D.R., Leslie, F.M. and Zhou, Q. (2002) Prokineticin 2 transmits the behavioural circadian rhythm of the suprachiasmatic nucleus. *Nature* 417, 405–410.

Clark, B.R. and Price, E.O. (1981) Domestication effects on sexual maturation and fecundity of Norway rats (*Rattus norvegicus*). *Journal of Reproduction and Fertility* 63, 215–220.

Clark, M.M. and Galef, B.G., Jr (1977) The role of the physical rearing environment in the domestication of the Mongolian gerbil (*Meriones unguiculatus*). *Animal Behaviour* 25, 298–316.

Clark, M.M., Stiver, K., Teall, T. and Galef, B.G., Jr (2006) Nursing one litter of Mongolian gerbils while pregnant with another: effects on daughters' mate attachment and fecundity. *Animal Behaviour* 71, 235–241.

Clegg, M.T., Beamer, W. and Bermant, G. (1969) Copulatory behaviour of the ram, *Ovis aries*. III: Effects of pre- and postpubertal castration and androgen replacement therapy. *Animal Behaviour* 17, 712–717.

Clutton-Brock, T.H. and McComb, K. (1993) Experimental tests of copying and mate choice in fallow deer. *Behavioral Ecology* 4, 191–193.

Clutton-Brock, T.H., Deutsch, J.C. and Nefdt, R.J.C. (1993) The evolution of leks. *Animal Behaviour* 46, 1121–1138.

Coblentz, B. (1976) Functions of scent-urination in ungulates with special reference to feral goats (*Capra hircus* L.) *American Naturalist* 110, 549–557.

Codenotti, T.L. and Alvarez, F. (2001) Mating behavior of the male Greater Rhea. *Wilson Bulletin* 113, 85–89.

Cole, D.D. and Schafer, J.B. (1966) A study of social dominance in cats. *Behaviour* 27, 39–53.

Collias, N.E. and Collias, E.C. (1996) Social organization of a red junglefowl, *Gallus gallus*, population related to evolution theory. *Animal Behaviour* 51, 1337–1354.

Collins, R. (1964) Inheritance of avoidance conditioning in mice: a diallel (sp) study. *Science* 143, 1188–1190.

Conboy, H.S. (1992) Training the novice stallion for artificial breeding. In: Beeman, G.M. (ed) *Stallion Management. Veterinary Clinics of North America: Equine Practice* Vol. 8, No. 1, 101–109.

Cook, W.T. and Siegel, P.B. (1974) Social variables and divergent selection for mating behaviour of male chickens. *Animal Behaviour* 22, 390–396.

Coppinger, L. and Coppinger, R. (2007) Dogs for herding and guarding livestock. In: Grandin, T. (ed) *Livestock Handling and Transport*, 3rd edn. CAB International, Wallingford, UK, pp. 199–213.

Coppinger, R., Glendinning, J., Torop, E., Matthay, C., Sutherland, M. and Smith, C. (1987) Degree of behavioral neoteny differentiates canid polymorphs. *Ethology* 75, 89–108.

Coquelin, A. and Bronson, F.H. (1979) Release of luteinizing hormone in male mice during exposure to females: habituation of the response. *Science* 206, 1099–1101.

Cornetto, T. and Estevez, I. (2001) Influence of vertical panels on use of space by domestic fowl. *Applied Animal Behaviour Science* 71, 141–153.

Cornetto, T., Estevez, I. and Douglass, L.W. (2002) Using artificial cover to reduce aggression and disturbances in domestic fowl. *Applied Animal Behaviour Science* 75, 325–336.

Cottle, C.A. and Price, E.O. (1987) Effect of the nonagouti pelage-color allele on the behavior of captive wild Norway rats (*Rattus norvegicus*). *Journal of Comparative Psychology* 101, 390–394.

Craig, J.V. (1981) *Domestic Animal Behavior*. Prentice-Hall, Englewood Cliffs, New Jersey.

Creel, S., Creel, N.M. and Monfort, S.L. (1996) Social stress and dominance. *Nature* 379, 212.

Cuomo-Benzo, M., Price, E.O. and Hartenstein, R. (1977) Catecholamine levels in whole brain of stressed and control domestic and wild rats (*Rattus norvegicus*). *Behavioural Processes* 2, 33–40.

Dantzer, R. and Morméde, P. (1981) Pituitary–adrenal consequences of adjunctive activities in pigs. *Hormones and Behavior* 15, 386–395.

Darwin, C. (1872) *The Expression of the Emotions in Man and Animals*. HarperCollins, London (1998 edition).

Davenport, R.J. (2001) A sniff in time scents mines. *Science* 291, 2072.

Davis, H. and Taylor, A. (2001) Discrimination between individual humans by domestic fowl (*Gallus gallus domesticus*). *British Poultry Science* 42, 276–279.

De Araujo, J.W., Borgwardt, R.E., Sween, M.L., Yelich, J.V. and Price, E.O. (2003) Incidence of repeat-breeding among Angus bulls (*Bos taurus*) differing in sexual performance. *Applied Animal Behaviour Science* 81, 89–98.

D'Eath, R.B. and Keeling, L.J. (2003) Social discrimination and aggression by laying hens in large groups: from pock orders to social tolerance. *Applied Animal Behaviour Science* 84, 197–212.

Dehnhard, M. and Claus, R. (1988) Reliability criteria of a bioassay using rats trained to detect estrus-specific odor in cow urine. *Theriogenology* 30, 1127–1138.

De Jong, I.C., Wolthuis-Fillerup, M. and van Reenen, C.G. (2007) Strength of preference for dustbathing and foraging substrates in laying hens. *Applied Animal Behaviour Science* 104, 24–36.

De Passillé, A.M. (2001) Sucking motivation and related problems in calves. *Applied Animal Behaviour Science* 72, 175–187.

De Passillé, A.M. and Rushen, J. (1997) Motivational and physiological analysis of the causes and consequences of non-nutritive sucking by calves. *Applied Animal Behaviour Science* 53, 15–31.

De Passillé, A.M.B., Rushen, J. and Hartsock, T.G. (1988) Ontogeny of teat fidelity in pigs and its relation to competition at suckling. *Canadian Journal of Animal Science* 68, 325–338.

De Passillé, A.M.B., Christopherson, R. and Rushen, J. (1993) Nonnutritive sucking by the calf and postprandial secretion of insulin, CCK and gastrin. *Physiology and Behavior* 54, 1069–1073.

Dewsbury, D.A. (1973) Copulatory behavior of montane voles (*Microtus montanus*). *Behaviour* 44, 186–202.

Dewsbury, D.A. (1981) Effects of novelty on copulatory behavior: The Coolidge Effect and related phenomena. *Psychological Bulletin* 89, 464–482.

Dilger, W.C. (1962) The behavior of lovebirds. *Scientific American* 206, 88–98.

D'Occhio, M.J. and Brooks, D.E. (1982) Threshold of plasma testosterone required for normal mating activity in male sheep. *Hormones and Behavior* 16, 383–394.

Dodd, G.H. and Squirrel, D.J. (1980) Structure and mechanism in the mammalian olfactory system. *Symposia of the Zoological Society of London* 45, 35–36.

Döhler, K.-D., Coquelin, A., Davis, F., Hines, M., Shryne, A.E. and Gorski, R.A. (1984) Pre- and post-natal influence of testosterone proprionate and diethylstilbestrol on differentiation of the sexually dimorphic nucleus of the preoptic area in male and female rats. *Brain Research* 302, 291–295.

Dorries, K.M., Adkins-Regan, E. and Halpern, B.P. (1991) Sex difference in olfactory sensitivity to the boar chemosignal, androstenone, in the domestic pig. *Animal Behaviour* 42, 403–411.

Doutrelant, C. and McGregor, P.K. (2000) Eavesdropping and mate choice in female fighting fish. *Behaviour* 137, 1655–1669.

Drickamer, L.C. (1982) Delay and acceleration of puberty in female mice by urinary chemosignals from other females. *Developmental Psychobiology* 15, 433–442.

Dudink, S., Simonse, H., Marks, I., de Jonge, F.H. and Spruijt, B.M. (2006) Announcing the arrival of enrichment increases play behaviour and reduces weaning-stress-induced behaviours of piglets directly after weaning. *Applied Animal Behaviour Science* 101, 86–101.

Duncan, A.J., Ginane, C., Elston, D.A., Kunaver, A. and Gordon, I.J. (2006) How do herbivores trade-off the positive and negative consequences of diet selection decisions? *Animal Behaviour* 71, 93–99.

Dunn, G.C., Price, E.O. and Katz, L.S. (1987) Fostering calves by odor transfer. *Applied Animal Behaviour Science* 17, 33–39.

Ellis, D.H., Gee, G.F., Hereford, S.G., Olsen, G.H., Chisholm, T.D., Nicolich, J.M., Sullivan, K.A., Thomas, N.J., Nagendran, M. and Hatfield, J.S. (2000) Post-release survival of hand-reared and parent-reared Mississippi sandhill cranes. *The Condor* 102, 104–112.

Ellis, D.H., Sladen, W.J.L., Lishman, W.A., Clegg, K.R., Duff, J.W., Gee, G.F. and Lewis, J.C. (2003) Motorized migrations: the future or mere fantasy. *BioScience* 53, 260–264.

Emlen, S.T. and Oring, L.W. (1977) Ecology, sexual selection, and the evolution of mating systems. *Science* 197, 215–223.

Enders, R.K. (1945) Induced changes in the breeding habits of foxes. *Sociometry* 8, 53–55.

Erhard, H.W., Price, E.O. and Dally, M.R. (1998) Competitive ability of rams selected for high and low levels of sexual performance. *Animal Science* 66, 403–408.

Estep, D.Q., Price, E.O., Wallach, S.J.R. and Dally, M.R. (1989) Social preferences of domestic ewes for rams (*Ovis aries*). *Applied Animal Behaviour Science* 24, 287–300.

Evans, C.S., Evans, L. and Marler, P. (1993) On the meaning of alarm calls: functional reference in an avian vocal system. *Animal Behaviour* 46, 23–38.

Ewbank, R.J. and Meese, G.B. (1971) Aggressive behaviour in groups of domesticated pigs on removal and return of individuals. *Animal Production* 13, 685–693.

Eysenck, H.J. and Broadhurst, P.L. (1964) Experiments with animals: Introduction. In: Eysenck, H.J. (ed) *Experiments in Motivation*. Macmillan, New York, pp. 285–291.

Faivre, B., Grégoire, A., Préault, M., Cézilly, F. and Sorci, G. (2003) Immune activation rapidly mirrored in a secondary sexual trait. *Science* 300, 103.

Falewee, C., Gaultier, E., Lafont, C., Bougrat, L. and Pageat, P. (2006) Effect of a synthetic equine maternal pheromone during a controlled fear-eliciting situation. *Applied Animal Behaviour Science* 101, 144–153.

Fatjó, J., Feddersen-Petersen, D., Luis, J., de la Torre, R., Amat, M., Mets, M., Braus, B. and Manteca, X. (2007) Ambivalent signals during agonistic interactions in a captive wolf pack. *Applied Animal Behaviour Science* 105, 274–285.

Febrer, K., Jones, T.A., Donnelly, C.A. and Dawkins, M.S. (2006) Forced to crowd or choosing to cluster? Spatial distribution indicates social attraction in broiler chickens. *Animal Behaviour* 72, 1291–1300.

Feist, J.D. and McCullough, D.R. (1976) Behavior patterns and communication in feral horses. *Zeitshrift für Tierpsychologie* 41, 337–371.

Festing, M.G.W. (1979) Inbred strains. In: Baker, H.J., Lindsey, J.R. and Weisbroth, S.H. (eds) *The Laboratory Rat*: Vol. 1. *Biology and Diseases*. Academic Press, New York, pp. 55–72.

Festing, M.F.W. (1989) Inbred strains of mice. In: Lyon, M.F. and Searle, A.G. (eds) *Genetic Variants and Strains of the Laboratory Mouse*, 2nd edn. Oxford University Press, New York, pp. 636–648.

Fillion, T.J. and Blass, E.M. (1986) Infantile behavioural reactivity to oestrous chemostimuli in Norway rats. *Animal Behaviour* 34, 123–133.

Fisher, H.S. and Rosenthal, G.G. (2006) Female swordtail fish use chemical cues to select well-fed mates. *Animal Behaviour* 72, 721–725.

Foote, R.H. (1975) Estrus detection and estrus detection aids. *Journal of Dairy Science* 58, 248–256.

Fox, R.A. and Millam, J.R. (2007) Novelty and individual differences influence neophobia in orange-winged Amazon parrots (*Amazona amazonica*). *Applied Animal Behaviour Science* 104, 107–115.

Francis, C. (1991) Prairie warriors. *Dog World* 76(7), 10–12.

Frank, D.F., Erb, H.N. and Houpt, K.A. (1999) Urine spraying in cats: presence of concurrent disease and effects of pheromone treatment. *Applied Animal Behaviour Science* 61, 263–272.

Fraser, D. (1975) The 'teat order' of piglets. II. Fighting during suckling and the effects of clipping the eye teeth. *Journal of Agricultural Science, Cambridge* 84, 393–399.

Fraser, D. (1987) Mineral-deficient diets and the pig's attraction to blood: implications for tail biting. *Canadian Journal of Animal Science* 67, 909–918.

Freeman, N.C.G. and Rosenblatt, J.S. (1978) The interrelationship between thermal and olfactory stimulation in the development of home orientation in newborn kittens. *Developmental Psychobiology* 11, 437–457.

Froy, O., Gotter, A.L., Casselman, A.L. and Reppert, S.M. (2003) Illuminating the circadian clock in monarch butterfly migration. *Science* 300, 1303–1305.

Fulkerson, W.J., Sawyer, G.J. and Crothers, I. (1983) The accuracy of several aids in detecting oestrus in dairy cattle. *Applied Animal Ethology* 10, 199–208.

Fuller, J.L. and Thompson, W.R. (1960) *Behavior Genetics*. Wiley, New York.

Galef, B.G. Jr (1970) Aggression and timidity: responses to novelty in feral Norway rats. *Journal of Comparative and Physiological Psychology* 70, 370–381.

Galef, B.G. Jr and White, D.J. (1998) Mate-choice copying in Japanese quail, *Coturnix coturnix japonica*. *Animal Behaviour* 55, 545–552.

Galef, B.G. Jr, Pretty, S. and Whiskin, E.E. (2006) Failure to find aversive marking of toxic foods by Norway rats. *Animal Behaviour* 72, 1427–1436.

Gallup, G.G. Jr, Povinelli, D.J., Suarez, S.D., Anderson, J.R., Lethmate, J. and Menzel, E.W. Jr (1995) Further reflections on self-recognition in primates. *Animal Behaviour* 50, 1525–1532.

Garcia, J. and Koelling, R.A. (1966) Relation of cue to consequence in avoidance learning. *Psychonomic Science* 4, 123–124.

Garcia, J., Ervin, F.R. and Koelling, R.A. (1966) Learning with prolonged delay of reinforcement. *Psychonomic Science* 5, 121–122.

Garcia, J., Hankins, W.G. and Rusiniak, K.W. (1976) Flavor aversion studies. *Science* 192, 265–266.

Gazzano, A., Mariti, C., Notari, L., Sighieri, C. and McBride, E.A. (2008) Effects of early gentling and early environment on emotional development of puppies. *Applied Animal Behaviour Science* 110, 294–304.

Gebhardt-Henrich, S.G., Vonlanthen, E.M. and Steiger, A. (2005) How does the running wheel affect the behaviour and reproduction of golden hamsters kept as pets? *Applied Animal Behaviour Science* 95, 199–203.

Gelez, H., Archer, E., Chesneau, D., Campan, R. and Fabre-Nys, C. (2004) Importance of learning in the response of ewes to male odor. *Chemical Senses* 29, 555–563.

Gerlach, G. and Lysiak, N. (2006) Kin recognition and inbreeding avoidance in zebrafish, *Danio rerio*, is based on phenotype matching. *Animal Behaviour* 71, 1371–1377.

Gibbon, J., Morrell, M. and Silver, R. (1984) Two kinds of timing in circadian incubation rhythm of ring doves. *American Journal of Physiology – Regulatory Integrative Comparative Physiology* 247, 1083–1087.

Gibbs, H.L., Weatherhead, R.J., Boag, P.T., White, B.N., Tabak, L.M. and Hoysak, D.J. (1990) Realized reproductive success of polygynous red-winged blackbirds revealed by DNA markers. *Science* 250, 1394–1397.

Ginsburg, B.E. and Hiestand, L. (1992) Humanity's 'best friend': the origins of our inevitable bond with dogs. In: Davis, H. and Balfour, D. (eds) *The Inevitable Bond: Examining Scientist–Animal Interactions*. Cambridge University Press, New York, pp. 93–108.

Gonzales-Mariscal, G. and Rosenblatt, J.S. (1996) Maternal behavior in rabbits. In: Rosenblatt, J.S. and Snowdon, C.T. (eds) *Parental Care: Evolution, Mechanisms, and Adaptive Significance. Advances in the Study of Behavior*. Vol. 25. Academic Press, New York, pp. 333–360.

Gorman, M.R., Goldman, B.D. and Zucker, I. (2001) Mammalian photoperiodism. In: Takahashi, J.S., Turek, F.W. and Moore, R.Y. (eds) *Handbook of Behavioral Neurobiology*, Vol. 12, *Circadian Clocks*. Kluwer Academic/Plenum Publishers, New York, pp. 481–508.

Gottlieb, G. (1968) Prenatal behavior of birds. *Quarterly Review of Biology* 43, 148–174.

Gottlieb, G. (1971) Ontogenesis of sensory function in birds and mammals. In: Tobach, E., Aronson, L.R. and Shaw, E. (eds) *The Biopsychology of Development*. Academic Press, New York, pp. 67–128.

Gottlieb, G. (1982) Development of species identification in ducklings: IX. The necessity of experiencing normal variations in embryonic auditory stimulation. *Developmental Psychobiology* 15, 507–517.

Gould, J.L. and Marler, P. (1987) Learning by instinct. *Scientific American* 256, 74–85.

Gowaty, P.A. and Karlin, A.A. (1984) Multiple maternity and paternity in single broods of apparently monogamous eastern bluebirds (*Sialia sialis*). *Behavioral Ecology and Sociobiology* 15, 91–95.

Graham, J.M. and Desjardins, C. (1980) Classical conditioning: induction of luteinizing hormone and testosterone secretion in anticipation of sexual activity. *Science* 210, 1039–1041.

Grandin, T. (1993) Behavioral agitation during handling of cattle is persistent over time. *Applied Animal Behaviour Science* 36, 1–9.

Grandin, T. (1998) Reducing handling stress improves both productivity and welfare. *The Professional Animal Scientist* 14, 1–10.

Grandin, T. (2007a) *Livestock Handling and Transport*, 3rd edn. CAB International, Wallingford, UK.

Grandin, T. (2007b) Behavioural principles of handling cattle and other grazing animals under extensive conditions. In: Grandin, T. (ed) *Livestock Handling and Transport*, 3rd edn. CAB International, Wallingford, UK, pp. 44–64.

Grandin, T. and Bruning, J. (1992) Boar presence reduces fighting in mixed slaughter-weight pigs. *Applied Animal Behaviour Science* 33, 273–276.

Gray, G.D. and Dewsbury, D.A. (1973) A quantitative description of copulatory behavior in prairie voles (*Microtus ochrogaster*). *Brain, Behavior and Evolution* 8, 437–452.

Greer, N.L., Bartolome, J.V. and Schanberg, S.M. (1991) Further evidence for the hypothesis that beta-endorphin mediates maternal deprivation effects. *Life Sciences* 48, 643–648.

Gubernick, D.J. (1980) Maternal 'imprinting' or maternal 'labeling' in goats. *Animal Behaviour* 28, 124–129.

Gubernick, D.J. (1981) Mechanisms of maternal 'labeling' in goats. *Animal Behaviour* 29, 305–306.

Gubernick, D.J. and Alberts, J.R. (1987) The biparental care system of the California mouse. *Journal of Comparative Psychology* 101, 169–177.

Guhl, A.M. (1953) Social behavior of the domestic fowl. *Technical Bulletin Number 73, Kansas Agriculture Experiment Station*, Manhattan, New York, pp. 3–48.

Guhl, A.M. and Allee, W.C. (1944) Some measureable effects of social organization in flocks of hens. *Physiological Zoology* 17, 320–347.

Guhl, A.M., Collias, N.E. and Allee, W.C. (1945) Mating behavior and the social hierarchy in small flocks of White Leghorns. *Physiological Zoology* 18, 365–390.

Guhl, A.M., Craig, J.V. and Mueller, C.D. (1960) Selective breeding for aggressiveness in chickens. *Poultry Science* 39, 970–980.

Hadley, C., Hadley, B., Ephraim, S., Yang, M. and Lewis, M.H. (2006) Spontaneous stereotypy and environmental enrichment in deermice (*P. maniculatus*): reversibility of experience. *Applied Animal Behaviour Science* 97, 312–322.

Hale, E.B. (1953) Social factors in sexual behavior of turkeys. *Pennsylvania Agricultural Experiment Station Progress Report*, No. 108, cited in Hale, E.B., Schleidt, W.M. and Schein, M.W. (1969) The behaviour of turkeys. In: Hafez, E.S.E. (ed) *The Behaviour of Domestic Animals*, 2nd edn. Williams & Wilkins, Co., Baltimore, Maryland.

Hale, E.B. (1966) Visual stimuli and reproductive behavior in bulls. *Journal of Animal Science* 25 (Supplement), 36–44.

Hale, E.B. and Almquist, J.O. (1960) Relation of sexual behavior to germ cell output in farm animals. *Journal of Dairy Science* 43 (Supplement), 145–169.

Haley, D.B., Bailey, D.W. and Stookey, J.M. (2005) The effects of weaning beef calves in two stages on their behavior and growth rate. *Journal of Animal Science* 83, 2205–2214.

Hall, F.S., Wilkinson, L.S., Humby, T., Inglis, W., Kendall, D.A., Marsden, C.A. and Robbins, T.W. (1998) Isolation rearing in rats: pre- and postsynaptic changes in striatal dopaminergic systems. *Pharmacology, Biochemistry and Behavior* 59, 859–872.

Hall, F.S., Wilkinson, L.S., Humby, T. and Robbins, T.W. (1999) Maternal deprivation of neonatal rats produces enduring changes in dopamine function. *Synapse* 32, 37–43.

Hansen, S.W. and Jensen, M.B. (2006) Quantitative evaluation of the motivation to access a running wheel or a water bath in farm mink. *Applied Animal Behaviour Science* 98, 127–144.

Harlander-Matauschek, A., Baes, C. and Bessei, W. (2006) The demand of laying hens for feathers and wood shavings. *Applied Animal Behaviour Science* 101, 102–110.

Harri, M., Mononen, J., Rekilä, T., Korhonen, H. and Niemelä, P. (1998) Effects of top nest boxes on growth, fur quality and behaviour of blue foxes during their growing season. *Acta Agriculturae Scandinavica, Section A, Animal Science* 48, 184–191.

Hart, B.L. (1985) *The Behavior of Domestic Animals*. W.H. Freeman Co., New York.

Hart, B.L. and Barrett, R.E. (1973) Effects of castration on fighting, roaming, and urine marking in adult male cats. *Journal of the American Veterinary Medical Association* 103, 290–292.

Hart, B.L. and Cooper, L.J. (1984) Factors relating to fighting and urine spraying in prepubertally gonadectomized male and female cats. *Journal of the American Veterinary Medical Association* 184, 1255–1258.

Hart, B.L., Hart, L.A. and Bain, M.J. (2006) *Canine and Feline Behavior Therapy*, 2nd edn. Blackwell Publishing, Ames, Iowa.

Hartsock, T.G. and Graves, H.B. (1976) Neonatal behavior and nutrition-related mortality in domestic swine. *Journal of Animal Science* 42, 235–241.

Hartsock, T.G., Graves, H.B. and Baumgardt, B.R. (1977) Agonistic behavior and the nursing order in suckling piglets: relationships with survival, growth and body composition. *Journal of Animal Science* 44, 320–330.

Hasler, A.D., Scholtz, A.T. and Horrall, R.M. (1987) Olfactory imprinting and homing in salmon. *American Scientist* 66, 347–355.

Hawley, D.M. (2006) Asymmetric effects of experimental manipulations of social status on individual immune response. *Animal Behaviour* 71, 1431–1438.

Hayssen, V. (1997) Effects of the nonagouti coat-color allele on behavior of deer mice (*Peromyscus maniculatus*): a comparison with Norway rats (*Rattus norvegicus*). *Journal of Comparative Psychology* 111, 419–423.

Hazary, R.C., Staines, H.J. and Wishart, G.J. (2001) Assessing the effect of mating ratio on broiler breeder performance by quantifying sperm:egg interaction. *Journal of Applied Poultry Research* 10, 1–4.

Hediger, H. (1950) *Wildtiere in Gefangenschaft-ein Grundriss der Tiergartenbiologie*. Benno Schwabe, Basle. (Reprinted as *Wild Animals in Captivity: An Outline of the Biology of Zoological Gardens*. Translated by Sircom, G., Butterworth, London.)

Hemmings, A., McBride, S.D. and Hale, C.E. (2007) Perseverative responding and the aetiology of equine oral stereotypy. *Applied Animal Behaviour Science* 104, 143–150.

Hemsworth, P.H. (2003) Human–animal interactions in livestock production. *Applied Animal Behaviour Science* 81, 185–198.

Hemsworth, P.H., Beilharz, R.G. and Balloway, D.B. (1977) Influence of social conditions during rearing on the sexual behavior of the domestic boar. *Animal Production* 24, 245–251.

Hemsworth, P.H., Beilharz, R.G. and Brown, W.J. (1978) The importance of the courting behaviour of the boar on the success of natural and artificial matings. *Applied Animal Ethology* 4, 341–347.

Hemsworth, P.H., Barnett, J.L. and Hansen, C. (1981a) The influence of handling by humans on the behavior, growth and corticosteroids in the juvenile female pig. *Hormones and Behavior* 15, 396–403.

Hemsworth, P.H., Brand, A. and Willems, P. (1981b) The behavioural response of sows to the presence of human beings and its relation to productivity. *Livestock Production Science* 8, 67–74.

Hemsworth, P.H., Gonyou, H.W. and Dziuk, P.J. (1986) Human communication with pigs: the behavioural response of pigs to specific human signals. *Applied Animal Behaviour Science* 15, 45–54.

Hemsworth, P.H., Barnett, J.L., Coleman, G.J. and Hansen, C. (1989) A study of the relationships between the attitudinal and behavioural profiles of stockpeople and level of fear of humans and the reproductive performance of commercial pigs. *Applied Animal Behaviour Science* 23, 301–314.

Hemsworth, P.H., Barnett, J.L., Treacy, D. and Magdwick, P. (1990) The heritability of the trait fear of humans and the association between this trait and subsequent reproductive performance of gilts. *Applied Animal Behaviour Science* 25, 85–95.

Hemsworth, P.H., Coleman, G.J., Barnett, J.L. and Jones, R.B. (1994) Behavioural responses of humans and the productivity of commercial broiler chickens. *Applied Animal Behaviour Science* 41, 101–114.

Hemsworth, P.H., Coleman, G.J., Barnett, J.L., Borg, S. and Dowling, S. (2002) The effects of cognitive behavioral intervention on the attitude and behavior of stockpersons and the behavior and productivity of commercial dairy cows. *Journal of Animal Science* 80, 68–78.

Henry, S., Hemery, D., Richard, M.-A. and Hausberger, M. (2005) Human–mare relationships and behaviour of foals toward humans. *Applied Animal Behaviour Science* 93, 341–362.

Hepper, P.G. (1988) Adaptive fetal learning: prenatal exposure to garlic affects postnatal preference. *Animal Behaviour* 36, 935–936.

Herlin, A.H. and Frank, B. (2007) Effect of protective gates at concentrate feed stations on behaviour and production in dairy cows: a note. *Applied Animal Behaviour Science* 103, 167–173.

Herman, L.M. (2006) Intelligence and rational behaviour in the bottlenose dolphin. In: Hurley, S. and Nudds, M. (eds) *Rational Animals?* Oxford University Press, Oxford, UK, pp. 439–467.

Hess, E. (1959) Imprinting. *Science* 130, 133–141.

Hill, D.L. and Przekop, P.R. Jr (1988) Influences of dietary sodium on functional taste receptor development: a sensitive period. *Science* 241, 1826–1828.

Hill, D. and Robertson, P. (1988) *The Pheasant: Ecology, Management and Conservation.* BSP Professional Books, London.

Hinde, R.A. (1954) Factors governing the changes in strength of a partially inborn response; as shown by the mobbing behaviour of the chaffinch (*Fringilla coelebs*). *Proceedings of the Royal Society, Series B*, 142, 306–331; 331–358.

Hinton, S.C. and Meck, W.H. (1997) The 'internal clocks' of circadian and interval timing. *Endeavour* 21, 3–8.

Hogue, M.-E., Beaugrand, J.P. and Laguë, P.C. (1996) Coherent use of information by hens observing their former dominant defeating or being defeated by a stranger. *Behavioural Processes* 38, 241–252.

Holmes, L.N., Song, G.K. and Price, E.O. (1987) Head partitions facilitate feeding by subordinate horses in the presence of dominant pen-mates. *Applied Animal Behaviour Science* 19, 179–182.

Holy, T.E., Dulac, C. and Meister, M. (2000) Responses of vomeronasal neurons to natural stimuli. *Science* 289, 1569–1572.

Hudson, S.J. and Mullord, M.M. (1977) Investigations of maternal bonding in dairy cattle. *Applied Animal Ethology* 3, 271–276.

Hughes, B.O., Carmichael, N.L., Walker, A.W. and Grigor, P.N. (1997) Low incidence of aggression in large flocks of laying hens. *Applied Animal Behaviour Science* 54, 215–234.

Hulet, C.V., Ercanbrack, S.K., Blackwell, R.L., Price, D.A. and Wilson, L.O. (1962) Mating behavior of the ram in the multi-sire pen. *Journal of Animal Science* 21, 865–869.

Hulet, C.V., Shelton, M., Gallagher, J.R. and Price, D.A. (1974) Effects of origin and environment on reproductive phenomena in Rambouillet ewes. I. Breeding season and ovulation. *Journal of Animal Science* 38, 1210–1217.

Hurnik, J.F., King, G.J. and Robertson, H.A. (1975) Estrous and related behaviour in postpartum Holstein cows. *Applied Animal Ethology* 2, 55–68.

Hurnik, J.G. (1987) Sexual behavior of female domestic animals. In: Price, E.O. (ed) *Farm Animal Behavior. Veterinary Clinics of North America: Food Animal Practice* Vol. 3, No. 2, pp. 423–461.

Hyman, J. and Hughes, M. (2006) Territory owners discriminate between aggressive and nonaggressive neighbours. *Animal Behaviour* 72, 209–215.

Ionata, L.M., Anderson, T.M., Pickett, B.W., Heird, J.C. and Squires, E.L. (1991) Effect of supplementary sexual preparation on semen characteristics of stallions. *Theriogenology* 36, 923–937.

Iwata, E., Kikusui, T., Tacheuchi, Y. and Mori, Y. (2003) Substances derived from 4-ethyl octanoic acid account for primer pheromone activity for the 'male effect' in male goats. *Journal of Veterinary Medical Science* 65, 1019–1021.

Jacobs, G.H. (1993) The distribution and nature of colour vision among the mammals. *Biological Review* 68, 413–471.

Jacobs, G.H. and Deegan, J.F. II (1994) Spectral sensitivity, photopigments, and color vision in the guinea pig (*Cavia porcellus*). *Behavioral Neuroscience* 108, 993–1004.

Jacobs, G.H., Deegan, J.F. II, Crognale, M.A. and Fenwick, J.A. (1993) Photopigments of dogs and foxes and their implications for canid vision. *Visual Neuroscience* 10, 173–180.

Jacobs, G.H., Deegan, J.F. II, Neitz, J., Murphy, B.P., Miller, K.V. and Marchinton, R.L. (1994) Electrophysiological measurements of spectral mechanisms in the retinas of two cervids, white-tailed deer (*Odocoileus virginianus*) and fallow deer (*Dama dama*). *Journal of Comparative Physiology* (A) 174, 551–557.

Jacobs, G.H., Deegan, J.F. II and Neitz, J. (1998) Photopigment basis for dichromatic color vision in cows, goats and sheep. *Visual Neuroscience* 15, 581–584.

Jacobs, G.H., Fenwick, J.A. and Williams, G.A. (2001) Cone-based vision of rats for ultraviolet and visible lights. *Journal of Experimental Biology* 204 (Pt. 14), 2439–2446.

Jacobs, G.H., Williams, G.A. and Fenwick, J.A. (2004) Influence of cone pigment coexpression on spectral sensitivity and color vision in the mouse. *Vision Research* 44, 1615–1622.

Jenni, D.A. (1974) Evolution of polyandry in birds. *American Zoologist* 14, 129–141.

Jones, A.C. and Gosling, S.D. (2005) Temperament and personality in dogs (*Canis familiaris*): a review and evaluation of past research. *Applied Animal Behaviour Science* 95, 1–53.

Jones, R.B. (1993) Reduction of the domestic chick's fear of human beings by regular handling and related treatments. *Animal Behaviour* 46, 991–998.

Katz, L.S. and Price, E.O. (1986) The role of penile stimulation and ejaculatory experience on the development and maintenance of reproductive behavior in the bull (*Bos taurus*). *Developmental Psychobiology* 19, 197–209.

Katz, L.S., Price, E.O., Wallach, S.J.R. and Zenchak, J.J. (1988) Sexual performance of rams reared with or without females after weaning. *Journal of Animal Science* 66, 1166–1173.

Keeler, C.E. (1942) The association of the black (non-agouti) gene with behavior. *Journal of Heredity* 33, 371–384.

Keeling, L.J. and Duncan, I.J.H. (1989) Inter-individual distances and orientation in laying hens housed in groups of three in two different-sized enclosures. *Applied Animal Behaviour Science* 24, 325–342.

Kendrick, K.M. (1994) Neurobiological correlates of visual and olfactory recognition in sheep. *Behavioural Processes* 33, 89–111.

Kendrick, K.M., Lévy, F. and Keverne, E.B. (1992) Changes in the sensory processing of olfactory signals induced by birth in sheep. *Science* 256, 833–836.

Kendrick, K.M., Hinton, M.R., Atkins, K., Haupt, M.A. and Skinner, J.D. (1998) Mothers determine sexual preferences. *Nature* 395, 229–230.

Kenward, B., Rutz, C., Weir, A.A.S. and Kacelnik, A. (2006) Development of tool use in New Caledonian crows: inherited action patterns and social influences. *Animal Behaviour* 72, 1329–1343.

Keverne, E.B., Lévy, F., Poindron, P. and Lindsay, D.R. (1983) Vaginal stimulation: an important determinant of maternal bonding in sheep. *Science* 219, 81–83.

Kiddy, C.A. (1977) Variation in physical activity as an indication of estrus in dairy cows. *Journal of Dairy Science* 60, 235–243.

Kiddy, C.A., Mitchell, D.S., Bolt, D.J. and Hawk, M.W. (1978) Detection of estrus-related odors in cows by trained dogs. *Biology of Reproduction* 19, 389–395.

Kiddy, C.A., Mitchell, D.S. and Hawk, M.W. (1984) Estrus-related odors in body fluids of dairy cows. *Journal of Dairy Science* 67, 388–391.

King, M.G. (1965) Disruptions of the pecking order of cockerels concomitant with degrees of accessibility to feed. *Animal Behaviour* 13, 504–506.

King, G.J., Hurnik, J.F. and Robertson, H.A. (1976) Ovarian function and estrus in dairy cows during early lactation. *Journal of Animal Science* 42, 688–692.

Kirkwood, R.N., Forbes, J.M. and Hughes, P.E. (1981) Influence of boar on attainment of puberty in gilts after removal of the olfactory bulbs. *Journal of Reproduction and Fertility* 61, 193–196.

Klemann, N. and Pelz, H.J. (2006) The feeding pattern of the Norway rat (*Rattus norvegicus*) in two differently structured habitats on a farm. *Applied Animal Behaviour Science* 97, 293–302.

Klemm, W.R., Hawkins, G.N. and De Los Santos, E. (1987) Identification of compounds in bovine cervico-vaginal mucus extracts that evoke male sexual behavior. *Chemical Senses* 12, 77–87.

Klinghammer, E. and Laidlaw, L. (1979) Analysis of 23 months of daily howl records in a captive grey wolf pack (*Canis lupus*). In: Klinghammer, E. (ed) *The Behaviour and Ecology of Wolves.* Garland STPM Press, New York, pp. 153–181.

Klopfer, P.H. and Gamble, J. (1966) Maternal 'imprinting' in goats: the role of the chemical senses. *Zeitschrift für Tierpsychology* 23, 588–592.

Klopfer, P.H., Adams, D.K. and Klopfer, M.S. (1964) Maternal 'imprinting' in goats. *Proceedings of the National Academy of Sciences* 52, 911–914.

Knight, T. (1983) Ram induced stimulation of ovarian and oestrous activity in anoestrous ewes – a review. *Proceedings of the New Zealand Society of Animal Production* 43, 7–11.

Knight, T.W. and Lynch, P.R. (1980) Source of ram pheromones that stimulate ovulation in the ewe. *Animal Reproduction Science* 3, 133–136.

Knight, T.W., McMillan, W.H., Kilgour, R., Hockey, H.-U.P. and Hall, D.R. (1989) Effect of slope of lambing site on lambs slipping and lamb mortality. *New Zealand Journal of Agricultural Research* 32, 199–206.

Konishi, M. (1963) The role of auditory feedback in the vocal behavior of the domestic fowl. *Zeitschrift für Tierpsychologie* 20, 349–367.

Konishi, M. (1965) The role of auditory feedback in the control of vocalization in the white-crowned sparrow. *Zeitschrift für Tierpsychologie* 22, 770–783.

Konopka, R.J. and Benzer, S. (1971) Clock mutants of *Drosophila melanogaster. Proceedings of the National Academy of Sciences* 68, 2112–2116.

Koski, S.E. and Sterck, E.H.M. (2007) Triadic postconflict affiliation in captive chimpanzees: does consolation console? *Animal Behaviour* 73, 133–142.

Kruijt, J.P. (1985) On the development of social attachments in birds. *Netherlands Journal of Zoology* 35, 45–62.

Kuhn, C.M. and Schanberg, S.M. (1998) Responses to maternal separation: mechanisms and mediators. *International Journal of Developmental Neuroscience* 16, 261–270.

Kuhn, C.M., Butler, S.R. and Schanberg, S.M. (1978) Selective suppression of serum growth hormone during maternal deprivation in rat pups. *Science* 201, 1034–1036.

Kuo, Z.-Y. (1932) Ontogeny of embryonic behavior in Aves. IV. The influence of embryonic movements upon behavior after hatching. *Journal of Comparative Psychology* 14, 109–122.

Ladewig, J., Price, E.O. and Hart, B.L. (1980) Flehmen in male goats: role in sexual behavior. *Behavioral and Neural Biology* 30, 312–322.

Lagerweij, E., Nelis, P.C., Weigant, V.M. and Van Ree, J.M. (1984) The twitch in horses: a variant of acupuncture. *Science* 225, 1172–1174.

Lane, A.J.P. and Wathes, D.C. (1998) An electronic nose to detect changes in perineal odors associated with estrus in the cow. *Journal of Dairy Science* 81, 2145–2150.

Lankin, V.S. (1997) Factors of diversity of domestic behaviour in sheep. *Genetics, Selection and Evolution* 29, 73–92.

Lansade, L., Bertrand, M. and Bouissou, M.-F. (2005) Effects of neonatal handling on subsequent manageability, reactivity and learning ability of foals. *Applied Animal Behaviour Science* 92, 143–158.

Lapidge, K., Oldroyd, B. and Spivak, M. (2002) Seven suggestive quantitative trait loci influence hygienic behavior of honeybees. *Naturwissenschaften* 89, 565–568.

Le Boeuf, B.J. (1970) Copulatory and aggressive behavior in the prepuberally castrated dog. *Hormones and Behavior* 1, 127–136.

Lehrman, D.S. (1961) Hormonal regulation of parental behavior in birds and infrahuman mammals. In: Young, W.C. (ed) *Sex and Internal Secretions*, Vol. 2. Williams and Wilkins, Baltimore, Maryland, pp. 1268–1382.

Le Neindre, P., Trillat, G., Sapa, J., Ménissier, F., Bonnet, J.N. and Chupin, J.M. (1995) Individual differences in docility in Limousin cattle. *Journal of Animal Science* 73, 2249–2253.

Lensink, B.J., Fernandez, X., Cozzi, G., Florand, L. and Veissier, I. (2001) The influence of farmers' behavior on calves' reactions to transport and quality of veal meat. *Journal of Animal Science* 79, 642–652.

Leon, M. (1974) Maternal pheromone. *Physiology and Behavior* 13, 441–453.

Leon, M. and Moltz, H. (1972) The development of the pheromonal bond in the albino rat. *Physiology and Behavior* 8, 683–686.

Leopold, A.S. (1944) The nature of heritable wildness in turkeys. *Condor* 46, 133–197.

Leuthold, W. (1977) *African Ungulates: A Comparative Review of their Ethology and Behavioral Ecology*, Springer-Verlag, New York.

Levine, E.D., Ramos, D. and Mills, D.S. (2007) A prospective study of two self-help CD based desensitization and counter-conditioning programmes with the use of Dog Appeasing Pheromone for the treatment of firework fears in dogs (*Canis familiaris*). *Applied Animal Behaviour Science* 105, 311–329.

Lévy, F., Poindron, P. and Le Neindre, P. (1983) Attraction and repulsion by amniotic fluids and their olfactory control in the ewe around parturition. *Physiology and Behavior* 31, 687–692.

Lévy, F., Keverne, E.B., Piketty, V. and Poindron, P. (1990) Physiological determination of olfactory attraction for amniotic fluid in sheep. In: MacDonald, D.W., Müller-Schwarze, D. and Natynczuk, S.E. (eds) *Chemical Signals in Vertebrates*, Vol. 5. Oxford University Press, New York, pp. 162–165.

Lévy, F., Gervais, R., Kindermann, U., Litterio, M., Poindron, P. and Porter, R. (1991) Effects of early post-partum separation on maintenance of maternal responsiveness and selectivity in parturient ewes. *Applied Animal Behaviour Science* 31, 101–110.

Lévy, F., Porter, R.H., Kendrick, K.M., Keverne, E.B. and Romeyer, A. (1996) Physiological, sensory, and experiential factors of parental care in sheep. In: Rosenblatt, J.S. and Snowdon, C.T. (eds) *Parental Care: Evolution, Mechanisms, and Adaptive Significance. Advances in the Study of Behavior*, Vol. 25. Academic Press, New York, pp. 385–422.

Leyhausen, P. (1979) *Cat Behavior: The Predatory and Social Behavior of Domestic and Wild Cats*. Garland STPM Press, New York.

Li, L.-L., Keverne, E.B., Aparicio, S.A., Ishino, F., Barton, S.C. and Surani, M.A. (1999) Regulation of maternal behavior and offspring growth by paternally expressed *Peg3*. *Science* 284, 330–333.

Lickliter, R. (1982) Effects of a post-partum separation on maternal responsiveness in primiparous and multiparous domestic goats. *Applied Animal Ethology* 8, 537–542.

Lickliter, R. and Stoumbos, J. (1992) Modification of prenatal auditory experience alters postnatal auditory preferences of bobwhite quail chicks. *Quarterly Journal of Experimental Psychology, B*, 44, 197–214.

Lidfors, L. (1997) Behavioural effects of environmental enrichment for individually caged rabbits. *Applied Animal Behaviour Science* 52, 157–169.

Lincoln, G.A. (1992) Photoperiod-pineal-hypothalamic relay in sheep. *Animal Reproduction Science* 28, 203–217.

Lincoln, G.A. and Davidson, W. (1977) The relationship between sexual and aggressive behavior, and pituitary and testicular activity during the seasonal sexual cycle of rams, and the influence of photoperiod. *Journal of Reproduction and Fertility* 49, 267–276.

Lindell, E.M. (1997) Diagnosis and treatment of destructive behavior in dogs. In: Houpt, K.A. (ed) *Progress in Companion Animal Behavior. Veterinary Clinics of North America: Small Animal Practice*, Vol. 27, No. 3, 533–547.

Lindsay, D.R. and Fletcher, I.C. (1972) Ram-seeking activity associated with estrous behavior in ewes. *Animal Behaviour* 20, 452–456.

Lindsay, D.R., Dunsmore, D.G., Williams, J.D. and Syme, G.J. (1976) Audience effects on mating behaviour of rams. *Animal Behaviour* 24, 818–821.

Line, S.W., Hart, B.L. and Sanders, L. (1985) Prepubertal versus postpubertal castration of male horses: effect on sexual and aggressive behavior. *Journal of the American Veterinary Medical Association* 186, 249–251.

Liu, D., Diorio, J., Tannenbaum, B., Caldji, C., Francis, D., Freedman, A., Sharma, S., Pearson, D., Plotsky, P.M. and Meaney, M.J. (1997) Maternal care, hippocampal glucocorticoid receptors, and hypothalamic–pituitary–adrenal responses to stress. *Science* 277, 1659–1662.

Loberg, J.M., Hernandez, C.E., Thierfelder, T., Jensen, M.B., Berg, C. and Lidfors, L. (2007) Reaction of foster cows to prevention of suckling from and separation from four calves simultaneously or in two steps. *Journal of Animal Science* 85, 1522–1529.

Lockwood, R. (1995) The ethology and epidemiology of canine aggression. In: Serpell, J. (ed) *The Domestic Dog. Its Evolution, Behaviour and Interactions with People.* Cambridge University Press, New York, pp. 131–138.

Lorenz, K. (1981) *The Foundations of Ethology.* Springer-Verlag, New York.

Luo, M., Fee, M. and Katz, L.C. (2003) Encoding pheromonal signals in the accessory olfactory bulb of behaving mice. *Science* 299, 1196–1201.

Lynch, J.J., Hinch, G.N. and Adams, D.B. (1992) *The Behaviour of Sheep.* CAB International, Wallingford, UK.

Lyons, D.M., Price, E.O. and Moberg, G.P. (1988a) Individual differences in temperament of domestic dairy goats: constancy and change. *Animal Behaviour* 36, 1323–1333.

Lyons, D.M., Price, E.O. and Moberg, G.P. (1988b) Social modulation of pituitary–adrenal responsiveness and individual differences in behavior of young domestic goats. *Physiology and Behavior* 43, 451–458.

Machado, L.C.P., Hurnick, J.F. and King, G.J. (1997) Timing of the attraction towards the placenta and amniotic fluid by the parturient cow. *Applied Animal Behaviour Science* 53, 183–192.

Macrides, F., Bartke, A. and Dalterio, S. (1975) Strange females increase plasma testosterone levels in male mice. *Science* 189, 1104–1106.

Mader, D.R. and Price, E.O. (1980) Discrimination learning in horses: effects of breed, age and social dominance. *Journal of Animal Science* 50, 962–965.

Mader, D.R. and Price, E.O. (1984) The effects of sexual stimulation on the sexual performance of Hereford bulls. *Journal of Animal Science* 59, 294–300.

Malmkvist, J. and Hansen, S.W. (2001) The welfare of farmed mink (*Mustela vison*) in relation to behavioural selection: a review. *Animal Welfare* 10, 41–52.

Mankovich, N.J. and Banks, E.M. (1982) An analysis of social orientation and the use of space in a flock of domestic fowl. *Applied Animal Ethology* 9, 177–193.

Marchlewska-Koj, A., Pochroń, E. and Śliwowska, A. (1990) Salivary glands and preputial glands of males as source of estrus-stimulating pheromone in female mice. *Journal of Chemical Ecology* 16, 2817–2822.

Marler, P. (1956) Studies of fighting in chaffinches. (3) Proximity as a cause of aggression. *British Journal of Animal Behaviour* 4, 23–30.

Marler, P. and Slabbekoorn, H. (eds) (2004) *Nature's Music.* Elsevier, Inc., New York.

Markowitz, T.M., Dally, M.R., Gursky, K. and Price, E.O. (1998) Early handling increases lamb affinity for humans. *Animal Behaviour* 55, 573–587.

Martin, N.L., Price, E.O., Wallach, S.J.R. and Dally, M.R. (1987) Fostering lambs by odor transfer: the add-on experiment. *Journal of Animal Science* 64, 1378–1383.

Mason, G. and Rushen, J. (eds) (2006) *Stereotypic Animal Behaviour.* CAB International, Wallingford, UK.

Mason, G., Clubb, R., Latham, N. and Vickery, S. (2007) Why and how should we use environmental enrichment to tackle stereotyped behaviour? *Applied Animal Behaviour Science* 102, 163–188.

Mateos, C. and Carranza, J. (1995) Female choice for morphological features of male ring-necked pheasants. *Animal Behaviour* 49, 737–748.

Matthews, K., Dalley, J.W., Matthews, C., Tsai, T.H. and Robbins, T.W. (2001) Periodic maternal separation of neonatal rats produces region- and gender-specific effects on biogenic amine content in postmortem adult brain. *Synapse* 40, 1–10.

McAdie, T.M., Keeling, L.J., Blokhuis, H.J. and Jones, R.B. (2005) Reduction in feather pecking and improvement in feather condition with the presentation of a string device to chickens. *Applied Animal Behaviour Science* 93, 67–80.

McBride, G. (1964) A general theory of social organization and behaviour. *University of Queensland Papers, Faculty of Veterinary Science* 1, 75–110.

McBride, G. (1971) Theories of animal spacing: the role of flight, fight and social distance. In: Esser, A.H. (ed) *The Social Use of Space in Animals and Man.* Plenum Press, New York, pp. 53–68.

McBride, G., James, J.W. and Shoffner, R.N. (1963) Social forces determining spacing and head orientation in a flock of domestic hens. *Nature* 197, 1272–1273.

McClearn, G.E. (1963) The inheritance of behavior. In: Postman, L. (ed) *Psychology in the Making.* Knopf, New York, pp. 144–252.

McConnell, P.B. (1991) Lessons from animal trainers: the effect of acoustic structure on an animal's response. In: Bateson, P.P.G. and Klopfer, P.H. (eds) *Perspectives in Ethology*, Vol. 9, *Human Understanding and Animal Awareness.* Plenum Press, New York, pp. 165–187.

McCort, W.D. and Graves, H.B. (1982) Social dominance relationships and spacing behaviour of swine. *Behavioural Processes* 7, 169–178.

McDonnell, S. (1986) Reproductive behavior of the stallion. In: Crowell-Davis, S.L. and Houpt, K.A. (eds) *Behavior. Veterinary Clinics of North America: Equine Practice* Vol. 2, No. 3, 535–555.

McGlone, J.J. and Anderson, D.L. (2002) Synthetic maternal pheromone stimulates feeding behavior and weight gain in weaned pigs. *Journal of Animal Science* 80, 3179–3183.

McGlone, J.J. and Morrow, J.L. (1988) Reduction of pig agonistic behavior by androstenone. *Journal of Animal Science* 66, 880–884.

Mech, L.D. (1970) *The Wolf: The Ecology and Behaviour of an Endangered Species.* Natural History Press, New York.

Mech, L.D. (1999) Alpha status, dominance, and division of labor in wolf packs. *Canadian Journal of Zoology* 77, 1196–1203.

Meehan, C.L. and Mench, J.A. (2007) The challenge of challenge: can problem solving opportunities enhance animal welfare? *Applied Animal Behaviour Science* 102, 246–261.

Meese, G.E. and Baldwin, B.A. (1975) Effects of olfactory bulb ablation on maternal behaviour in sows. *Applied Animal Ethology* 1, 379–386.

Mellen, J.D. (1992) Effects of early rearing experience on subsequent adult sexual behavior using domestic cats (*Felis catus*) as a model for exotic small felids. *Zoo Biology* 11, 17–32.

Melrose, D.R., Reed, H.C.B., and Patterson, R.L.S. (1971) Androgen steroids associated with boar odour as an aid to the detection of oestrus in pig artificial insemination. *British Veterinary Journal* 127, 497–501.

Mendl, M., Zanella, A.J. and Broom, D.M. (1992) Physiological and reproductive correlates of behavioural strategies in female domestic pigs. *Animal Behaviour* 44, 1107–1121.

Metzler, J.A., Price, E.O., Kitchell, R.L. and BonDurant, R.H. (1988) Sexual behavior of male dairy goats: effects of deafferentation of the genitalia. *Physiology and Behavior* 43, 207–212.

Miller, K.A. and Mench, J.A. (2005) The differential effects of four types of environmental enrichment on the activity budgets, fearfulness and social proximity preference of Japanese quail. *Applied Animal Behaviour Science* 95, 169–187.

Mills, D.S. and Riezebos, M. (2005) The role of the image of a conspecific in the regulation of stereotypic head movements in the horse. *Applied Animal Behaviour Science* 91, 155–165.

Moberg, G.P., Anderson, C.O. and Underwood, T.R. (1980) Ontogeny of the adrenal and behavioral responses of lambs to emotional stress. *Journal of Animal Science* 51, 138–142.

Mononen, J., Kasanen, S., Harri, M., Sepponen, J. and Rekilä, T. (2001) The effects of elevated platforms and concealment screens on the welfare of blue foxes. *Animal Welfare* 10, 373–385.

Moore, A.S., Gonyou, H.W. and Ghent, A.W. (1993) Integration of newly introduced and resident sows following grouping. *Applied Animal Behaviour Science* 38, 257–267.

Moritz, R.F.A. (1988) A re-evaluation of the two-locus model for hygienic behavior in honeybees (*Apis mellifera* L.). *Journal of Heredity* 79, 257–262.

Morrow, D.A. (1969) Estrous behavior and ovarian activity in prepuberal and postpuberal dairy heifers. *Journal of Dairy Science* 52, 224–227.

Morrow-Tesch, J. and McGlone, J.J. (1990a) Sources of maternal odors and the development of odor preferences in baby pigs. *Journal of Animal Science* 68, 3563–3571.

Morrow-Tesch, J. and McGlone, J.J. (1990b) Sensory systems and nipple attachment behavior in neonatal pigs. *Physiology and Behavior* 47, 1–4.

Mounier, L., Veissier, I., Andansen, S., Delval, E. and Boissy, A. (2006) Mixing at the beginning of fattening moderates social buffering in beef bulls. *Applied Animal Behaviour Science* 96, 185–200.

Müller, R. and von Keyserlingk, M.A.G. (2006) Consistency of flight speed and its correlation to productivity and to personality in *Bos taurus* beef cattle. *Applied Animal Behaviour Science* 99, 193–204.

Mundinger, P.C. (1995) Behaviour–genetic analysis of canary song: inter-strain differences in sensory learning, and epigenetic rules. *Animal Behaviour* 50, 1491–1511.

Mykytowycz, R. (1968) Territorial marking by rabbits. *Scientific American* 218(5), 116–126.

Mykytowycz, R. (1970) The role of skin glands in mammalian communication. In: Johnston, J.W. Jr, Moulton, D.G. and Turk, A. (eds) (1970) *Advances in Chemoreception*, Vol. 1, *Communication by Chemical Signals*. Appleton Century Crofts, New York, pp. 327–360.

Natynczuk, S., Bradshaw, J.W.S. and McDonald, D.W. (1989) Chemical constituents of the anal sacs of domestic dogs. *Biochemical Systematics and Ecology* 17, 83–87.

Neitz, J. and Jacobs, G.H. (1989) Spectral sensitivity of cones in an ungulate. *Visual Neuroscience* 2, 97–100.

Neitz, J., Geist, T. and Jacobs, G.H. (1989) Color vision in the dog. *Visual Neuroscience* 3, 119–125.

Nicol, C. (1999) Understanding equine stereotypies. *Equine Veterinary Journal* (Supplement) 28, 20–25.

Nikoletseas, M. and Lore, R. (1981) Aggression in domesticated rats reared in a burrow-digging environment. *Aggressive Behavior* 7, 245–252.

Nikulina, E.M. (1990) Brain catecholamines during domestication of the silver fox (*Vulpes vulpes*). *Journal of Evolution, Biochemistry and Physiology* 26, 118–121.

Norbury, G., O'Connor, C. and Byrom, A. (2005) Conditioned food aversion to eggs in captive-reared ferrets, *Mustela furo*: a test of seven potential compounds. *Applied Animal Behaviour Science* 93, 111–121.

Nottebohm, F. (1989) From bird song to neurogenesis. *Scientific American* 260, 74–79.

Novotny, M., Harvey, S., Jemiolo, B. and Alberts, J. (1985) Synthetic pheromones that promote inter-male aggression in mice. *Proceedings of the National Academy of Sciences* 82, 2059–2061.

Odend'Hal, S., Miller, K.V. and Demarais, S. (1996) Preputial glands in Artiodactyla. *Journal of Mammalogy* 77, 417–421.

O'Keefe, T.R., Graves, H.B. and Siegel, H.S. (1988) Social organization in caged layers: the peck order revisited. *Poultry Science* 67, 1008–1014.

Oliveira, R.F., Lopes, M., Carneiro, L.A. and Canário, A.V.M. (2001) Watching fights raises fish hormone levels. *Nature* 409, 475.

Olmstead, C.E., Villablanca, J.R., Torbiner, M. and Rhodes, D. (1979) Development of thermo-regulation in the kitten. *Physiology and Behavior* 23, 489–495.

Oppenheim, R.W. (1974) The ontogeny of behavior in the chick embryo. In: Lehrman, D.S., Rosenblatt, J.S., Hinde, R.A. and Shaw, E. (eds) *Advances in the Study of Behavior*, Vol. 5. Academic Press, New York, pp. 133–172.

Osborn, D.A., Miller, K.V., Hoffman, D.M., Dickerson, W.H., Gassett, J.W. and Quist, C.F. (2000) Morphology of the white-tailed deer tarsal gland. *Acta Theriologica* 45, 117–122.

Osborne, K.A., Robichon, A., Burgess, E., Butland, S., Shaw, R.A., Coulthard, A., Pereira, H.S., Greenspan, R.J. and Sokolowski, M.B. (1997) Natural behavior polymorphism due to a cGMP-dependent protein kinase of *Drosophila*. *Science* 277, 834–836.

Packer, C., Collins, D.A., Sindimwo, A. and Goodall, J. (1995) Reproductive constraints on aggressive competition in female baboons. *Nature* 373, 60–63.

Page, R.E., Waddington, K.D., Hunt, G.J. and Fondrk, M.K. (1995) Genetic determinants of honey bee foraging behaviour. *Animal Behaviour* 50, 1617–1625.

Palen, G.F. and Goddard, G.V. (1966) Catnip and oestrous behaviour in the cat. *Animal Behaviour* 14, 372–377.

Palmer, E. and Guillaume, D. (1998) Some mechanisms involved in the response of mares to photoperiodic stimulation of reproductive activity. *Reproduction in Domestic Animals* 33, 205–208.

Pedersen, P.E., Williams, C.L. and Blass, E.M. (1982) Activation and odor conditioning of suckling behavior in 3-day-old albino rats. *Journal of Experimental Psychology: Animal Behavior Processes* 8, 329–341.

Pelletier, F. and Festa-Bianchet, M. (2006) Sexual selection and social rank in bighorn rams. *Animal Behaviour* 71, 649–655.

Pengelley, E.T. and Asmundson, S.J. (1974) Circannual rhythmicity in hibernating mammals. In: Pengelley, E.T. (ed) *Circannual Clocks*. Academic Press, New York, pp. 95–160.

Pengelley, E.T., Asmundson, S.J., Barnes, B. and Aloia, R.C. (1976) Relationship of light intensity and photoperiod to circannual rhythmicity in the hibernating ground squirrel, *Citellus lateralis*. *Comparative Biochemistry and Physiology Part A: Physiology* 53, 273–277.

Pennisi, E. (2000) The snarls and sneers that keep violence at bay. *Science* 289, 576–577.

Pepelko, W.E. and Clegg, M.T. (1965) Studies of mating behaviour and some factors influenc-ing the sexual response in the male sheep, *Ovis aries*. *Animal Behaviour* 13, 249–258.

Pepperberg, I.M. (2006) Cognitive and communicative abilities of Grey parrots. *Applied Animal Behaviour Science* 100, 77–86.

Pérez-Guisado, J., Rodríguez, R. and Muñoz-Serrano, A. (2006) Heritability of dominant-aggressive behavior in English Cocker Spaniels. *Applied Animal Behaviour Science* 100, 219–227.

Perkins, A., Fitzgerald, J.A. and Price, E.O. (1992a) Sexual performance of rams in serving capacity tests predicts success in pen breeding. *Journal of Animal Science* 70, 2722–2725.

Perkins, A., Fitzgerald, J.A. and Price, E.O. (1992b) Luteinizing hormone and testosterone response of sexually active and inactive rams. *Journal of Animal Science* 70, 2086–3093.

Peters, R.P. and Mech, L.D. (1975) Scent marking in wolves. *American Scientist* 63, 628–637.

Pickerel, T.M., Crowell-Davis, S.L, Caudle, A.B. and Estep, D.Q. (1993) Sexual preference of mares (*Equus caballus*) for individual stallions. *Applied Animal Behaviour Science* 38, 1–13.

Platt, D.M. and Novak, M.A. (1997) Videostimulation as enrichment for captive rhesus monkeys (*Macaca mulatta*). *Applied Animal Behaviour Science* 52, 139–155.

Poindron, P. and Carrick, M. (1976) Hearing recognition of the lamb by its mother. *Animal Behaviour* 24, 600–602.

Polsky, R.H. (1995) Wolf hybrids: are they suitable as pets? *Veterinary Medicine* 90, 1122–1124.

Pongrácz, P., Molnár, C. and Miklósi, Á. (2006) Acoustic parameters of dog barks carry emotional information for humans. *Applied Animal Behaviour Science* 100, 228–240.

Popova, N.K., Voitenko, N.N., Kulikov, A.V. and Avgustinovich, D.F. (1991a) Evidence for the involvement of central serotonin in mechanism of domestication of silver foxes. *Pharmacology, Biochemistry and Behaviour* 40, 751–756.

Popova, N.K., Kulikov, A.V., Nikulina, E.M., Kozlachkova, E.Y. and Maslova, G.B. (1991b) Serotonin metabolism and serotonergic receptors in Norway rats selected for low aggressiveness to man. *Aggressive Behavior* 17, 207–213.

Post, T.B., Christensen, H.R. and Seifert, G.W. (1987) Reproductive performance and productive traits of beef bulls selected for different levels of testosterone response to GnRH. *Theriogenology* 27, 317–328.

Prayitno, D.S., Phillips, C.J.C. and Omed, H. (1997) The effects of color of lighting on the behavior and production of meat chickens. *Poultry Science* 76, 452–457.

Price, E.O. (1966) Influence of light on reproduction in *Peromyscus maniculatus gracilis*. *Journal of Mammalogy* 47, 343–344.

Price, E.O. (1972) Domestication and early experience effects on escape conditioning in the Norway rat. *Journal of Comparative and Physiological Psychology* 79, 51–55.

Price, E.O. (1978) Genotype versus experience effects on aggression in wild and domestic Norway rats. *Behaviour* 64, 340–353.

Price, E.O. (1980) Sexual behaviour and reproductive competition in male wild and domestic Norway rats. *Animal Behaviour* 28, 657–667.

Price, E.O. (1987) Male sexual behavior. In: Price, E.O. (ed) *Farm Animal Behavior. Veterinary Clinics of North America: Food Animal Practice*, Vol. 3, No. 2, 405–422.

Price, E.O. (2002) *Animal Domestication and Behavior*. CAB International, Wallingford, UK.

Price, E.O. and Wallach, S.J.R. (1990a) Physical isolation of hand-reared Hereford bulls increases their aggressiveness toward humans. *Applied Animal Behaviour Science* 27, 263–267.

Price, E.O. and Wallach, S.J.R. (1990b) Rearing bulls with females fails to enhance sexual performance. *Applied Animal Behaviour Science* 26, 339–347.

Price, E.O. and Wallach, S.J.R. (1991) Inability to predict the adult sexual performance of bulls by prepuberal sexual behaviors. *Journal of Animal Science* 69, 1041–1046.

Price, E.O., Dunbar, J.M. and Dally, M.R. (1984a) Behavior of ewes and lambs subjected to restraint fostering. *Journal of Animal Science* 58, 1084–1089.

Price, E.O., Dunn, G.C., Talbot, J.A. and Dally, M.R. (1984b) Fostering lambs by odor transfer: the substitution experiment. *Journal of Animal Science* 59, 301–307.

Price, E.O., Smith, V.M. and Katz, L.S. (1984/85) Sexual stimulation of male dairy goats. *Applied Animal Behaviour Science* 13, 83–92.

Price, E.O., Katz, L.S., Moberg, G.P. and Wallach, S.J.R. (1986) Inability to predict sexual and aggressive behaviors by plasma concentrations of testosterone and luteinizing hormone in Hereford bulls. *Journal of Animal Science* 62, 613–617.

Price, E.O., Katz, L.S., Wallach, S.J.R. and Zenchak, J.J. (1988) The relationship of male–male mounting to the sexual preferences of young rams. *Applied Animal Behaviour Science* 21, 347–355.

Price, E.O., Wallach, S.J.R. and Silver, G.V. (1990) The effects of long-term individual vs. group housing on the sexual behaviour of beef bulls. *Applied Animal Behaviour Science* 27, 277–285.

Price, E.O., Erhard, H., Borgwardt, R. and Dally, M.R. (1992) Measures of libido and their relation to serving capacity in the ram. *Journal of Animal Science* 70, 3376–3380.

Price, E.O., Hutson, G.D., Price, M.I. and Borgwardt, R. (1994) Fostering in swine as affected by age of offspring. *Journal of Animal Science* 72, 1697–1701.

Price, E.O., Borgwardt, R., Dally, M.R. and Hemsworth, P.H. (1996) Repeated matings with individual ewes by rams differing in sexual performance. *Journal of Animal Science* 74, 542–544.

Price, E.O., Borgwardt, R., Orihuela, A. and Dally, M.R. (1998a) Sexual stimulation in male sheep and goats. *Applied Animal Behaviour Science* 59, 317–322.

Price, E.O., Dally, M., Erhard, H., Gerzevske, M., Kelly, M., Moore, N., Schultze, A. and Topper, C. (1998b) Manipulating odor cues facilitates add-on fostering in sheep. *Journal of Animal Science* 76, 961–964.

Price, E.O., Borgwardt, R.E. and Dally, M.R. (2001) Male–male competition fails to sexually stimulate domestic rams. *Applied Animal Behaviour Science* 74, 217–222.

Price, E.O., Adams, T.E., Huxsoll, C.C. and Borgwardt, R.E. (2003a) Aggressive behavior is reduced in bulls actively immunized against gonadotropin-releasing hormone. *Journal of Animal Science* 81, 411–415.

Price, E.O., Harris, J.E., Borgwardt, R.E., Sween, M.L. and Connor, J.M. (2003b) Fenceline contact of beef calves with their dams at weaning reduces the negative effects of separation on behavior and growth rate. *Journal of Animal Science* 81, 116–121.

Ralls, K., Brugger, K. and Ballou, J. (1979) Inbreeding and juvenile mortality in small populations of ungulates. *Science* 206, 1101–1103.

Ralls, K., Lundrigan, B. and Kranz, K. (1987) Mother–young relationships in captive ungulates: behavioral changes over time. *Ethology* 75, 1–14.

Ralph, M.R. and Menaker, M. (1988) A mutation of the circadian system in golden hamsters. *Science* 241, 1225–1227.

Ralph, M.R., Foster, R.G., Davis, F.C. and Menaker, M. (1990) Transplanted suprachiasmatic nucleus determines circadian period. *Science* 247, 975978.

Ramírez, A., Quiles, A., Hevia, M.L., Sotillo, F. and Ramírez, M.C. (1996) Effect of immediate and early post-partum separation on maintenance of maternal responsiveness in parturient multiparous goats. *Applied Animal Behaviour Science* 48, 215–224.

Rauw, W. (2006) A note on behavioural response to a novel arena in lactating mice highly selected for litter size. *Applied Animal Behaviour Science* 99, 357–365.

Reinhardt, V. (1991) Training adult rhesus monkeys to actively cooperate during in-homecage venipuncture. *Animal Technology* 42, 11–17.

Reinhardt, V. (1996) Refining the blood collection procedure for macaques. *Laboratory Animals* 25, 32–35.

Reiss, D. and Marino, L. (2001) Mirror self-recognition in the bottlenose dolphin: a case of cognitive convergence. *Proceedings of the National Academy of Sciences* 98, 5937–5942.

Robinson, R. (1965) *Genetics of the Norway Rat.* Pergamon Press, Oxford, UK.

Romeyer, A., Poindron, P. and Orgeur, P. (1994) Olfaction mediates the establishment of selective bonding in goats. *Physiology and Behavior* 56, 693–700.

Rosenblatt, J.S. and Aronson, L.R. (1958) The influence of experience on the behavioural effects of androgen in prepuberally castrated male cats. *Animal Behaviour* 6, 171–182.

Roselli, C.E., Stormshak, F., Stellflug, J.N. and Resko, J.A. (2002) Relationship of serum testosterone concentrations to mate preferences in rams. *Biology of Reproduction* 67, 263–268.

Roselli, C.E., Larkin, K., Resko, J.A., Stellflug, J.N. and Stormshak, F. (2004) The volume of a sexually dimorphic nucleus in the ovine medial preoptic area/anterior hypothalamus varies with sexual partner preference. *Endocrinology* 145, 478–483.

Roselli, C.E., Stadelman, H., Reeve, R., Bishop, C.V. and Stormshak, F. (2007) The ovine sexually dimorphic nucleus of the medial preoptic area is organized prenatally by testosterone. *Endocrinology* 148, 4450–4457.

Rothenbuhler, W.C. (1964a) Behaviour genetics of nest cleaning in honeybees. I. Responses of four inbred lines to disease-killed brood. *Animal Behaviour* 12, 578–583.

Rothenbuhler, W.C. (1964b) Behavior genetics of nest cleaning in honeybees. IV. Responses of F1 and backcross generations to disease-killed brood. *American Zoologist* 4, 111–123.

Rottman, S.J. and Snowden, C.T. (1972) Demonstration and analysis of an alarm pheromone in mice. *Journal of Comparative and Physiological Psychology* 81, 483–490.

Rouquier, S., Blancher, A. and Giorgi, D. (2000) The olfactory receptor gene repertoire in primates and mouse: evidence for reduction of the functional fraction in primates. *Proceedings of the National Academy of Sciences* 97, 2870–2874.

Rusak, B. (1977) The role of the suprachiasmatic nuclei in the generation of circadian rhythms in the golden hamster, *Mesocricetus auratus*. *Journal of Comparative Physiology* 118, 145–164.

Rushen, J. (1986) Aversion of sheep for handling treatments: paired-choice studies. *Applied Animal Behaviour Science* 16, 363–370.

Rushen, J. (1987) A difference in weight reduces fighting when unacquainted newly weaned pigs first meet. *Canadian Journal of Animal Science* 67, 951–960.

Rushen, J. and de Passillé, A.M. (1995) The motivation of non-nutritive sucking in calves, *Bos taurus*. *Animal Behaviour* 49, 1503–1510.

Rushen, J., Munksgaard, L., de Passillé, A.M.B., Jensen, M.B. and Thodberg, K. (1998) Location of handling and dairy cows' response to people. *Applied Animal Behaviour Science* 55, 259–267.

Rushen, J., Taylor, A.A. and de Passillé, A.M. (1999) Domestic animals' fear of humans and its effect on their welfare. *Applied Animal Behaviour Science* 65, 285–303.

Ryan, B.C. and Vandenbergh, J.G. (2002) Intrauterine position effects. *Neuroscience and Biobehavioral Reviews* 26, 665–678.

Rybarczyk, P., Koba, Y., Rushen, J., Tanida, H. and de Passillé, A.M. (2001) Can cows discriminate people by their faces? *Applied Animal Behaviour Science* 74, 175–189.

Ryner, L.C., Goodwin, S.F., Castrillon, D.H., Anand, A., Villella, A., Baker, B.S., Hall, J.C., Taylor, B.J. and Wasserman, S.A. (1996) Control of male sexual behavior and sexual orientation in Drosophila by the *fruitless* gene. *Cell* 87, 1079–1089.

Sapolsky, R.M. (1992) Cortisol concentrations and the social significance of rank instability among wild baboons. *Psychoneuroendocrinology* 17, 701–709.

Sapolsky, R. (2004) Social status and health in humans and other animals. *Annual Review of Anthropology* 33, 393–418.

Schalke, E., Stichnoth, J., Ott, S. and Jones-Baade, R. (2007) Clinical signs caused by the use of electric training collars on dogs in everyday life situations. *Applied Animal Behaviour Science* 105, 369–380.

Schanberg, S.M. and Field, T.M. (1987) Sensory deprivation stress and supplemental stimulation in the rat pup and preterm human neonates. *Child Development* 58, 1431–1437.

Schenkel, R. (1948) Ausdrucks-studien an Wölfen: Gefangenschafts-beobachtungen. *Behaviour* I, 81–129.

Schenkel, R. (1967) Submission: its features and function in the wolf and dog. *American Zoologist* 7, 319–329.

Schichowski, C., Moors, E. and Gauly, M. (2008) Effects of weaning lambs in two stages or by abrupt separation on their behavior and growth rate. *Journal of Animal Science* 86, 220–225.

Schjelderup-Ebbe, T. (1922) Beiträge zur Sozialpsychologie des Haushuhns. *Zeitschrift für Psychologie* 88, 225–252. English translation by Schleidt, M. and Schleidt, W. (1975) Contributions to the social psychology of the domestic chicken. In: Schein, M.W. (ed) *Social Hierarchy. Benchmark Papers in Animal Behavior*. Vol. 3. Halsted Press, Division of John Wiley & Sons, New York, pp. 35–49.

Schjelderup-Ebbe, T. (1935) Social behavior in birds. In: Murchison, C.A. (ed) *A Handbook of Social Psychology*. Clark University Press, Worcester, UK, pp. 947–972.

Scott, J.P. (1958) *Animal Behavior*. University of Chicago Press, Chicago, Illinois.

Scott, J.P. (1962) Critical periods in behavioral development. *Science* 138, 949–958.

Scott, J.P. and Fuller, J.L. (1965) *Genetics and the Social Behavior of the Dog.* University of Chicago, Chicago, Illinois.

Scraba, S.T. and Ginther, O.J. (1985) Effect of lighting programs on the ovulatory season in mares. *Theriogenology* 24, 667–679.

Seabrook, M.F. (1972) A study to determine the influence of the herdsman's personality on milk yield. *Journal of Agricultural Labour Science* 1, 44–59.

Searby, A. and Jouventin, P. (2003) Mother–lamb acoustic recognition in sheep: a frequency coding. *Proceedings, Biological Sciences, Royal Society of London* 270, 1765–1771.

Searle, L.V. (1949) The organization of hereditary maze-brightness and maze-dullness. *Genetic Psychology Monographs* 39, 279–325.

Séguin, M.J., Friendship, R.M., Kirkwood, R.N., Zanella, A.J. and Widowski, T.M. (2006) Effects of boar presence on agonistic behavior, shoulder scratches, and stress response of bred sows at mixing. *Journal of Animal Science* 84, 1227–1237.

Semsar, K., Kandel, F.L.M. and Godwin, J. (2001) Manipulations of the AVT system shift social status and related courtship and aggressive behavior in the bluehead wrasse. *Hormones and Behavior* 40, 21–31.

Sevcik, R.A. and Savage-Rumbaugh, E.S. (1994) Language comprehension and use by great apes. *Language and Communication* 14, 37–58.

Shapiro, L.E. and Dewsbury, D.A. (1990) Differences in affiliative behavior, pair bonding, and vaginal cytology in two species of vole (*Microtus ochrogaster* and *M. montanus*). *Journal of Comparative Psychology* 104, 268–274.

Sharp, D.C., Kooistra, L. and Ginther, O.J. (1975) Effect of artificial light on the oestrus cycle of the mare. *Journal of Reproduction and Fertility* 34, 241–246.

Sheppard, G. and Mills, D.S. (2003) Evaluation of dog-appeasing pheromone as a potential treatment for dogs fearful of fireworks. *Veterinary Record* 152, 432–436.

Shipka, M.P. and Ellis, L.C. (1998) No effects of bull exposure on expression of estrous behavior in high-producing dairy cows. *Applied Animal Behaviour Science* 57, 1–7.

Siegel, P.B. (1972) Genetic analysis of male mating behaviour in chickens (*Gallus domesticus*). I. Artificial selection. *Animal Behaviour* 20, 564–570.

Signoret, J.P. (1971) The reproductive behaviour of pigs in relation to fertility. *Veterinary Record* 88, 34–38.

Signoret, J.P. (1990) The influence of the ram effect on the breeding activity of ewes and its underlying physiology. In: Oldham, C.M., Martin, G.B. and Purvis, I.W. (eds) *Reproductive Physiology of Merino Sheep.* University of Western Australia, Perth, Australia, pp. 59–70.

Silver, G.V. and Price, E.O. (1986) Effects of individual vs. group rearing on the sexual behavior of prepuberal beef bulls: mount orientation and sexual responsiveness. *Applied Animal Behaviour Science* 15, 287–294.

Silver, R. (1990) Biological timing mechanism with special emphasis on the parental behavior of doves. In: Dewsbury, D.A. (ed.) *Contemporary Issues in Comparative Psychology,* Sinauer Associates, Sunderland, MA, pp. 252–277.

Smith, F.V., Van-Toller, C. and Boyes, T. (1966) The 'critical period' in the attachment of lambs and ewes. *Animal Behaviour* 14, 120–125.

Smith, M.E., Linnell, J.D.C., Odden, J. and Swenson, J.E. (2000) Review of methods to reduce livestock depredation: I. Guardian animals. *Acta Agriculturæ Scandinavica, Section A, Animal Science* 50, 279–290.

Snowder, G.D., Stellflug, J.N. and Van Vleck, L.D. (2002) Heritability and repeatability of sexual performance scores of rams. *Journal of Animal Science* 80, 1508–1511.

Sparks, J. (1982) *The Discovery of Animal Behaviour.* Little, Brown and Co., Boston, Massachusetts.

Spoon, T.R., Millam, J.R. and Owings, D.H. (2006) The importance of mate behavioural compatibility in parenting and reproductive success by cockatiels, *Nymphicus hollandicus. Animal Behaviour* 71, 315–326.

Staats, J. (1981) List of inbred strains. In: Green, M.C. (ed) *Genetic Variants and Strains of the Laboratory Mouse*. G. Fischer Verlag, New York, pp. 373–376.

Stellflug, J.N., Cockett, N.E. and Lewis, G.S. (2006) Relationship between sexual behavior classification of rams and lambs sired in a competitive breeding environment. *Journal of Animal Science* 84, 463–468.

Stevenson, J.S. (1996) Strategies for getting high-producing cows pregnant. Presentation given at University of California Cooperative Extension, South Valley Dairy Days, California.

Stone, C.P. (1932) Wildness and savageness in rats of different strains. In: Lashley, K.S. (ed) *Studies in the Dynamics of Behavior*. University of Chicago Press, Chicago, Illinois, pp. 1–55.

Stookey, J.M. and Gonyou, H.W. (1994) The effects of regrouping on behavioral and production parameters in finishing swine. *Journal of Animal Science* 72, 2804-2811.

Stookey, J.M. and Gonyou, H.W. (1998) Recognition in swine: recognition through familiarity or genetic relatedness? *Applied Animal Behaviour Science* 55, 291–305.

Stowers, L., Holy, T.E., Meister, M., Dulac, C. and Koentges, G. (2002) Loss of sex discrimination and male–male aggression in mice deficient for TRP2. *Science* 295, 1493–1500.

Subiaul, F., Cantlon, J.F., Holloway, R.L. and Terrace, H.S. (2004) Cognitive imitation in Rhesus macaques. *Science* 305, 407–410.

Sugiyama, T., Sasada, H., Masaki, F. and Yamashita, K. (1981) Unusual fatty acids with specific odor from mature male goats. *Agricultural Biological Chemistry* 45, 2655–2658.

Svartberg, K. (2006) Breed-typical behaviour in dogs – historical remnants or recent constructs? *Applied Animal Behaviour Science* 96, 293–313.

Swaddle, J.P., Cathey, M.G., Correll, M. and Hodkinson, B.P. (2005) Socially transmitted mate preferences in a monogamous bird: a non-genetic mechanism of sexual selection. *Proceedings of the Royal Society B. Biological Sciences* 272, 1053–1058.

Takahashi, J.S. and Hoffman, M. (1995) Molecular biological clocks. *American Scientist* 83, 158–165.

Tang-Martinez, Z. (2001) The mechanisms of kin discrimination and the evolution of kin recognition in vertebrates: a critical re-evaluation. *Behavioural Processes* 53, 21–40.

Tomlinson, K.A. and Price, E.O. (1980) The establishment and reversibility of species affinities in domestic sheep and goats. *Animal Behaviour* 28, 325–330.

Topál, J., Gácsi, M., Miklósi, Á., Virányi, Z., Kubinyi, E. and Csányi, V. (2005) Attachment to humans: a comparative study on hand-reared wolves and differently socialized dog puppies. *Animal Behaviour* 70, 1367–1375.

Tribukait, B. (1956) Die Aktivitätsperiodik der Weissen Maus im Kunsttag von 16-29 Stunden Länge. *Zeitschrift für vergleichende Physiologie* 38, 479–490.

Trivers, R.L. (1972) Parental investment and sexual selection. In: Campbell, B. (ed) *Sexual Selection and the Descent of Man, 1871–1971*. Heinemann, London, pp. 136–179.

Trut, L.N. (1999) Early canid domestication: the farm-fox experiment. *American Scientist* 87, 160–169.

Tryon, R.C. (1940) Genetic differences in maze-learning ability in rats. *Thirty-ninth Yearbook of the National Society for the Study of Education*. Public School Publishing, Bloomington, Illinois, pp. 111–119.

Tsutsumi, H., Morikawa, N., Niki, R. and Tanigawa, M. (2001) Acclimatization and response of minipigs toward humans. *Laboratory Animals* 35, 236–242.

Turner, D.C. and Bateson, P. (2000) *The Domestic Cat: The Biology of its Behaviour*, 2nd edn. Cambridge University Press, New York.

Turner, S.P., Horgan, G.W. and Edwards, S.A. (2001) Effect of social group size on aggressive behaviour between unacquainted domestic pigs. *Applied Animal Behaviour Science* 74, 203–215.

Ulyan, M.J., Burrows, A.E., Buzzell, C.A., Raghanti, M.A., Marcinkiewicz, J.L. and Phillips, K.A. (2006) The effect of predictable and unpredictable feeding schedules on the behavior and physiology of captive brown capuchins (*Cebus apella*). *Applied Animal Behaviour Science* 101, 154–160.

Underwood, H. (2001) Circadian organization in nonmammalian vertebrates. In: Takahashi, J.S., Turek, F.W. and Moore, R.Y. (eds) (2001) *Handbook of Neurobiology*, Vol. 12, *Circadian Clocks*. Kluwer Academic/Plenum Publishers, New York, pp. 111–140.

Ungerfeld, R. and Silva, L. (2005) The presence of normal vaginal flora is necessary for normal sexual attractiveness of ewes. *Applied Animal Behaviour Science* 93, 245–250.

Ungerfeld, R., Forsberg, M. and Rubianes, E. (2004) Overview of the response of anoestrous ewes to the ram effect. *Reproduction, Fertility and Development* 16, 479–490.

Valenstein, E.S. and Young, W.C. (1955) An experiential factor influencing the effectiveness of testosterone proprionate in eliciting sexual behavior in male guinea pigs. *Endocrinology* 56, 173–177.

Valenta, J.G. and Rigby, M.K. (1968) Discrimination of the odor of stressed rats. *Science* 161, 599–601.

Van den Berg, C.L., Hol, T., Van Ree, J.M., Spruijt, B.M., Everts, H. and Koolhaas, J.M. (1999) Play is indispensable for an adequate development of coping with social challenges in the rat. *Developmental Psychobiology* 34, 129–138.

Vandenbergh, J.G. (1969) Male odor accelerates female sexual maturation in mice. *Endocrinology* 84, 658–660.

Van de Weerd, H.A., Docking, C.M., Day, J.E.A., Breuer, K. and Edwards, S.A. (2006) Effects of species-relevant environmental enrichment on the behaviour and productivity of finishing pigs. *Applied Animal Behaviour Science* 99, 230–247.

Van Schaik, C.P., Ancrenaz, M., Borgen, G., Galdikas, B., Knott, C.D., Singleton, I., Suzuki, A., Utami, S.S. and Merrill, M. (2003) Orang-utan cultures and the evolution of material culture. *Science* 299, 102–105.

Varela, F.J., Palacios, A.G. and Goldsmith, T.H. (1993) Color vision of birds. In: Zeigler, H.P. and Bischof, H.J. (eds) *Vision, Brain and Behavior in Birds*. Massachusetts Institute of Technology Press, Cambridge, Massachusetts, pp. 77–98.

Vas, J., Topál, J., Gácsi, M., Miklósi, Á. and Csányi, V. (2005) A friend or an enemy? Dogs' reaction to an unfamiliar person showing behavioural cues of threat and friendliness at different times. *Applied Animal Behaviour Science* 94, 99–115.

Veissier, I. (1993) Observational learning in cattle. *Applied Animal Behaviour Science* 35, 235–243.

Veissier, I., de Passillé, A.M., Després, G., Rushen, J., Charpentier, I., Ramirez de la Fe, A.R. and Pradel, P. (2002) Does nutritive and non-nutritive sucking reduce other oral behaviors and stimulate rest in calves? *Journal of Animal Science* 80, 2574–2587.

Vidal, J.-M. (1980) The relations between filial and sexual imprinting in the domestic fowl: effects of age and social experience. *Animal Behaviour* 28, 880–891.

Vieira, A. de P., Guesdon, V., de Passillé, A.M., von Keyserlingk, M.A.G., and Weary, D.M. (2008) Behavioural indicators of hunger in dairy calves. *Applied Animal Behaviour Science* 109, 180–189.

Villalba, J.J. and Provenza, F.D. (1999) Effects of food structure and nutritional quality and animal nutritional state on intake behaviour and food preferences of sheep. *Applied Animal Behaviour Science* 63, 145–163.

Vince, M.A. (1973) Effects of external stimulation on the onset of lung ventilation and the time of hatching in the fowl, duck and goose. *British Poultry Science* 14, 389–401.

Vince, M.A. (1993) Newborn lambs and their dams: the interaction that leads to sucking. In: Slater, P.B., Rosenblatt, J.S., Snowdon, C.T. and Milinski, M. (eds) *Advances in the Study of Behavior*, Vol. 22. Academic Press, New York, pp. 239–268.

Von Uexküll, J. (1934) *A Stroll Through the World of Animals and Men: A Picture Book of Invisible Worlds*. Springer, Berlin. Translation by Schiller, C.H. In: Schiller, C.H. (ed) (1957) *Instinctive Behavior: The Development of a Modern Concept*. International Universities Press, New York, pp. 5–80.

Waddington, C.H. (1966) *Principles of Development and Differentiation*. Macmillan, New York.

Wagner, A.R., Siegel, S., Thomas, E. and Ellison, G.D. (1964) Reinforcement history and the extinction of a conditioned salivary response. *Journal of Comparative and Physiological Psychology* 58, 354–358.

Walker, D.B., Walker, J.C., Cavnar, P.J., Taylor, J.L., Pickel, D.H., Hall, S.B. and Suarez, J.C. (2006) Naturalistic quantification of canine olfactory sensitivity. *Applied Animal Behaviour Science* 97, 241–254.

Wallach, S.J.R. and Price, E.O. (1988) Bulls fail to show preference for estrous females in serving capacity tests. *Journal of Animal Science* 66, 1174–1178.

Walther, F.R. (1974) Some reflections on expressive behaviour in combats and courtship of certain horned ungulates. In: Geist, V. and Walther, F. (eds) *The Behaviour of Ungulates and its Relation to Management*. IUCN Publications, No. 24, International Union for Conservation of Nature and Natural Resources, Morges, Switzerland, pp. 56–106.

Waran, N.K. and Henderson, J. (1998) Stable vices: what are they and can we prevent them? *Equine Practice* 20, 6–8.

Waran, N.K., Robertson, V., Cuddeford, D., Kokoszko, A. and Marlin, D.J. (1996) Effects of transporting horses facing either forwards or backwards on their behaviour and heart rate. *Veterinary Record* 139, 7–11.

Watson, A. and Jenkins, D. (1968) Experiments on population control by territorial behaviour in red grouse. *Journal of Animal Ecology* 37, 595–614.

Weatherhead, P.J. (1995) Effects of female reproductive success of familiarity and experience among male red-winged blackbirds. *Animal Behaviour* 49, 967–976.

Wecker, S.C. (1963) The role of early experience in habitat selection by the prairie deer-mouse, *Peromyscus maniculatus bairdii*. *Ecological Monographs* 33, 307–325.

Welch, A.M., Semlitsch, R.D. and Gerhardt, H.C. (1998) Call duration as an indicator of genetic quality in male gray tree frogs. *Science* 280, 1928–1930.

Wells, D.L. and Hepper, P.G. (2006) Prenatal olfactory learning in the domestic dog. *Animal Behaviour* 72, 681–686.

Wesley, F. (1967) Stereotypy and teat selection in pigs. *Zeitschrift für Saugetlerkunde* 32, 362–366.

West-Eberhard, M.J. (2003) *Developmental Plasticity and Evolution*. Oxford University Press, New York.

Whalen, R.E. and Edwards, D.A. (1967) Hormonal determinants of the development of masculine and feminine behavior in male and female rats. *Anatomical Record* 157, 173–180.

Wheelwright, N.T., Freeman-Gallant, C.R. and Mauck, R.A. (2006) Asymmetrical incest avoidance in the choice of social and genetic mates. *Animal Behaviour* 71, 631–639.

White, D.J. and Galef, B.G. Jr (1999) Social effects on mate choices of male Japanese quail, *Coturnix coturnix*. *Animal Behaviour* 57, 1005–1012.

White, F.J., Wettemann, R.P., Looper, M.L., Prado, T.M. and Morgan, G.L. (2002) Seasonal effects on estrous behavior and time of ovulation in nonlactating beef cows. *Journal of Animal Science* 80, 3053–3059.

Whitten, W.K. (1956) Modification of the estrous cycle of the mouse by external stimuli associated with the male. *Journal of Endocrinology* 13, 399–404.

Wiley, R.H. (1991) Lekking in birds and mammals: behavioral and evolutionary issues. *Advances in the Study of Behavior* 20, 201–291.

Willard, J.G., Willard, J.C., Wolfram, S.A. and Baker, J.P. (1977) Effect of diet on cecal pH and feeding behavior of horses. *Journal of Animal Science* 45, 87–93.

Wilson, E.O. (1975) *Sociobiology*. Harvard University Press, Cambridge, Massachusetts.

Winskill, L.C., Young, R.J., Channing, C.E., Hurley, J. and Waran, N.K. (1996) The effect of a foraging device (the modified 'Edinburgh Foodball') on the behaviour of the stabled horse. *Applied Animal Behaviour Science* 48, 25–35.

Wisenden, B.D., Vollbrecht, K.A. and Brown, J.L. (2004) Is there a fish alarm cue? Affirming evidence from a wild study. *Animal Behaviour* 67, 59–67.

Würbel, H. and Stauffacher, M. (1997) Age and weight at weaning affect corticosterone level and development of stereotypy in ICR-mice. *Animal Behaviour* 53, 891–900.

Würbel, H. and Stauffacher, M. (1998) Physical condition at weaning affects exploratory behaviour and stereotypy development in laboratory mice. *Behavioural Processes* 43, 61–69.

Wyatt, T.D. (2003) *Pheromones and Animal Behaviour.* Cambridge University Press, New York.

Wyeth, G.S.F. and McBride, G. (1964) Social behaviour of domestic animals. V. Note on suckling behaviour in young pigs. *Animal Production* 6, 245–247.

Wysocki, C.J., Bean, N.J. and Beauchamp, G.K. (1986) The mammalian vomeronasal system: its role in learning and social behavior. In: Duvall, D., Müller-Schwarze, D. and Silverstein, R.M. (eds) *Chemical Signals in Vertebrates.* Vol. 4, *Ecology, Evolution and Comparative Biology.* Plenum Press, New York, pp. 471–485.

Xu, Z.Z., McKnight, D.J., Vishwanath, R., Pitt, C.J. and Burton, L.J. (1998) Estrus detection using radiotelemetry or visual observation and tail painting for dairy cows on pasture. *Journal of Dairy Science* 81, 2890–2896.

Yin, S. (2002) A new perspective on barking in dogs. *Journal of Comparative Psychology* 116, 189–193.

Ylander, D.M. and Craig, J.V. (1980) Inhibition of agonistic acts between domestic hens by a dominant third party. *Applied Animal Ethology* 6, 63–69.

Yoshihara, Y., Nagao, H. and Mori, K. (2001) Sniffing out odors with multiple dendrites. *Science* 291, 835–837.

Zucker, I. (1980) Light, behavior and biologic rhythms. In: Krieger, D.T. and Hughes, J.C. (eds) *Neuroendocrinology,* Sinauer Associates, Sunderland, MA, pp. 93–101.

Index

Hierarchy, social (*see* Social organization)
Home range 214
Hormones, effects on behavior 21, 30, 52–53, 99, 101–104, 115, 122–124, 126, 130–131, 147, 154, 159, 180, 193, 197–199, 221, 226
Housing, effects on behavior 196, 211, 221, 222–224, 225, 238
Human-animal interactions
  approach humans, tendency to 28–29, 33, 48, 227–231, 233, 234–239, 245–246, 282
  avoid humans, tendency to 25, 33, 227–231, 233, 234–239, 245–246, 247, 250, 252–253, 255–257, 282
  context of the interaction (e.g. place) 240, 256
  fear, of humans 23–25, 33, 195, 227–231, 233, 234–246, 247, 250 252–253, 255–257, 282
  flight initiation distance 227, 231, 237, 239–240, 242, 252–253
  flight zone 227, 250, 252–253, 255
  genetic effects, on ease of handling, tameability and tameness 23–25, 228–233, 236, 245–246
  human behavior and posture 238–243, 247, 250, 252–254, 255, 265–266, 282, 283–284
  moving animals 174, 247, 248–258, 260–261, 263–267
  productivity of animals 240–243, 247–248
  recognition of individual humans 236, 240
  reproductive success of animals 241–242, 244, 246, 247–248
  tameability, tameness and taming, defined 228
  tameness, undesirable aspects 44–45, 244–245, 282
  worker training 241, 243
  training animals, techniques 174, 195, 245, 256–258, 259, 260, 261–262, 265–266, 267
  (*see also* Tameness)
Hunger
  motivation 4–5, 11, 20, 157–158, 191–192, 199, 278, 280, 282, 283–284
  physiological basis 11, 158

Hybridization, effects on behavior (*see* Genetic effects)
Hypothalamus 21, 75, 101, 117, 122–123, 126, 147, 179, 193, 245–246

Immobility, significance for sexual behavior (*see* Sexual behavior)
Imprinting
  auditory imprinting 39–40
  filial imprinting 35–36, 156–157
  habitat imprinting 16–17, 39
  olfactory imprinting 38–39, 42–43, 179–181
  sensitive periods in development 34–40, 42
  species imprinting 36–38
Individual distance (*see* Inter-individual distance)
Individual recognition
  dominance hierarchy, role in stability of 206–209
  dominant-subordinate relationships, role in 199–201, 206–208
  (*see also* Communication, function, Recognition, Social organization)
Ingestive behavior (*see* Feeding behavior)
Innate
  defined 15
  (*see also* Genetic effects)
Insight learning (*see* Learning, abstract thinking)
Instinct
  abuse and subsequent disfavor in use of term 12–15
  criteria for use of term by the early ethologists 12–15
  historical relevance 12
  (*see also* Innate)
Intention movements 63, 132–133
Inter-individual distance (*see also* Flight initiation distance) 201, 202, 220–221, 226, 248–249
Interpreting function, mistakes 6–7
Inter-sucking (*see* Atypical behavior, non-nutritive sucking)
Invisible fence 259–260
Isolation, prior to parturition 142
Isolation rearing, effects of (*see* Early experience)
Isolation, and stress 41–42, 254, 272, 277, 280, 283–284

amniotic fluid, attraction and
   repulsion 144, 152, 154–155
audition, role in offspring
   discrimination 151
bonding (social attachment) to
   offspring 147–148, 149–150,
   152–156
brain systems involvement 21, 147, 149
breed differences 28–29, 145–146
cats 4, 143, 146–147, 157
cattle 129, 148, 154, 157, 162
discrimination (identification), of
   young 147–148, 152–154, 180
dogs 143, 157
evolutionary considerations 152,
   155–156
fostering 29, 38, 151–156, 229, 235
goats 148, 157
hormone involvement 21, 83–84, 143,
   147, 154, 159
horses 157
huddling (crouching), over
   offspring 21, 83, 146–147
isolation, prior to parturition 143
labor and expulsion of fetus 21,
   143–145
licking (grooming) offspring 4, 41,
   143–146, 148, 150, 246
mice 21
milk ejection (release), stimulus
   control 21, 52, 159
nestbuilding 21, 99, 102, 103, 104, 143
nursing 4, 129, 146, 156–157, 159
odor transfer (fostering
   technique) 152–154
olfaction, role in offspring
   discrimination 149–150,
   152–155, 180
pigs 143, 149, 155, 157, 159, 162
preparturient behavior 143
primiparous vs multiparous
   mothers 148
rabbits 158–159, 180
rats 41, 143
rejection of young 145–146, 147–148,
   149–151, 180
responsiveness to *any* young, sensitive
   period 147–148
restraint fostering 152–153
retrieval, of young to nest 21
sensitive periods, in expression
   of 147–148

separation, from young 145–146, 148,
   254–255
sheep 143–146, 148, 149–155, 157, 180
skin grafting (fostering
   technique) 152–153
slime grafting (fostering
   technique) 152
stealing young neonates 145
uterus/cervical/vaginal stimulation
   (fostering technique) 154–155
visual cues, role in offspring
   discrimination 150–151
vocalizations, role of 157, 159
weaning (*see also* Weaning) 162–163
Maternal effect, in development of offspring
   behavior 4, 28–29, 42–43, 229,
   231, 235, 244
Mating systems (*see* Reproduction)
Maze learning (*see* Learning)
Meat quality, stress effects 242–243, 257
Melatonin 77
Metestrus (*see* Sexual behavior)
Milk ejection (let-down) (*see* Maternal
   behavior)
Milk production aggression, effects on milk
   production 209
   artificial selection for 211
   dominance, lack of correlation 211
   human-animal interactions, effects on
      milk production 242–243
   (*see also* Lactation)
Milk replacer, use of 158, 164
Mitral cells, in olfaction 150, 178
Monogamy (*see also* Reproduction) 83–85
Mounting (*see* Sexual behavior)
Moving (animals)
   cattle 248–256, 266–267
   chutes and alleyways 248–252
   dogs, use in moving livestock 49, 253,
      255, 264–267
   facility design 248–252
   fear-inducing stimuli, effects 255–256
   goats 254
   handlers, position and movement 250,
      252–254, 266
   horses 254, 256, 257–258, 260–261,
      263
   pastures, moving livestock in and out
      of 253–254, 264–267
   pens, moving livestock in and out
      of 252–253
   pigs 248–250, 251, 263–264

THE SECOND MYSTERY IS ABOUT ALIEN CREATURES

Many people have seen human-like creatures climbing into or out of UFOs which have landed on Earth. But often people only see the UFO. At other times some people see alien creatures on their own.

This photograph appeared in a German newspaper in the 1950s. It was a trick shot.

The first UFO was seen in 1947. Films about **aliens** were very popular at this time, too. Many people thought that UFOs were spaceships carrying aliens. But nobody knows if the strange creatures and UFOs are really linked.

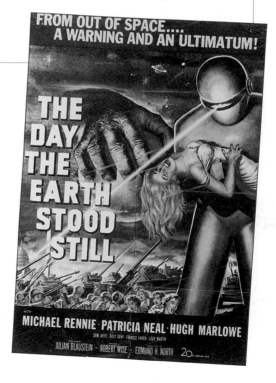

FROM OUT OF SPACE....
A WARNING AND AN ULTIMATUM!

THE DAY THE EARTH STOOD STILL

MICHAEL RENNIE · PATRICIA NEAL · HUGH MARLOWE

SAM JAFFE · BILLY GRAY · FRANCES BAVIER · LOCK MARTIN

JULIAN BLAUSTEIN · ROBERT WISE · EDMUND H. NORTH

UFOs

Most people who see UFOs describe them in detail. Although UFOs can be very different, they share certain features.

WHAT DO UFOs LOOK LIKE?

UFOs are usually round or oval in shape with smooth outlines. Most UFOs don't have doors or windows. People often say a UFO is blurred or looks out of focus. UFOs may be yellow or white, or glow orange or red. Sometimes they throw out sparks and flashes.

Contents

Introduction

This book looks at two different mysteries.

THE FIRST MYSTERY IS ABOUT UFOS

The letters **UFO** stand for Unidentified Flying Object. Pilots and air traffic controllers first used the term 'UFO' to describe any object in the sky which they did not recognize. People now use the term UFO to describe any object which flies and lands on Earth and cannot be explained. Thousands of people have seen UFOs but nobody knows what they are.

UFO photographed by Ella Louise Fortune in New Mexico City, USA on 16 October 1957.

UFO photographed by Paul Villa in New Mexico City, USA on 16 June 1963.

WHAT HAPPENS WHEN UFOS ARE AROUND?

UFOs can also have a strange effect on things. Witnesses often say that electrical systems stop working when a UFO is close. Radios stop working. **Compasses** and other magnetic equipment may do strange things.

HOW DO UFOS MOVE?

Most UFOs wobble, bounce or skip along. Sometimes they move very quickly. They can also hover or move very slowly. UFOs may make a noise when they move. Some people have said UFOs buzz or whine. These noises can be very loud.

Flying saucers

Until 1947 nobody had heard of **flying saucers** or UFOs.
Then on 4 June 1947 they became world famous.

An American called Kenneth Arnold was flying his
private aircraft towards Yakima, in Washington State,
USA. As he flew near Mount Rainier, Arnold saw a
flash of light and six large round silver objects. He

Kenneth Arnold standing with his plane.

*Kenneth Arnold with a drawing of
the UFO he saw.*

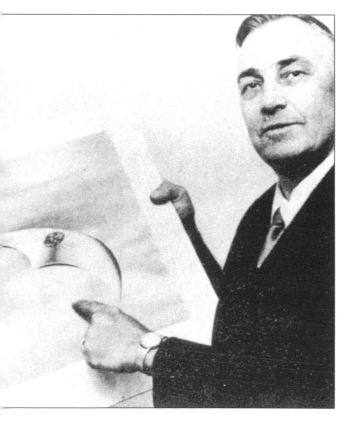

*Kenneth Arnold wrote
a book about his adventure.*

guessed they were about twice the size of a bus and
were moving at about 1,600 kilometres per hour.
That's much faster than any jet could fly in those days.

Later that day, Arnold told a reporter that the
objects moved "like giant saucers being skipped on
water". The newspapers called the objects flying
saucers. The name is still used today.

Flashes at night

On 30 November 1973 a pilot called Ricardo Marano saw a UFO in Turin, Italy. He was trying to land his aircraft at Turin airport, but he couldn't. There was a strange object blocking his path.

This bright object is similar to the one Marano saw.

Marano looked round and saw flashes of light coming from the object. It became dimmer, then brighter. It changed colour from green to red to yellow. Then it began to move. Marano chased it, but his aircraft was too slow. The object headed south-east at 900 kilometres per hour.

When Marano finally landed at Turin, he told everyone what he had seen. Another pilot had also seen the object, and a Colonel of the Italian Air Force had seen the UFO on his **radar** set. He said it looked about as big as a passenger jet.

In 1954 Fate magazine told a similar story to Marano's.

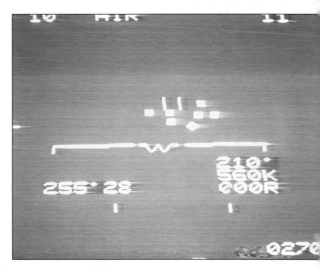

This radar screen has detected a UFO in the same way that Marano's did.

Air crash

A UFO may have caused
an air crash in Australia in 1978.
Frederick Valentich flew to King
Island, Australia, to collect some
shellfish but never arrived.

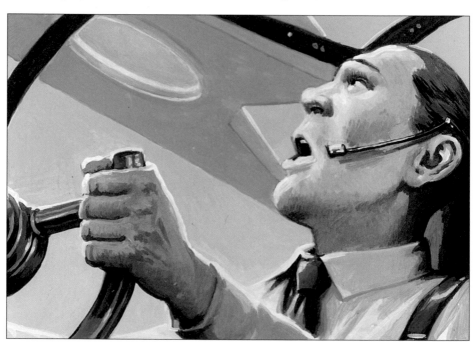

At 7.06 pm Valentich radioed
the air base to report a large
aircraft above him with four bright lights. A few
minutes later, Valentich radioed again, "It's due east of

me. It seems to be playing some sort of game ... It's orbiting above me and has some sort of metallic light on the outside." Then the object flew off and vanished.

Six minutes later Valentich radioed once more. "The engine is coughing. The unknown aircraft is now on top of me." Then there was silence.

Valentich never arrived at King Island. There was a large oil slick on the sea where he vanished.

Some people think Valentich was attacked by the UFO. Others think the UFO made his engine fail. A few people think Valentich went mad and saw things that were not really there, and then crashed.

UFOs on film

One of the most famous UFO sightings was on 30 December 1978 in New Zealand. A plane was flying from Wellington to Blenheim with a film crew on board.

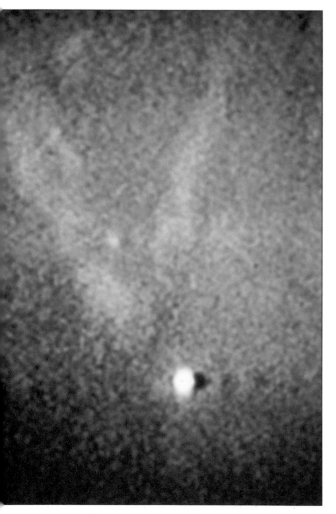

This is what the cameraman saw.

A radar station in Wellington radioed the aircraft to say, "Eleven strange craft are following you." The Captain of the aircraft looked back to see several bright lights tailing his aircraft. The cameraman on board saw the light too, got his camera and filmed the UFOs for 30 seconds. Then they became too faint to see. The radar monitored the UFOs for another 20 minutes.

Some people think the cameraman filmed the planet Venus by mistake. Other people think the lights were a reflection of the Moon. Nobody can explain the radar trace.

Did the cameraman really see Venus?

Or was it a reflection of the Moon?

Radar

Radar works by bouncing radio beams off solid objects. Any solid object will produce a radar trace. This shows the radar operator where the object is and how large it is. If a UFO shows on radar, it is probably solid.

Under attack

On 21 August 1955 alien creatures attacked a farm in Kentucky, USA.

Luck Sutton saw a UFO land near his farm. He ran home to tell his family. A few minutes later the dogs started barking. Luck went outside to see what was wrong. He saw a creature with big ears running towards him. Luck grabbed a shotgun and shot the creature. The creature fell down as if dead. Then it got up and ran away.

Luck's brother came out to see what had happened. As he stepped out of the house a creature sitting on the roof pulled his hair. He turned and shot the creature with a rifle. It fell to the ground but then ran off. Several creatures then ran towards the farm. Then the Sutton family got into their car and fled.

They came back with the local police four hours later. The farm was a mess. All the windows were smashed and the doors were broken. The creatures had gone.

Kidnapped

On 19 September 1961 Betty and Barney Hill were
driving across open countryside in New Hampshire,
USA. Suddenly, they saw a bright light flying through

the air. The light landed on the road in front of them.
The next thing they knew, it was two hours later and
they were 50 kilometres away. They did not know how
they got there or what had happened.

Later, the Hills went to a hypnotist to try to remember what had happened. Under **hypnosis**, the Hills remembered being taken aboard a large orange UFO. The alien creatures were tall and looked like humans. They took pieces of hair and finger nails from Betty and Barney. Barney was then tied to a large table. The alien creatures did some experiments which hurt a lot.

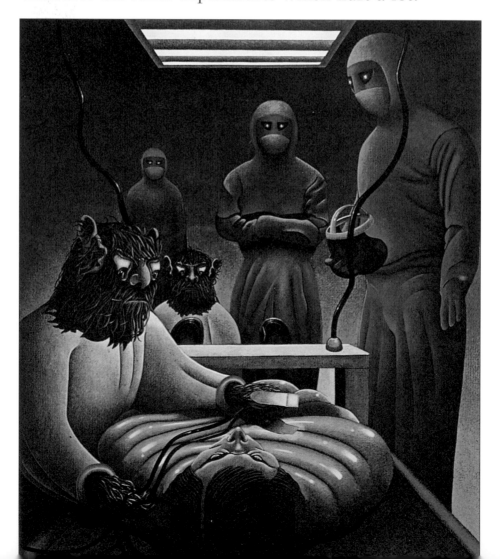

After this the alien creatures hypnotized Betty and Barney so that they would forget what happened. Then the alien creatures put Betty and Barney back in their car and left.

Betty and Barney Hill had to be hypnotized many times before they were able to remember everything that had happened.

Barney and Betty Hill with a drawing of what they saw.

A few years before, Fate magazine described a similar story to Barney and Betty's.

Hypnosis

Under hypnosis people may remember things they have forgotten. Some people can remember right back to when they were babies.

UFO landing

On 24 April 1964 a policeman called Lonnie Zamora was driving his patrol car near the town of Socorro, in New Mexico, USA. Suddenly, he heard a loud explosion. He drove up a dirt track to the top of a hill to find out what was going on.

Lonnie Zamora

Zamora saw an object about three metres tall with a symbol on the side. It stood on four metal legs. Two small men were standing by the object looking at some plants. Zamora radioed for help. Then he got out of the car and walked towards the object. The two men had vanished.

When Zamora was about 40 metres from the object, there was a loud roar. The object lifted into the air and threw out an orange flame. He dived for cover behind a rock as the object flew away.

A few minutes later, another policeman arrived. He and Zamora searched the site. They found four imprints where Zamora had seen the object. They looked like marks left by the legs of the strange object.

Marks left by the UFO.

What do alien creatures look like?

People are often frightened when they see alien creatures. They are not always clear about what they have seen. There seem to be three types of these creatures.

1 ▶ The first type is short, ugly and vicious. The creatures which attacked the Suttons in Kentucky were like this.

2 ▶ The second type looks very like a human. These creatures are about the same size as humans. They have arms, legs and a head. But their faces are often quite odd. They may not have a nose or their eyes may be very large. The creatures that took Betty and Barney Hill on board their UFO were like this.

ET was a small and shy alien in the film ET.

3 ▶ The third type of alien creature is very small and shy. These creatures often have big heads and large eyes. Many people see them picking flowers or looking at plants. They run away when they see humans. The aliens that Zamora saw were like this.

UFO groups

Some people who say they have met alien creatures believe they have been chosen by the aliens for a special mission. These people often want others to believe in their mission. Sometimes people form groups based on their beliefs.

Sir George King, Founder of the Aetherius Society

The Aetherius Society began in 1954 when a master of thought power, George King, was met by an alien being. The alien told him that he would get messages from other planets. Since then George King received thousands of messages. One of the messages tells us that we are going into a 'New Age'.

In August 1954 many people believed a woman called Margaret Keech. She said that she had met an alien creature in Lake City, Utah, USA. Mrs Keech said that the alien creature had told her that the USA would be destroyed by a great flood on 21 December 1954.

Mrs Keech said that people who were with her on the day would be saved from disaster by a fleet of UFOs. Hundreds of people came to Lake City with her, but nothing happened. There was no flood and there were no aliens.

What are UFOs?

Lots of people have tried to explain what UFOs could be. There are different theories.

Is this a UFO or Venus?

THEORY 1

There are no UFOs at all

People who think they see a UFO are really seeing a normal object under odd conditions. For example, the planet Venus can look very bright, like a UFO, if the air is clear. Or the UFO might be a **mirage**.

Is this a UFO or a distorted image of the Moon?

THEORY 2

UFOs are time-travel ships

Perhaps UFOs are sent by scientists in the future who want to study humans in their past. Nobody has invented a **time-travel ship,** but perhaps somebody might one day.

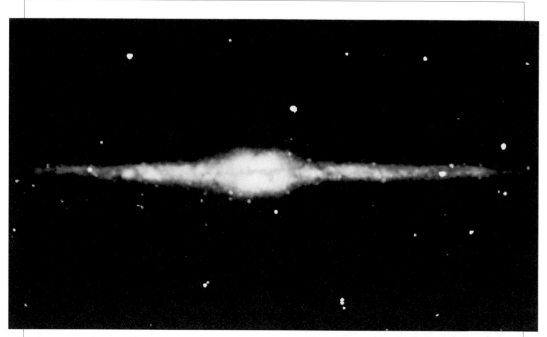

Is this a UFO or the Milky Way?

THEORY 3

UFOs are natural objects

UFOs usually glow, bounce and affect electrical equipment. Lightning is a powerful burst of natural electricity. Perhaps there are some conditions which make balls of energy that act like UFOs.

Is this a UFO or a comet?

THEORY 4

UFOs are alien spaceships

Scientists do not think there could be aliens on any of the planets near our star, the Sun. So if there are alien spaceships around, they must come from planets near other stars. The nearest stars to Earth are millions of kilometres away. Scientists on Earth do not think it is possible to travel so far, but perhaps some clever aliens know a way.

Travelling in space

MARS AND VENUS

The Earth is a planet which goes round the Sun. Life on Earth needs heat, oxygen and water. If any other planet has life like that on Earth, it must have these things. They are most likely to be found on a planet near to Earth.

Mars

The two planets nearest to Earth are Mars and Venus. In the past, people thought that alien beings might live on these two planets. People also thought that UFOs were spacecraft from these planets.

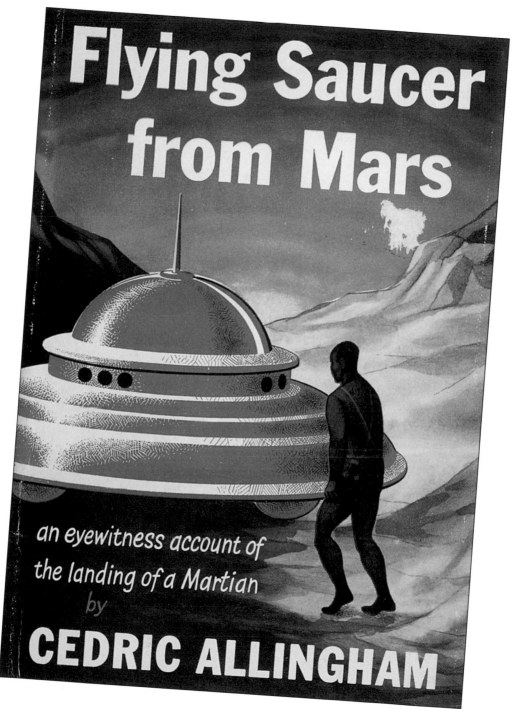

This book was published in 1955.

We now know much more about these two planets.
Space probes have flown from Earth to Mars and Venus.

Mars is further from the Sun than the Earth is. It is
slightly smaller than the Earth. The surface is made up
of red rock and dust. There are mountains, valleys and

volcanoes. The **atmosphere** is made up of different gases. Humans could not live on Mars because there is not enough oxygen. Some scientists now believe they have found traces of simple plants in rocks from Mars. These plants may have lived millions of years ago but there is no life on Mars now.

Venus is closer to the Sun than the Earth is. It is about the same size as Earth. Venus is very hot with no oxygen. No humans could live on Venus. Space probes have not found any sign of life on Venus.

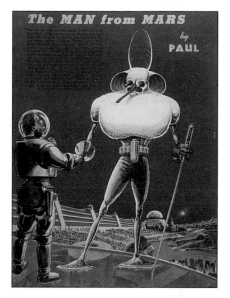

Creatures could only exist on Mars if they could cope with very hot and cold weather.

Creatures could only exist on Venus if they could live on the land and water.

THE OUTER PLANETS

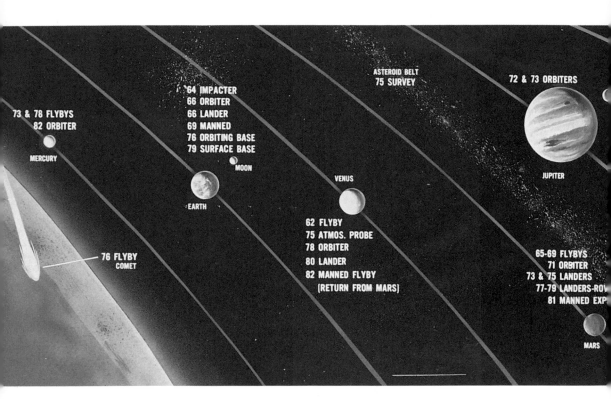

73 & 78 FLYBYS
82 ORBITER

MERCURY

64 IMPACTER
66 ORBITER
66 LANDER
69 MANNED
76 ORBITING BASE
79 SURFACE BASE

MOON

EARTH

76 FLYBY
COMET

ASTEROID BELT
75 SURVEY

VENUS

62 FLYBY
75 ATMOS. PROBE
78 ORBITER
80 LANDER
82 MANNED FLYBY
(RETURN FROM MARS)

72 & 73 ORBITERS

JUPITER

65-69 FLYBYS
71 ORBITER
73 & 75 LANDERS
77-79 LANDERS-ROV
81 MANNED EXP

MARS

There are nine planets which go around the Sun.
Scientists now know that life does not exist on Mars or
Venus. What about the other six planets?

The closest planet to the Sun is Mercury. It is
burning hot in the day but freezing cold at night. There
is no oxygen or water on the surface. Life like that on
Earth could not exist there.

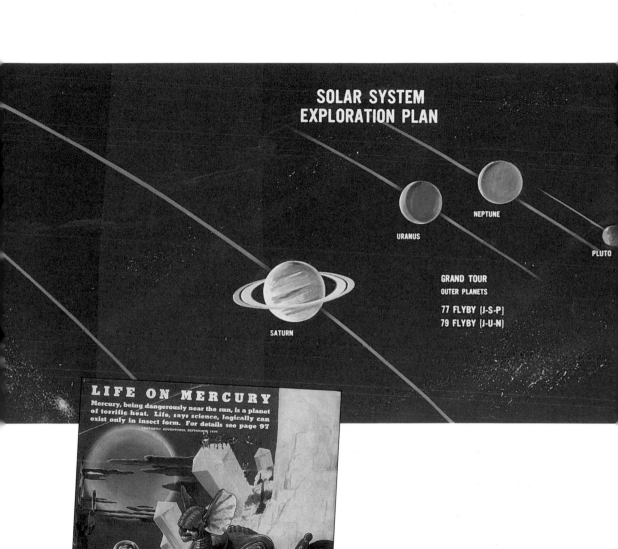

SOLAR SYSTEM EXPLORATION PLAN

URANUS

NEPTUNE

PLUTO

SATURN

GRAND TOUR
OUTER PLANETS

77 FLYBY (J-S-P)
79 FLYBY (J-U-N)

LIFE ON MERCURY

Mercury, being dangerously near the sun, is a planet of terrific heat. Life, says science, logically can exist only in insect form. For details see page 97

Creatures could only exist on Mercury if they could endure extreme heat, like insects.

The planets Jupiter, Saturn, Uranus and Neptune are called the gas giants. They are huge balls of gas with a rocky centre. But there is no oxygen. This means that life like that on Earth is impossible on these planets.

LIFE ON JUPITER

Jupiter's inhabitants would need to be massive, of tremendous strength to cope with the enormous gravity of this giant world. They would probably be forced to a clumsy means of locomotion, since long legs would be impossible. An Earthman would need a tractor car to get about.

For complete details, see page 97
FANTASTIC ADVENTURES, JANUARY, 1940

Creatures could only exist on Jupiter if they could cope with the strong force of gravity.

LIFE ON SATURN Life on Saturn would evolve along insect lines, with light body, capable of walking spider-like across its swampy, unstable surface. See page 97 for details

Creatures could only exist on Saturn if they could walk across the swampy ground, like a spider.

LIFE ON PLUTO — This world of cold and eternal twilight would most likely be inhabited by winged bat-people with heavy protecting fur. Details on page 97

FANTASTIC ADVENTURES, FEBRUARY, 1940

Creatures could only exist on Pluto if they could cope with the cold and dark, like bats.

Pluto is small, rocky and icy. It is so far from the Sun that temperatures never rise above -230°C. That is far too cold for Earth-like life to exist.

Scientists do not think any of these planets have life on them.

SPACESHIPS

Scientists have tried to work out how to travel to other stars. They have come up with three ideas.

1 ▶ A giant spaceship could be built. The crew of this ship would not live long enough to reach another star. But, they could have children who would learn to fly the ship. Then their children would have children and so on. It would take hundreds of years for the ship to arrive. Not many astronauts like this idea.

2 ▶ The crew of a spaceship could be kept asleep for the journey. They would only wake up once the ship arrived. Nobody has found a way to keep humans asleep like this.

3 ▶ A spaceship could be built that could travel fast enough for a human to survive the journey. A nuclear bomb exploded behind the ship would push the ship forward with huge force. It is thought that 30,000 bombs would push a ship to the nearest star in about 130 years.

In science-fiction books and films spaceships use warp drives, ion drives and sub-space drives. These power giant ships at fantastic speeds. All the drives are make-believe. So far nobody knows how to make them really work.

Movie space travel

The science fiction series Star Trek began as a television series in the 1960s, but has grown to include several movies, television shows, toys and games. The shows and movies include space travel, time shifts and lots of different aliens. Some people think that stories like this help people to believe in aliens.

BE A UFO DETECTIVE

It is easy to become a **UFO detective.** It is an interesting hobby and may last all your life. You could start by collecting reports of UFO sightings from newspapers and the TV. You could ask at your local library for old copies of local newspapers. Make a record of UFO sightings in previous years.

UFO photographed by Mrs Trent in Oregon, USA on 11 May 1950.

UFO photographed by Paul Villa in New Mexico, USA on 18 April 1965.

UFO photographed by an unknown photographer in Natal, South Africa on 29 May 1973.

Once you have started your collection, you could catalogue the sightings. You could arrange the sightings in a scrap book according to the type of sighting. Once you have collected a lot of data, you may find that a pattern emerges. Perhaps you could make a great UFO discovery.

UFO photographed by Hannah McRoberts in British Columbia, Canada in October 1981.

Time line

4 June 1947
US businessman Kenneth Arnold sees six large round objects flying over Washington State. He calls them 'flying saucers'.

August 1954
An American woman named Marian Keech is told by alien creatures that the USA will be destroyed by floods in December. Nothing happens.

1947

1954

1969

1950

1961

1964

June 1950
US Marine Captain Donald Keyhoe publishes a book called *The Flying Saucers are Real*. He has the idea that UFOs are spaceships from an alien civilization.

19 September 1961
An American couple, Betty and Barney Hill, say they have been kidnapped by alien creatures.

24 April 1964
US policeman, Lonnie Zamora, sees alien creatures looking at plants near their UFO in New Mexico.

1969
US government publishes the Condon Report. It says that there is no evidence to suggest alien spaceships exist. UFO detectives say the government is covering up vital evidence.

21 October 1978
An Australian aircraft vanishes after the pilot sees a UFO coming towards him at high speed. No trace of the aircraft is ever found.

30 December 1978
A New Zealand TV crew film a UFO near Wellington. Radar also picked up a strange object.

July 1996
The film Independence Day tells the fictional story of an alien invasion of Earth. The film is based on UFO reports.

August 1996
Scientists studying rocks from Mars say that they have found traces of simple plants. These may be the first signs of life on another planet ever found.

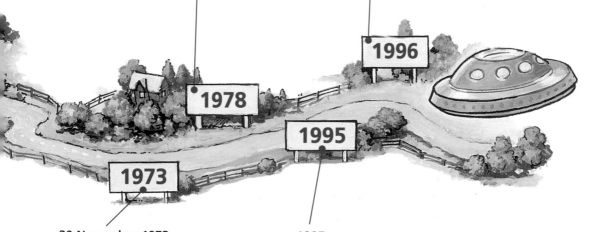

30 November 1973
UFO seen by pilots and shown on radar at Turin, Italy.

1995
US government scientist says he has worked on a captured UFO at a secret US government research base in Utah.

45

Glossary

alien A creature or person from another planet. Nobody has proved that aliens exist.

atmosphere A blanket of gases which surrounds a planet.

compass An instrument with a needle that always points towards north.

flying saucer A UFO that is shaped like a saucer. This term is sometimes used to mean all types of UFO.

hypnosis A method of mind control. People under hypnosis fall fast asleep. They can often remember things that they can't remember when they are awake.

radar A way of finding objects by bouncing radio beams off them. Any solid object gives a radar trace. The radar trace shows where an object is, how big it is and how fast it is moving.

space probe A craft with no people in it that travels into space to explore planets.

time-travel ship A craft which can carry people or aliens through time. Nobody has invented a time-travel ship yet.

UFO (Unidentified Flying Object) Any unknown object in the sky. The term is used to mean objects which cannot be explained by science.

UFO detective A person who studies UFOs.

Index